International Conference on Recent Advances in Mechanical and Automation Engineering (ICRAMAE-2021)

IOP Conference Series: Materials Science and Engineering
Volume 1224

Jaipur, India
29 – 30 November 2021

Editors:

Mohd Mukhtar Alam
Vijay Kumar Pandey

ISBN: 978-1-7138-4682-6
ISSN: 1757-8981

Printed from e-media with permission by:

Curran Associates, Inc.
57 Morehouse Lane
Red Hook, NY 12571

CURRAN ASSOCIATES INC. proceedings .com

Some format issues inherent in the e-media version may also appear in this print version.

Institute of Physics
Dirac House, Temple Back
Bristol BS1 6BE UK

Phone: 44 1 17 929 7481
Fax: 44 1 17 920 0979

techtracking@iop.org

Additional copies of this publication are available from:

Curran Associates, Inc.
57 Morehouse Lane
Red Hook, NY 12571 USA
Phone: 845-758-0400
Fax: 845-758-2633
Email: curran@proceedings.com
Web: www.proceedings.com

TABLE OF CONTENTS

Preface

International Conference on Recent Advances in Mechanical and Automation Engineering (ICRAMAE-2k21), Department of Mechanical Engineering, Vivekananda Global University, Jaipur, Rajasthan, India. 29th -30th November, 2021.

Department of Mechanical Engineering, Vivekananda Global University, Jaipur, Rajasthan, India organized International conference during 29th -30th November, 2021 and is pleased to publish the selected papers in IOP Conference Series: Materials Science and Engineering. The conference was interdisciplinary and covered several scientific thematic areas from Mechanical, Robotics and Automation Engineering.

Virtual Disclaimer for ICRAMAE-2k21: Due to the uncertainty associated with COVID-19 pandemic and travel restrictions across the globe and considering the safety of the participants it was decided to conduct the conference in hybrid mode using the offline/Meet jitsi platform. The conference was not postponed to a later date, as the scientific community would have been facing a delay in publishing their research work.

The organizers were located at Department of Mechanical Engineering, Vivekananda Global University, Jaipur, Rajasthan, India and the participants across the globe presented their research works through oral presentation. Each registered participant was allotted 12 minutes including questionnaire. The conference was conducted smoothly without any major bottlenecks and the organizers would like to express their gratitude to all the people associated especially the invited keynote speakers, session chairs and the participants who supported this conference. Also, the thanks go to the IOP Science for agreeing to publish the all presented papers in its Conference Series: Materials Science and Engineering journal.

Day 1 (Monday, 29-11-2021):

The conference was inaugurated by the Chief Guest Dr. Amar Patnaik, Associate Professor, MNIT Jaipur, Guest of Honor Prof. (Dr.) S.N. Vijay Vargiya, Technological Director, Genus, Jaipur; Dr. K.R. Bagaria, Vice Chairperson, Prof. (Dr.) Vijay Vir Sing, President, Dr. Pramod Kumar, HOD Mechanical Engineering, Dr. Mohd Mukhtar Alam, Organizing Secretary and Dr. Vijay Kumar Pandey, Convener, ICRAMAE-2k21. After the inaugural function keynote speech was delivered by Prof. (Dr.) Juan M.Carchado, Director BISITE, University of

IOP Conf. Series: Materials Science and Engineering 1224 (2022) 011001 doi:10.1088/1757-899X/1224/1/011001

Salamanca, Spain on Deployment of DeepTech AI Models in Engineering Solutions. After Keynote address, Technical Session-I (Theme: Manufacturing and Industrial Development) was conducted followed by oral presentation. In afternoon, another Keynote address was derived by Prof. (Dr) Rahul Goyal, Manipal University, Jaipur followed Technical session-II (Theme: Thermal Science and Automobile Engineering).

Day 2 (Tuesday, 30-11-2021):

Day 2 was started with Keynote address by Prof. (Dr.) Pradeep Kumar, IIT Roorkee on sustainability and Green Manufacturing, and then a invited talk by Dr. Sarvesh Mishra, Postdoc fellow, BCAM/UPV EHU-Spain on Lasers in advanced manufacturing followed by Technical session-III (Composite and Materials) followed by oral presentation. After lunch, another keynote address was delivered by Dr. Siddharth Bhardwaj, Postdoc fellow, IIT Gandhinagar, India followed by Technical session IV (Theme: Design, Ergonomics and Automation) followed by oral presentation. Finally, Valedictory function and certificate distribution was conducted.

List of editors

1. Dr. Mohd Mukhtar Alam
   Department of Mechanical Engineering
   Vivekananda Global University, Jaipur, Rajasthan, India, 302012
   mohd.mukhtaralam@vgu.ac.in
   Contact: +919602985309

2. Dr. Vijay Kumar Pandey
   Department of Mechanical Engineering
   Vivekananda Global University, Jaipur, Rajasthan, India, 302012
   vijay_pandey@vgu.ac.in
   Contact: +919828706888

IOP Conf. Series: Materials Science and Engineering 1224 (2022) 011001 doi:10.1088/1757-899X/1224/1/011001

List of committees

PATRONS

Dr. Lalit K.Panwar
Chairperson, VGU

Dr. K. R. Bagaria
Vice Chairperson, VGU

Er. Onkar Bagaria
CF & AO, VGU

Dr. K. Ram
Chief Patron, VGU

Dr. Vijay Vir Singh
President, VGU

Prof. (Dr.) Baldev Singh
Dean, Engineering, VGU

Organizing Secretary

Dr. Mohd Mukhtar Alam

Associate Professor, ME, VGU

Conveners

1. Dr Pramod Kumar

   Associate Professor, ME, VGU
2. Dr. Vijay Kumar Pandey

   Associate Professor, ME, VGU

Co-Conveners

1. Dr. Rupali Baghel

   Assistant Professor, ME, VGU
2. Mr. Satendra Singh

   Assistant Professor, ME, VGU

Organizing Committee

Dr. Meetu Nag	Asst. Prof. ECE, VGU
Dr. Ravikant Pareek	Asst. Prof. CE, VGU
Dr. Mala Mathur	HOD, First Year, VGU
Mr. Sita Ram Gupta	Asst. Prof. CSE, VGU
Mr. Parmeshwar Kumawat	Asst. Prof. ECE, VGU
Dr. Bhuvnesh Sharma	Asst. Prof. ME, VGU
Dr. Devendra Doda	Asst. Prof. ECE, VGU
Mr. Sanjay Sharma	Asst. Prof. ME, VGU
Mr. Surjeet Yadav	Asst. Prof. ME, VGU
Mr. Ganesh Kantak	Asst. Prof. ME, VGU
Mr. Vishal Sharma	Asst. Prof. ME, VGU
Mr. Sourav Kumar	Asst. Prof. ME, VGU

Mr. Mohit Jain	Asst. Prof. ECE, VGU
Mr. Sandeep Jain	Asst. Prof. CSE, VGU
Mr. Ravindra Kumar	Asst. Prof. CE, VGU
Mr. Dushhyant Singh	HOD, CSE, VGU
Dr. Mohammad Asif Iqbal	Asso. Prof. EE, VGU

Keynotes Speakers

1. Prof. (Dr.) Juan M. Corchado
   Director BISITE, University of Salamanca, Spain
2. Prof (Dr.) Pradeep Kumar
   Professor, Department of Mechanical Engineering
   Indian Institute of Technology Roorkee
3. Dr. Shiddharth Bhardwaj
   Post Doc Fellow, IIT Gandhi Nagar
4. Prof. (Dr.) Rahul Goyal
   Department of Mechanical Engineering
   Manipal University, Jaipur
5. Dr. Sarvesh Mishra
   Postdoc fellow, BCAM/UPV EHU-Spain

International Advisory Committee

1. Prof. (Dr.) Juan M. Corchado, University of Salamanca , Spain
2. Manjusri Misra, University of Guelph, Canada
3. Prof. Vikas Kumar, University of the West of England, Bristol, UK
4. Prof. Amit Bandyopadhyay, Washington State University, Pullman
5. Dr. Sanjeev Jakhar, Institute of Agriculture Engineering, Israel
6. Prof. Ratna Babu Chinnam, Wayne State University, USA
7. Prof. David Kruger, University of South Africa, South Africa
8. Prof. Orhan Korhan, Eastern Mediterranean University, Iran
9. Dr. Roberto Chavez, Swinburne University of Technology, Australia
10. Dr. Basharat Jamil, Universidad Rey Juan Carlos, Madrid, Spain
11. Dr. Abullellah M. Ali, Jazan University, Saudi Arabia
12. Dr. Bader A. Al-Ablani, Benha University, Cairo, Egypt
13. Dr. Marwa Mekky, Cairo University, Giza, Egypt

IOP Conf. Series: Materials Science and Engineering 1224 (2022) 011001 doi:10.1088/1757-899X/1224/1/011001

14. Dr. Noura A. Al-Ghimlas, Kuwait University, Kuwait
15. Dr. Masood Ashraf, Prince Sattam bin Abdulaziz University, Saudi Arabia

National Advisory Committee

1. Prof (Dr.) Pradeep Kumar, Indian Institute of Technology Roorkee,
2. Prof. Abid Ali Khan, Aligarh Muslim University, Aligarh, UP
3. Dr. Siddharth Bhardwaj, IIT Gandhi Nagar
4. Prof. Qasaim Murtuja, Delhi Technology University
5. Prof. G. S. Dangayach, MNIT, Jaipur
6. Dr. Jai Prakash Bhamu, Govt. Engineering College Bikaner
7. Dr. Dharmendra Singh, Govt. Engineering College Bikaner
8. Dr. Hari Lal Singh Mali, MNIT, Jaipur
9. Dr. Md Azhar, MMANTC
10. Prof. Kamal Kishor Khatri, LMNIT
11. Prof. Vasant Matsagar, IIT Delhi
12. Dr. Harpreet Singh Arora, Shiv Nadar University, Noida
13. Dr. Ashish Kumar Singh, VIT University
14. Dr. Sumit Gupta, Amity University, Noida
15. Prof. Prabir Mukhopadhyay, IIITDM, Jabalpur

IOP Conf. Series: Materials Science and Engineering 1224 (2022) 011002 doi:10.1088/1757-899X/1224/1/011002

Peer Review Statement

All papers published in this volume have been reviewed through processes administered by the Editors. Reviews were conducted by expert referees to the professional and scientific standards expected of a proceedings journal published by IOP Publishing.

- **Type of peer review:** Double Anonymous
- **Conference submission management system:** Morressier
- **Number of submissions received:** 45
- **Number of submissions sent for review:** 45
- **Number of submissions accepted:** 28
- **Acceptance Rate (Submissions Accepted / Submissions Received × 100):** 62.2
- **Average number of reviews per paper:** 2
- **Total number of reviewers involved:** 29
- **Contact person for queries:**
  Name: MOHD MUKHTAR ALAM
  Email: mohd.mukhtaralam@vgu.ac.in
  Affiliation: Vivekananda Global University

Performance Investigation of Wire EDM Process Parameters

Waddah Mohamedosman Elyass, Vishal Sharma, Ganesh Kantak, Mohd Mukhtar Alam

Department of Mechanical Engineering, Vivekananda Global University, Jaipur, Rajasthan, India

mohd.mukhtaralam@vgu.ac.in

Abstract. It is exceptionally difficult to create confusing geometry in such a material using traditional machining measurement. At present, companies use a non-regular machining strategy which uses energies like electricity, chemical, sound, light, etc. which are useful for machining hard materials. WEDM machines are used to cut complex and simple patterns and have the potential to achieve the maximum production capacity since the last few decades. This study was carried out in two parts. The first segment was consolidated with analyses of the Taguchi method depending on the signal-to-noise ratio (S / N) and regression modelling for the S / N ratio. In the second segment, the model condition was created using the RSM methodology. L16 orthogonal array was used with four levels of parameters namely, pulse on time, pulse off time, wire feed rate and peak current to find the effects on optimal response variables namely, cutting time, cutting rate, overcut and dimensional deviations. The results of the regression analysis showed that there was significant effect of pulse on/off and peak current on the cutting time and for cutting rate there was also significant effect of wire feed rated. Therefore, it can be concluded that all the four process parameters can affect the response variables.
Keywords: WEDM, Taguchi method, regression modelling, signal to noise ratio

1. Introduction

In today's production industry, processing measurement undertakes an important task, in which cost and quality are the main factors. Materials such as alloys, ceramics, and composite materials are created for clear purposes and have low machinability. It is very difficult to create confusing geometry in this material using traditional machining measurements. Conventional machining measurements use mechanical power to remove material, so expecting hard materials to be machined is absurd. Currently, the company uses an unconventional processing strategy that uses electricity, chemical, sound, light and other energy useful for processing hard materials.

Arikatla et al. [1] carried out an experiment on a titanium alloy (Ti-6Al-4V) and a brass wire as an electrode and selected five input parameters, i.e. pulse on time, pulse off time, servo voltage, input power and wire tension. They studied the influence of the input parameters on the MRR and the SR. They concluded that with the increase in the wire tension and the servo tension, the roughness of the surface decreases and improves the quality of the machined surface. Daneshmand et al. [2] used copper tools and deionized water to study the effect of EDM input parameters (pulse time, stop time, pulse current and voltage) on output parameters (MRR, surface roughness, relative electrode wear and tool wear rate) for the Ni-Ti alloy. The process took place in two stages. First, by increasing the pulse time and pulse current, which resulted in an increase in MRR until the part got closer to its final shape. Second, the workpiece has been machined with other tools to achieve a high quality surface finish.

In addition, Balasubarmanian and Ganapathy [3], studied to examine the effects of various factors, namely spark current, T_{on}, T_{off}, gap voltage, wire feed speed and wire voltage on the MRR and RS using gray regression analysis. They found that the optimal condition based on the method can give better surface quality. Puri and Deshpande [4] selected high chromium and high carbon matrix steel as the workpiece. They used Taguchi's method coupled with a fuzzy logic system for the optimization of multiple responses such as MRR and SR in the WEDM process. It has been observed that the approach has proven to be effective and efficient. Moreover, the result showed that the two parameters: MRR and SR were improved. In a study Alias et al. [5] investigated the effect of feed rate on wire cutting performance of titanium alloy Ti-Al-4V. Results of the study showed that the best combination of processing parameters for higher MRR and minimum SR is machine feed rate (4 mm/min), wire speed (8 m/min), wire tension (1.4kg) and voltage (60V). Shandilya et al. [6] studied the four input process parameters of WEDM, namely servo voltage (V), pulse on time, pulse off time and wire feed speed (WF)) to study the cutting width (kerf) and surface deal with. The results show that the pulse on time has no significant effect on the cutting width.

Wire EDM machine has been used to cut delicate and hard modern materials into any shape with greater precision in the end result, but the ideal activity of this machine was not accomplished at this stage, so in different parts of the world, different scientists are trying to improve more and accurate working capacity. In the present investigation, this machine was chosen to explore the measurement parameters of the WEDM machine. In the process, past distributed exploration work was reviewed to uncover the research gaps for the study.

2. Method

2.1. Design of Experiments

This study was carried out in two parts. The first segment was consolidated with analyzes of the Taguchi method depending on the signal-to-noise ratio (S / N) and regression modeling for the S / N ratio. In the second segment, the model condition was created using the RSM methodology. L16 orthogonal array was used with four levels of parameters namely, pulse on time, pulse off time, wire feed rate and peak current (Table 1 and Table 2) to find the effects on optimal response variables namely, cutting time, cutting rate, overcut and dimensional deviations. The trials were performed in a local industry in Jaipur, Rajasthan, India. In the present work, pilot tests were also performed to recognize the impact of cycle parameters on machine reactions. The analysis were carried out on a wire EDM machine (Maxicut-e WEDM) from Electronica Machine Tools Ltd.

Table 1 Process Parameters and their levels for Taguchi Method

Factor	**Parameter Name**	**Levels**			
		A	**B**	**C**	**D**
Ton	Pulse on time	4	6	8	**10**
Toff	Pulse off time	4	6	8	**10**
Feed	Wire Feed rate	3	5	7	**9**
Current	**Peak Current**	**2**	**4**	**6**	**8**

2.2. Work Piece Selection

Modern 75mm x 75mm x 0.75mm steel sheet (Figure 1) was used as the part material for the present analyses. This modern industrial steel is normally hot worked industrial steel with moderate hardness and strength properties. It is used for medium load conditions. The service life and dimensional accuracy of this steel and instruments can be improved with proper heat treatment.

Table 2 L_{16} Orthogonal array design of experiment

S. No.	T_{on}	T_{off}	Feed(mm/min)	Current
1	4	4	3	2
2	4	6	5	4
3	4	8	7	6
4	4	10	9	8
5	6	4	5	6
6	6	6	3	8
7	6	8	9	2
8	6	10	7	4
9	8	4	7	8
10	8	6	9	6
11	8	8	3	4
12	8	10	5	2
13	10	4	9	4
14	10	6	7	2
15	10	8	5	8
16	10	10	3	6

Figure 1 Design of cut for present experimental work

3. Data analysis

The L16 orthogonal array was used with four parameter levels, namely pulse on time, pulse stop time, wire feed rate and peak current, as shown in table 1. The parameters of the processes and their levels for the Taguchi method were shown in Table 2. Table 3 showed the experimental results of cutting time and cutting speed. Each experiment was performed three times to find differences between the selected responses. In addition, the experimental results of dimensional deviation and kerf are shown in Table 4. Table 5 showed the results for the cut time (Ct) and the cut speed (Cr). Table 6 shows the summary of the results for overcuts and dimensional deviations. Table 7 shows the rank of each factor for each response variable.

Table 3 Experimental results of cutting time and cutting rate

Trial	Cutting Time (sec)			S/N Ratio	Cutting Rate (mm/sec)			S/N Ratio
	T1	T2	T3		T1	T2	T3	
1	151.243	151.692	151.700	-43.61	0.375	0.376	0.383	**-8.45**
2	151.684	150.537	151.715	-43.59	0.380	0.375	0.385	**-8.40**
3	151.246	151.181	151.614	-43.59	0.379	0.381	0.385	**-8.36**
4	150.439	150.239	151.605	-43.56	0.380	0.380	0.389	**-8.33**
5	151.127	151.658	151.493	-43.60	0.376	0.375	0.377	**-8.49**
6	150.066	150.199	152.171	-43.56	0.372	0.370	0.383	**-8.52**
7	151.597	151.649	151.989	-43.62	0.385	0.384	0.391	**-8.25**
8	151.246	151.381	151.915	-43.60	0.384	0.381	0.390	**-8.29**
9	151.251	150.973	152.702	-43.61	0.373	0.372	0.385	**-8.48**
10	151.536	151.649	151.876	-43.61	0.379	0.376	0.385	**-8.40**
11	151.843	151.247	151.038	-43.60	0.381	0.382	0.389	**-8.31**
12	151.368	151.612	152.087	-43.61	0.381	0.382	0.395	**-8.27**
13	152.191	152.547	152.306	-43.65	0.376	0.378	0.386	**-8.40**
14	151.816	152.286	153.100	-43.65	0.383	0.382	0.385	**-8.32**
15	151.117	150.721	152.440	-43.60	0.376	0.378	0.381	**-8.44**
16	**150.317**	**151.040**	**152.384**	**-43.59**	**0.378**	**0.379**	**0.388**	**-8.36**

Table 4 Experimental results of dimensional deviation and kerf

Trial	Dimensional Deviation (DD)			S/NRatio	Overcut (OC)			S/N Ratio
	T1	T2	T3		T1	T2	T3	
1	0.933	0.446	0.657	3.00609	0.0028	0.0041	0.0023	**49.9807**
2	0.576	0.663	0.538	4.51411	0.0091	0.0091	0.0061	**41.6648**
3	1.009	1.032	0.837	0.32449	0.0089	0.0011	0.0036	**45.0833**
4	0.521	0.828	0.639	3.41806	0.0071	0.0071	0.0051	**43.7134**
5	0.482	0.133	0.777	5.45202	0.0100	0.0048	0.0070	**42.4114**
6	0.822	0.145	0.118	6.25235	0.0014	0.0044	0.0031	**49.8574**
7	0.959	0.823	0.845	1.13241	0.0059	0.0044	0.0054	**45.5716**
8	0.915	0.585	0.128	3.99314	0.0073	0.0038	0.0026	**46.0366**
9	0.280	0.554	1.045	3.07614	0.0063	0.0065	0.0062	**43.9789**
10	0.362	0.126	0.837	5.48692	0.0085	0.0023	0.0073	**43.5872**
11	0.803	0.753	0.437	3.30098	0.0020	0.0094	0.0063	**43.5340**
12	0.643	0.863	0.813	2.17139	0.0045	0.0016	0.0080	**45.4000**
13	0.655	0.691	0.957	2.16535	0.0033	0.0037	0.0037	**48.9702**
14	0.834	0.905	0.810	1.40396	0.0029	0.0044	0.0065	**46.2832**
15	0.826	0.763	0.684	2.38347	0.0059	0.0054	0.0048	**45.3474**
16	**0.939**	**0.767**	**0.731**	**1.75309**	**0.0060**	**0.0059**	**0.0027**	**45.8164**

4. Results and discussion

In the current experimental work, the effect of process parameters on the response parameters of the wire electric discharge machining process was investigated. A total of 4 process parameters were selected, each with 4 levels to investigate the effects on four response variables (CT, CR, OC, and DD). The results showed the optimal solution (A1B4C1D4) for the cutting time with predicted optimal value of 150.398 seconds; however the experiment value for the cutting time was 151.517 seconds. In addition, the optimal solution (A3B4C4D1) for cutting rate with the predicted optimal value was 0.389 mm/sec; however, the experiment value for the cutting rate was 0.381. Also the results showed decrease in over cut rate with the optimal solution (A4B1C3D1) with the predicted optimal value of 0.0035 as compared with the experimental result (0.0049). There was also decrease in dimensional deviation using the optimal solution (A2B2C1D4) with the predicted optimal value of 0.322 as compared to the experimental result (0.677).

Table 5 Summary of the results for cutting time (Ct) and cutting rate (Cr)

Run Order	Pt Type	Blocks	T_{on}	T_{off}	Feed	Current	Cutting Time	Cutting Rate
1	0	1	7	7	6	5	151.828	**0.326**
2	2	1	10	4	6	5	153.378	**0.337**
3	0	1	7	7	6	5	152.352	**0.328**
4	2	1	7	4	6	8	151.575	**0.344**
5	2	1	7	10	9	5	151.722	**0.342**
6	2	1	10	10	6	5	153.111	**0.347**
7	2	1	10	7	3	5	152.975	**0.323**
8	2	1	4	7	3	5	149.946	**0.361**
9	2	1	10	7	6	2	154.883	**0.379**
10	2	1	10	7	9	5	153.762	**0.342**
11	2	1	7	7	9	8	151.380	**0.356**
12	2	1	4	7	9	5	150.868	**0.342**
13	2	1	7	4	9	5	152.320	**0.324**
14	0	1	7	7	6	5	151.941	**0.344**
15	2	1	7	4	6	2	153.206	**0.365**
16	2	1	4	7	6	2	151.381	**0.349**
17	2	1	4	10	6	5	150.680	**0.319**
18	2	1	10	7	6	8	152.950	**0.342**
19	2	1	7	10	3	5	151.704	**0.312**
20	2	1	7	7	9	2	152.991	**0.342**
21	2	1	4	7	6	8	150.155	**0.364**
22	2	1	7	10	6	8	150.547	**0.354**
23	2	1	4	4	6	5	150.987	**0.360**
24	2	1	7	7	3	8	151.363	**0.353**
25	2	1	7	7	3	2	153.071	**0.340**
26	2	1	7	4	3	5	151.574	**0.330**
27	**2**	**1**	**7**	**10**	**6**	**2**	**153.060**	**0.358**

In addition, the model equations were generated for each response variable using RSM technique. Therefore, the regression coefficient (R2 =95.21%) and the equation for response cutting time (sec) was

$CT = 150.885 + 0.4734\ T_{on} - 0.0616\ T_{off} + 0.0670\ Feed - 0.595\ Current + 0.0300\ Current*Current$

In addition, the regression coefficient (R2 =98.88%) and the equation for response cutting rate (mm/sec) was

$CR = 0.32766 + 0.009460\ Ton - 0.00359\ T_{off} - 0.00147\ Feed - 0.000770\ Current + 0.000317\ T_{off}*T_{off} + 0.000188\ Feed*Feed$

Table 6 Summary of the results for over cut and dimensional deviations

Run Order	Pt Type	Blocks	T_{on}	T_{off}	Feed	Current	OC	DD
1	0	1	7	7	6	5	0.0058	**0.6302**
2	2	1	10	4	6	5	0.0065	**0.6173**
3	0	1	7	7	6	5	0.0061	**0.5668**
4	2	1	7	4	6	8	0.0057	**0.5220**
5	2	1	7	10	9	5	0.0063	**0.9046**
6	2	1	10	10	6	5	0.0075	**0.9575**
7	2	1	10	7	3	5	0.0075	**0.6206**
8	2	1	4	7	3	5	0.0050	**0.5133**
9	2	1	10	7	6	2	0.0063	**0.9056**
10	2	1	10	7	9	5	0.0064	**0.8064**
11	2	1	7	7	9	8	0.0063	**0.7332**
12	2	1	4	7	9	5	0.0046	**0.5842**
13	2	1	7	4	9	5	0.0050	**0.6672**
14	0	1	7	7	6	5	0.0060	**0.5998**
15	2	1	7	4	6	2	0.0045	**0.6506**
16	2	1	4	7	6	2	0.0041	**0.5216**
17	2	1	4	10	6	5	0.0054	**0.5953**
18	2	1	10	7	6	8	0.0073	**0.7573**
19	2	1	7	10	3	5	0.0066	**0.5752**
20	2	1	7	7	9	2	0.0053	**0.8110**
21	2	1	4	7	6	8	0.0053	**0.3915**
22	2	1	7	10	6	8	0.0068	**0.6696**
23	2	1	4	4	6	5	0.0040	**0.4865**
24	2	1	7	7	3	8	0.0067	**0.4386**
25	2	1	7	7	3	2	0.0059	**0.7299**
26	2	1	7	4	3	5	0.0058	**0.4511**
27	**2**	**1**	**7**	**10**	**6**	**2**	**0.0059**	**0.7974**

Table 7 Using S/N ratio analysis, rank of each factor for each response variable

Response	Ton	Toff	Feed	Current
CT	2	4	3	**1**
CR	4	1	3	**2**
OC	2	3	4	**1**
DD	**1**	**4**	**3**	**2**

Further, the regression coefficient (R2 =98.31%) and the equation for response overcut (mm) was

$$OC = -0.000641 + 0.000686\ T_{on} + 0.000397\ T_{off} + 0.000031\ Feed + 0.000296\ Current - 0.000015\ T_{on}*T_{on} - 0.000015\ T_{off}*T_{off} - 0.000012\ Current*Current - 0.000019\ T_{on}*Feed$$

However, the regression coefficient (R2 =89.58%) and the equation for response dimensional deviation (mm) was

$$DD = 0.550 - 0.0013\ T_{on} - 0.0143\ T_{off} + 0.0031\ Feed - 0.0607\ Current + 0.00643\ T_{on}*T_{off} + 0.00593\ Feed*Current$$

Therefore, the results of the regression analysis showed that there was significant effect of pulse on/off and peak current on the cutting time and for cutting rated wire feed rate there was also significant effect of wire feed rated. In addition, there was significant effect of all four process parameter on the overcut and the dimensional deviation. Moreover, using S/N ratio analysis, rank of each factor for each response variable were also presented in Table 8 in which pulse on time was ranked 1 for the dimensional deviation, pulse off time for the cutting rate and peak current for the cutting time and overcut. However, Shandilya et al. [6] studied the four input process parameters of WEDM, namely servo voltage (V), pulse on time, pulse off time and wire feed speed (WF)) to study the cutting width (kerf) and surface finish. The results show that the pulse on time has no significant effect on the cutting width. Therefore, it can be concluded that all the four process parameters can affect the response variables.

Conflict of Interest
None to report.

References

[1] S. P. Arikatla, K. T. Mannan and K. Arkanti , "Parametric Optimization in Wire Electrical Discharge Machining of Titanium Alloy Using Response Surface Methodology," in *5th International Conference of Materials Processing and Characterization*, 2017.

[2] D. Saeed, F. K. Ehsan, A. Esmail and M. M. Abdolhosseini, "Influence of Machining Parameters on Electro Discharge Machining of NiTi Shape Memory Alloys," *International Journal of electrochemical science,* vol. 8, pp. 3095-3104, 2013.

[3] S. Balasubramanian and S. Ganapathy, "Grey Relational Analysis to determine optimum Process parameters for Wire Electro Discharge Machining," *International Journal of Engineering Science and Technology,* vol. 3, no. 1, pp. 95-101, 2011.

[4] Y. M. Puri and N. V. Deshpande, "Simultaneous optimization of multiple quality characteristics of WEDM based on fuzzy logic and Taguchi technique," in *Fifth Asia Pacific Industrial Engineering and Management Systems Conference*, 2013.

[5] A. Alias, B. Abdullah and N. M. Abbas, "WEDM: Influence of machine feed rate In machining Titanium Ti-6al-4v using brass and constant current," *Procedia Engineering* , vol. 41, pp. 1812-1817, 2012.

[6] P. Shandilya, P. K. Jain and N. K. Jain, "Parametric optimization during wire electrical discharge machining using response surface methodology," *Procedia Engineering,* vol. 38, pp. 2371-2377, 2012.

Experimental investigation of Marble powder mixed micro-electric discharge machining

Rupali Baghel

Department of Mechanical Engineering, Vivekananda Global University, Jaipur

rupali.baghel@vgu.ac.in

Abstract. Micro-electrical discharge machining (EDM) is among the most important machining techniques for creating micro features, notably for cutting challenging materials such as superalloys, ceramics, and composite materials. To improve machining performance characteristics, micro-EDM technology was integrated with the mixing of marble powder in the dielectric fluid in this research. This work investigates the potential of marble powder mixed micro-EDM (PM- EDM) to manufacture high aspect ratio micro holes in a newly designed ceramic, TiN-Al2O3. SEM photos of the action of marble powder on the machined surface are also examined, with thermal cracking (Spalling) and melting being the most prominent mechanisms of material removal.

Keywords: EDM, marble powder, Ceramics, MRR, EWR.

1. Introduction

Advanced manufacturing technologies are mostly utilized to create miniaturized goods with dimensions ranging from 1 to 999 micrometers. Sensors turbine blades, a printed circuit board, injection nozzles, and other miniature goods have a wide range of applications in the automobile, aircraft, optical, and biomedical industries [1]. To resist wear and promote heat transmission, microdevices such as micro heat exchangers and microreactors with micro features such as micro holes, micro-channels, and fins are required [2]. Miniaturized advanced ceramics goods have low weight and great mechanical strength, which are required for certain applications in the current environment.

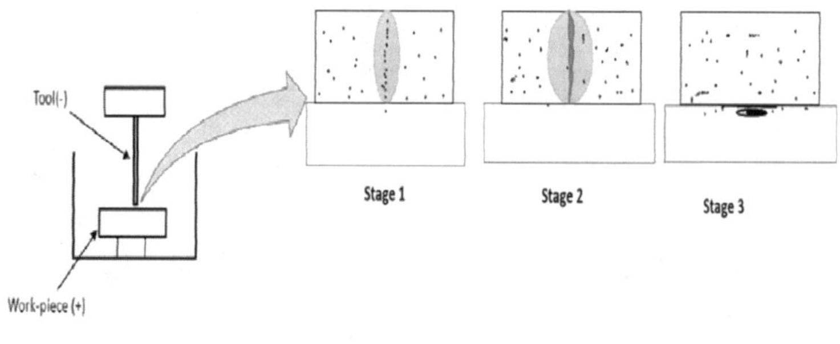

Fig. 1 Schematic representation of PM-μEDM

Ceramic matrix composites (CMCs) are appropriate for modern material requirements. CMCs have physiological qualities that are superior to metals, particularly high toughness, hot hardness, low density to weight ratio, and high stiffness [3]. In MMCs and CMCs, Micro-electro-discharge machining (-EDM) offers the distinct benefit of cutting complicated micro-features [4]. Although -EDM has drawbacks such as high energy consumption, low material removal rate (MRR), and high tool wear when cutting micro-features with high aspect ratios, there is always the issue of debris flushing [5]. By hybridizing the machining process, the limitations of innovative machining operations could be addressed. In a hybrid-EDM, two or more processes work together to achieve improved response parameters. [6] [7]. Baghel et al. [8] used vibration-aided -EDM to drill a 200 m diameter micro hole in Titanium Nitride Aluminium Oxide and found that MRR was enhanced by 1.98 times under the effect of low-frequency vibration. They also used diamond grinding aided EDM to machine Titanium Nitride Aluminium Oxide and could see a considerable increase in MRR [9]. Powder mixed EDM is a technique proposed by several researchers. By adding impurities into the dielectric fluid, the properties of the dielectric fluid are modified. Fig. 1 is showing the process in three stages, as machinability in EDM is highly dependent on dielectric properties [10]. When a potential difference is applied to the electrodes in powder mixed EDM, the particles polarise and align themselves as a bridge across the electrodes in the first stage, resulting in effortless dielectric fluid breakdown and persistent sparking. The frequency of discharge is increased by the presence of microparticles. Researchers incorporate a range of powders to improve EDM performance, including aluminum, chromium, copper, and others. The proposed paper is based on marble powder mixed electro-discharge machining of Titanium Nitride-Aluminum Oxide (TiN-Al_2O_3), an advanced ceramic-matrix composite. The foreign particle reduces the dielectric strength of the fluid, resulting in a persistent spark.

2. Experimentations

Titanium Nitride-Aluminium Oxide (TiN-Al_2O_3) is a newly discovered CMC with better mechanical qualities to metal, especially high toughness and chemical and mechanical wear resistance. High-temperature reaction sintering has been used to make this CMC. Traditional machining techniques have had trouble fabricating micro-features on these ceramics because they are exceptionally brittle and have a high hardness. In this study, an attempt is made to use PM-EDM to create micro-holes with a high aspect ratio, i.e. 12. All tests were conducted out using a Hybrid-EDM (DT-110i) machine. These experiments are designed using a Taguchi L16 orthogonal array. The parameters are determined by a review of the literature and the machine's capabilities. Table 1 shows the range of process parameters used in the trials. The response parameters are materials removal rate (MRR in mm^3/min) and electrode wear rate (EWR in mm^3/min). The MRR and EWR are calculated by dividing the volume loss by the machining time.

Table 1 Input parameters for PM-μEDM

Sn	Input parameter	Level of input parameters			
1	Speed (rpm)	1000	1200	1400	1600
2	Voltage (Volt)	85	100	115	130
3	Marble-Powder Concentration (g/l)	0	2	4	6

3. Results and Discussion

The Taguchi L16 orthogonal array design has been used in all of the experiments, with three control factors at four levels and two response parameters. For each response parameter in Table 2, the signal to noise (S/N) ratio is obtained. If the output parameter has to be maximized, the (S/N) ratio can be calculated by equation 1. For MRR (S/N) ratio follow the " Larger the Better" model.

$$\left(\frac{S}{N}\right) \text{ratio} = \log_{10}\frac{\sum_{i=1}^{n}\frac{1}{Y^2}}{n} \quad \text{.......................} (1)$$

n= number of replications

"Electrode wear" is the terminology for this type of damage. The change in the volume of the electrode in each machine time is used to determine the electrode wear rate (EWR). The value of electrode wear should be as low as possible, and if the value of the response parameter must be lowered, the signal-to-noise ratio can be determined using equation 2 and the principle "smaller is better."

$$\left(\frac{S}{N}\right) ratio = \log_{10}\frac{\sum_{i=1}^{n} Yi^2}{n} \quad \text{...............} (2)$$

n=number of replications

Each Response parameter is tested to an analysis of variance (ANOVA) test. Table 2 provides the experimental observations. Figure 2 depicts the main effect curves for MRR and EWR.

In Fig. 2(a), the MRR plot shows a high concentration of marble powder, with MRR being the largest. The high particle concentration represents a low discharge gap and high conductivity, resulting in a steady spark with a high MRR. EWR, on the other hand, is largest at high marble powder proportions, as illustrated in Fig. 2. (b). As mentioned earlier, a high concentration of marble powder causes persistent sparking, which causes the temperature of the tool-tip to rise.

The high temperature of the tooltip accelerates tool degradation, resulting in a larger EWR. It also shows that speed is the most major determinant, followed by voltage and focus. Consistent sparking affects the geometry and size of the electrode, resulting in increased EWR. Table 3 shows the ranking of several factors.

Table 2 Experimental design matrix with a set of process parameters and corresponding response parameters

speed	voltage	Conc.	MRR	EWR	S/N Ratio (MRR)	S/N Ratio (EWR)
RPM	Volt	g/l	mm^3/min	mm^3/min		
1000	85	0	0.022847	0.028730	-32.8234	32.5149
1000	100	2	0.024870	0.027540	-32.0865	30.8333
1000	115	4	0.026730	0.028750	-31.4600	31.2007
1000	130	6	0.034870	0.026780	-29.1510	30.8272
1200	85	2	0.027467	0.024856	-31.2238	31.4438
1200	100	0	0.030128	0.022145	-30.4206	32.0914
1200	115	6	0.036740	0.025735	-28.6972	33.0945
1200	130	4	0.027462	0.023780	-31.2254	31.7895
1400	85	4	0.026439	0.025483	-31.5551	32.4758
1400	100	6	0.033749	0.035820	-29.4348	31.8750
1400	115	0	0.025387	0.035629	-31.9078	28.9175
1400	130	2	0.028946	0.025473	-30.768 2	28.9639
1600	85	6	0.034779	0.024582	-29.1737	31.8784
1600	100	4	0.027888	0.039760	-31.0917	32.1877
1600	115	2	0.028638	0.034500	-30.8611	28.0111
1600	130	0	0.027894	0.028730	-31.0898	29.2436

Table 3 Response table for the signal to noise ratios

Response table for the signal to noise ratios							
MRR- Larger is better				EWR- Smaller is better			
Level	speed	voltage	Conc.	Level	Speed	Voltage	Conc.
1	-31.38	-31.19	-31.56	1	31.34	32.08	25.69
2	-30.39	-30.76	-31.23	2	32.10	31.75	24.81
3	-30.92	-30.73	-31.33	3	20.56	25.21	31.91
4	-30.55	-30.56	-29.11	4	30	25.21	31.92
Delta	0.99	0.64	2.45	Delta	11.55	6.87	7.11
Rank	2	3	1	Rank	1	2	3

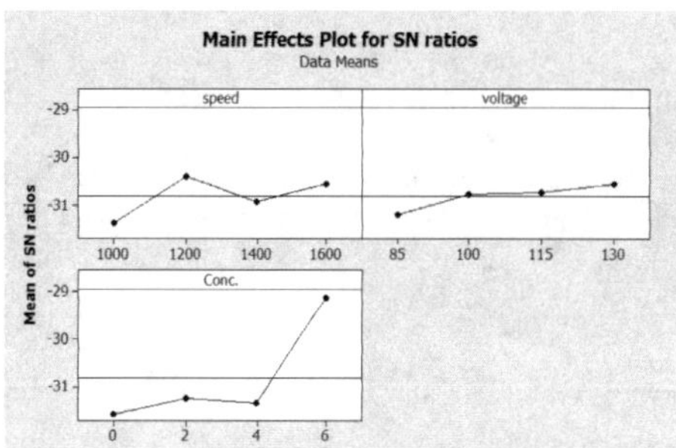

(a) For MRR

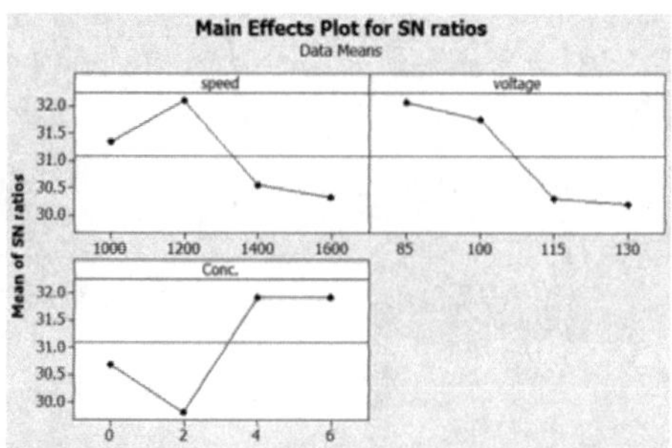

(b) For EWR

Fig. 2 Main effects plot SN ratio

Table 4 ANOVA table for Material removal rate

Analysis of Variance for MRR						
Source	DOF	Seq SS	Adj SS	Adj MS	F	P
Speed	3	0.0000226	0.0000226	0.0000075	2.98	0.118
Voltage	3	0.0000081	0.0000081	0.0000027	1.07	0.430
Conc.	3	0.0001926	0.0001926	0.0000642	25.34	0.001
Error	6	0.0000152	0.0000152	0.0000025		
Total	15	0.0002385				
S = 0.00159163		R-Sq = 93.63%	R-Sq(adj) = 84.07%			

Table 5 ANOVA table for Electrode wear rate

Analysis of Variance for EWR						
Source	DOF	Seq SS	Adj SS	Adj MS	F	P
speed	3	0.0000969	0.0000969	0.0000323	7.70	0.018
voltage	3	0.0001373	0.0001373	0.0000458	10.91	0.008
Conc.	3	0.0001519	0.0001519	0.0000506	12.07	0.006
Error	6	0.0000252	0.0000252	0.0000042		
Total	15	0.0004112				
S = 0.00204816		R-Sq = 93.88%	R-Sq(adj) = 84.70%			

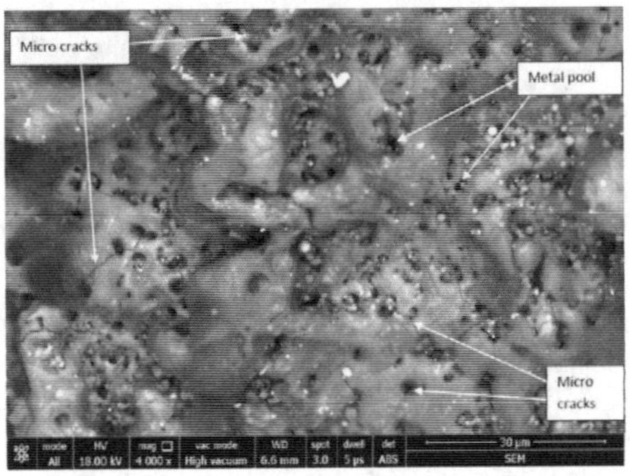

Fig. 3 SEM images of the machined surface by PM-μEDM

Tables 4 and 5 show the findings of the ANOVA. According to the ANOVA results, marble powder concentration has a significant effect on MRR, and all three parameters have a major effect on EWR. The model's goodness of fit is shown by the determination coefficient (R2). MRR has a determination coefficient of 0.9363 and EWR has a determination coefficient of 0.9388, indicating that the model only explains around 7 % of total variance for both MRR and EWR.

The adjusted determination coefficient (adjusted R2=0.8407 for MRR, adjusted R2=0.8470) is also high and close to the determination coefficient, suggesting that these models are highly significant. Scanned electrode microscopy (SEM) can be used to analyze the interior surface of micro-holes, as shown in Fig. 3. On the machined surface, there are pools (with debris) of melting metals. As the melted metal in it shrunk, micro-cracks appeared.

It has cracks and debris, indicating that it is melting and vaporizing in a machined zone. Thermal cracking and flakes removal, sometimes known as thermal spalling, are used to remove materials. The mechanisms by which materials are removed are melting, vaporization, and thermal shock failure, i.e. significant internal stresses induced by a fast temperature gradient causing the material to flake and shatter in advanced ceramics like TiN-Al_2O_3.

4. Conclusions

On a TiN-Al_2O_3 ceramic-composite workpiece, a powder mixed micro-electro-discharge machining method is applied to produce micro-holes. The following conclusions have been drawn from the analysis of the data:

1. The percentage of marble particles is the most affecting parameter for MRR during PM-EDM machining of TiN-Al_2O_3, followed by the applied voltage and electrode rotation speed.
2. The electrode wear rate has mostly been influenced by the powder percentage and the rotation speed of the electrode.
3. PMEDM has now been confirmed to be the recommended way for machining TiN-Al_2O_3 ceramic-composite because the material removal rate is much higher than EDM since the presence of micro pow der increases the frequency of discharge, resulting in continuous sparking.
4. Melting for metals, such as TiN, and thermal spalling for ceramics, such as Al_2O_3, are the mechanisms for material removal in composites.
5. PM-EDM is an effective process for machining difficult-to-cut materials such as TiN-Al_2O_3 ceramic-composite, based on the above-mentioned research observations.

REFERENCES

1. M. Hourmand, A. A.D. Sarhan, and N. Mohd Yusof, "Development of new fabrication and measurement techniques of micro-electrodes with high aspect ratio for micro EDM using typical EDM machine," Measurement, vol. 97, pp. 64-78, 2017.
2. S. Ravisubramanian and M. S. Shunmugam, "Investigations into peck drilling process for large aspect ratio micro-holes in aluminum 6061-T6," Materials and Manufacturing Processes, vol. 33(9), pp. 935-942, 2018.
3. R. Baghel, H. S. Mali and S. K. Biswas, "Parametric optimization and surface analysis of diamond grinding-assisted EDM of TiN-Al_2O_3 ceramic composite," The International Journal of Advanced Manufacturing Technology, vol. 1, pp. 1-10, 2018.
4. R. Kumar, A. Singh, and I. Singh, "Electric discharge hole grinding in hybrid metal matrix composite," Materials and Manufacturing Processes, vol. 32(7), pp. 127-134, 2017.
5. M. S. Rayat, S. S. Gill, R. Singh, and L. Sharma, "Fabrication and machining of ceramic composites-A review on the current scenario," Materials and manufacturing process, vol. 1, pp. 1-24, 2017.
6. R. Baghel and H. S. Mali, "A study on effects of discharge energy on geometric characteristics of high aspect ratio micro-holes on TiN-Al_2O_3 ceramics," in 8th International Conference on Materials Processing and Characterization (ICMPC), India, 2018.
7. R. Baghel and H. S. Mali, "An Experimental Study on Fabrication of Micro Channels in Titanium Nitride Alumina Composite Using Electro-discharge Milling," International Journal of Modern Manufacturing Technologies, vol. 10(2), pp. 24-29, 2018.
8. R. Baghel, H.S. Mali, & S.K. Biswas; "Micro hole fabrication in TiN-Al2O3 ceramic composite by SiC powder assisted micro- EDM", Engineering Research Express 2 (2020), 015028, https://doi.org/10.1088/2631-8695/ab6ce
9. R. Baghel, H. S. Mali and S. K. Biswas, "Parameter Optimization of Diamond Grinding Assisted EDM of TiN-Al2O3 Ceramics using Taguchi Method," in 6th International & 27th All India Manufacturing Technology, Design, and Research Conference, Coep, Pune, 2016.
10. G. S. Prihandana, M. Mahardika, M. Hamdi, Y. S. Wong and K. Mitsui, "Effect of micro-powder suspension and ultrasonic vibration of dielectric fluid in micro-EDM processes-Taguchi approach," International Journal of Machine Tools and Manufacture, vol. 49(12-13), pp. 1035-1041, 2009.

IOP Conf. Series: Materials Science and Engineering 1224 (2022) 012003 doi:10.1088/1757-899X/1224/1/012003

Machinability of Non-ConductiveCeramic by EDM: A Review

Rupali Baghel

Department of Mechanical Engineering, Vivekananda Global University Jaipur

rupali.baghel@vgu.ac.in

Abstract.
AdvancedCeramics are gaining a foothold in the lightweight aerospace, electronics, and structural engineering component markets. These ceramics could be extensively used in modern industry, such as ballistic body armour, ceramic carbon fibre composite automoti ve brakes, diesel particulate filters, prosthetic limbs, piezoelectric sensors, and computer memory products, due to their higher compressive strength, resistance to abrasion, lower thermal expansion coefficient, higher density, and chemical stability. Ceramics are notori ously difficult to handle due to the increased hardness and brittleness. Low electric-conductive ceramics, on the other hand, can be machined using the EDM technique, in which plasma energy is used to accurately remove the material by continuous sparking between the surface and the electrode submerged in dielectric. It is observed that EDM can be applied to the material having electrical resistivity below 100 Ω.cm. Most recently it has been observed that EDM could be applied to insulating ceramics too. An attempt has been made in this paper to critically review the machining of ceramics by the EDM process.

1. Introduction

Non-conductive ceramics, because of their exceptional mechanical, thermal, and physicochemical characteristics are likely to encounter a wide spectrum of uses in a variety of industries. They are likely to find a wide range of applications in a variety of industries. Aluminum oxide (Al_2O_3) and zirconium oxide (ZrO_2) are the most commonly used ceramics in machinery for purposes like grinding wheels grinding wheel, bearing, gas sensor, automobile part and medical tools. The biggest issue with ceramics is that they are very onerous to machine. The most prevalent method for machining ceramics is diamond grinding [1, 2], although this method is somewhat expensive and inefficient. Ceramics' greater hardness causes a higher grinding force, which causes diamond cutting edges to wear out quickly. Ceramics' strength is degraded during diamond grinding because of surface and subsurface fractures [3]. They are difficult to operate due to their machining characteristics, which include high cutting force, elevated heat, poor surface integrity, and poor tool life. In recent years, Electrical Discharge Machining (EDM) of advanced engineering ceramics has gotten a lot of attention. EDM provides for the flexible and precise machining of complicated shaped harden-conductive objects. Electrode discharge machining has the potential to be a cost-effective and efficient method for producing tools and parts from conducting ceramic blanks [4]. In 1943, B. R. Lazarenko and N. I. Lazarenko, two Russian scientists, were charged with figuring out how to keep tungsten electrical contacts from corroding due to sparking. They failed at this goal, but discovered that if the electrodes were submerged in a dielectric fluid, the erosion could be more precisely regulated. This led to the development of the EDM system, which could be used to process materials that are hard to machine, such as tungsten, super alloys, and composites [5].

The Lazarenko's machine is known as an R-C- type machine. It was also used to finish surgeries equipment, as well as aeronautical and automotive. To remove material from a component of this phase,

In the presence of a dielectric fluid, a series of successive electrical discharges between an elect rode and the workpiece are utilized. The electrodes are brought closer to the workpiece until there is a tiny enough distance between them for sparking to occur and an impression to be created. The electrode is placed closer to the work-piece until the distance between them is small enough for sparking to be happened & the impression to be created. The material is removed by the erosive effect of the electrical discharges from the tool-electrode and work-piece. During machining, EDM reduces mechanical tension chatter and vibration problems by avoiding direct interaction between the electrode and the work-piece. Any material can be cut regardless of its hardness as long as it can conduct electricity [6, 7]. Advanced ceramics may be machined by EDM if their electrical conductivity reachesa certain threshold value of the order of 0.01 S/cm (or resistivity less than 100 cm) [8, 9]. When machining advanced ceramics with EDM, the thermal spalling mechanismis used to extract the material [3]. EDM is proving to be a promising and appealing technology for machining ceramics, as long as the materials have a high adequate electrical conductivity. On improve processing parameters, an electrically conductive film must be added to non-conductive ceramics having Alumina, Silicon-Nitride, or Zirconia matrix [10].

2. EDM of Non-Conductive Ceramics

Metallic and non-metallic constituents make up non-conductive ceramics. Ceramics are significantly stronger than metals because of these types of bonds. Non-conductive ceramics have been used as electrical insulators and high-temperature resistance in vehicle spark plugs for many years. Ceramics are now widely regarded as the ideal material for diverse angles and perspectives manufacture of home, industrial, and building products, as well as aesthetic artefacts. Machining tools, self-sustainable bearings, engine components, diesel engines, heat exchangers, bulletproof shield, automotive brakes, diesel particulate filters, and a range of prosthetic implants are just a few categories [11]. Engineering ceramic micro-parts are applied in the biomedical industry to make femoral heads and ace-tabular cups for total hip replacement, dentures and remodels, bone fillers, and tissue engineering scaffolds [12].

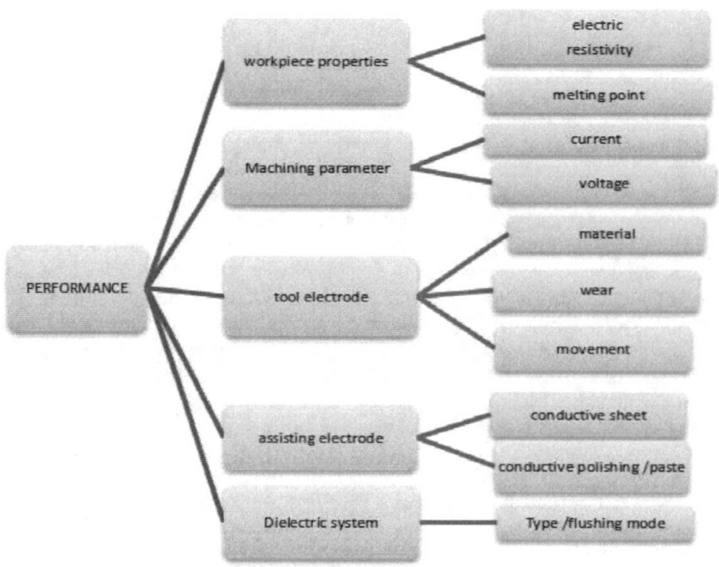

Fig. 1 A line diagram of the EDM process variables for ceramics

Based on the input parameters provided in Fig. 1, the machining of advanced-ceramics has been analyzed. Surface roughness number and material removal rate are used to demonstrate the EDM process' performance. The performance of conductive and non-conductive ceramics depends mainly on machining parameters like current, voltage, work-piece property like electric resistivity conductivity, assisting electrode properties, etc. These parameters are found to be affecting the performance of the process in various degrees. Electrical conductive workpiece is required for machining by EDM, researchers used a conductive layer between the workpiece and electrode as an assisting electrode. The pictorial view of the arrangement of assisting electrode is given in Fig. 2. Here a conductive material like copper or copper alloy is made to pass between workpiece and the tool electrode.

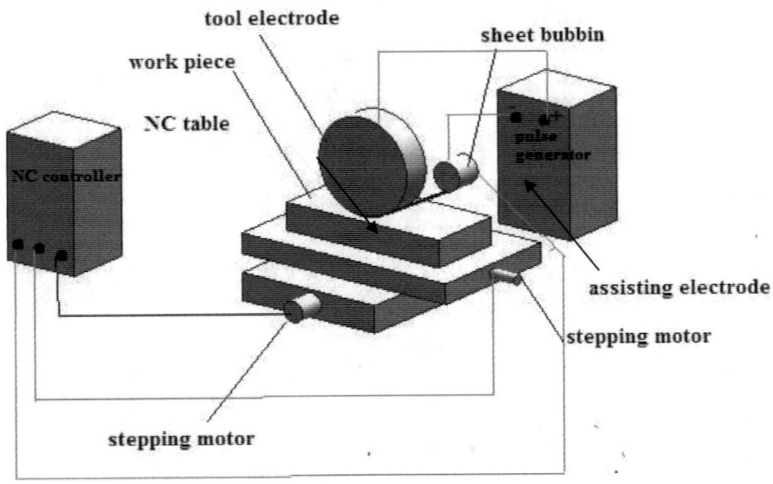

Fig. 2 EDM of insulating material with auxiliary electrode

The pulse generator's positive and negative sides are connected to the tool electrode and the auxiliary electrode, respectively. A steel wheel is mounted on a rotary spindle that is powered by an A.C. motor. A numerically controlled (NC) table provides the workpiece, which is an insulating ceramic blank. A thin coating of conductive material serves as an auxiliary electrode in this case. The tool electrode rotates at a high speed during machining, and the auxiliary electrode is fed along the surface of the insulating ceramic work towards the tool electrode.

The conductive layer allows for sparking, which contributes to material removal as the spark energy triggers a crack in the workpiece [12, 13]. This spark leads to the dissociation of the dielectric. On the machined surface, dissociated carbon and tool electrode particles rebuild a conductive layer. As a result, another EDM spark occurs. EDM comes in a variety of forms, the most common of which are wire EDM and die-sink EDM. A thin conductive material wire serves as the electrode in wire EDM, and the workpiece is placed on a CNC-controlled worktable. By controlling the movement in X-Y direction, the setup is able to cut complex 2 -dimensional shapes. While die sink EDM, a three-dimensional form conductive material tool is employed to machine a conductive material immersed in a dielectric tank. An electrical spark is produced for a very short time when a regulated voltage is applied between the electrode and the workpiece, and the material on the workpiece melts and evaporates locally. The appropriate cavity in the workpiece is formed by a series of successive sparks. In comparison, much less material is extracted from the tool material, allowing the tool to wear out over time.

Melting, vaporization, and disintegration, as well as fracturerelated spalling, are key phenomena t hat occur during EDM owing to spark erosion of ceramic materials, depending on the mechanical characteristics and EDM settings. Material removal mechanisms during EDM of ceramic composites have been carefully examined [13]. To acquire an insight into the mechanism, most researchers looked at microscopic scans of the machined part. The researchers find various models of the mechanism of material removal [14]. It is said the main mechanism of material removal is melting evaporation, dissociation, and micro-cracking .the various mechanisms observed by different researchers are shown in Fig. 3.

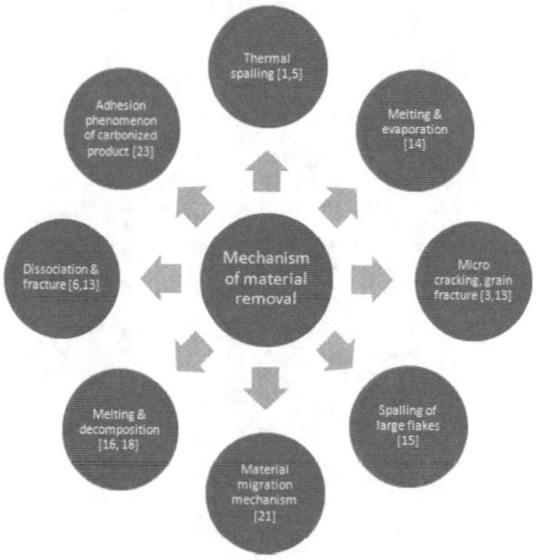

Fig. 3 Theoretical models available in literature for mechanism of material removal

3. Progress so far in the machining of insulated ceramics by EDM

Diamond grinding has typically been used to process exceptionally hard and challenging materials, however the problem with diamond grinding is that it is inefficient and expensive. Because EDM can process the material despite its hardness, it is thought to be a very viable technology for machining ceramics. EDM can be used to efficiently machine conductive ceramics. Patel et al. [15] explored whether melting, evaporation, and dissociation, or fracture-related spalling, are the predominant material removal mechanisms during electrical discharge machining of conductive Al_2O_3 composites. D.Hanaoka et al. [16] experimentally investigated that Insulating Si3N4 ceramics, and Si3N4/CNT and Si3N4/GNP can be efficiently machined .they also explain that surface roughness, wear rate properties are better for conductive ceramics and MRR obtain in insulated ceramic is better than conductive ceramics. Y.H. Liu. et. al. [17] have explained that using the copper sheet as an assisting electrode, the non-conductive ceramic can be easily machined in EDM milling. They also explained the importance of using water-based emulsion as adielectric medium. Abdus Sabur et al. [14] has experimentally shows that by using copper sheet around the insulated ceramics help to machine it by EDM machine. Apiwat Muttamara et al. [18] has shown experimentally that copper layer and carbon powder painted work well as an assisting electrode while micro-machining of non- conductive Si3N4 ceramics is done. They also shown that the pipe electrode is better than the solid type of electrode. A. Schubert et al. [19] have used silver varnish as assisting electrode for micro-milling of non-conductive ZrO_2 ceramics. The table given below shows the various non-conductive ceramics machined with the assisting electrode method and the process parameters with the outcomes of the processes. A classification based on various materials that have machined by EDM, parameters used during machining and theoutcomes of their research is given in Table 1.

IOP Conf. Series: Materials Science and Engineering 1224 (2022) 012003 doi:10.1088/1757-899X/1224/1/012003

Table 1 Classification based on various materials that have been machined by EDM

Sn	JOURNAL NAME	CERAMICS TO BE MACHINE		ASSISTING ELECTRODE TYPE	PROCESS PARAMETER		OUTCOMES
1[16]	Procedia CIRP 6 (2013) 95 – 100	Si_3N_4 ceramics (SN)		Pyrolytic carbon	Machining conditions		1: The insulating materials generated higher MRRs than the conductive materials. 2: Conductive materials have better EDM properties than insulating materials in terms of tool wear ratio and surface quality. 3:The aiding electrode method produced a better edge contour than the standard approach on a GNP composite with lower thermal conductivity.
		1	SN		Electrode Cu	(Φ8mm)	
		2	SN/0.9 CNT		Open circuit voltage (V)	100	
		3	SN/5.3CNT		Setting Current (A)	5.0	
		4	SN/11.3CNT		Discharge duration (μs)	1.0	
		5	SN/20.6CNT		Electrode Polarity	negative	
2[17]	International Journal of Machine Tools & Manufacture 48 (2008) 1030–1035	Al2O3		Thin copper sheet	-		1: There is no hazardous gas produced when a water-based emulsion is used as the machining fluid and the machinery is not corroded.
4[18]	Journal of Materials Processing Technology 140 (2003) 243–247	Si3N4		Copper sheet	**Machining conditions**		1. The material is extracted predominantly by spalling non-conductive ZrO2 ceramics in an EDM process. 2. A small amount of material is removed via melting and evaporating.
					Tool electrode	Copper	
					Tool polarity	-ve	
					Input power (KVA)	1.1, 1.2, 1.3, 1.4	

				Pulse-on-time	Ton (µs) 12	
				, Pulse-off-time	, Toff (µs) 12	
5[19]	Procedia CIRP 6 (2013) 297 – 302	zirconia ceramics	silver varnish with 45% silver	**Machining conditions**		1: Micro EDM-milling of Non conductive ZrO_2 ceramics could successfully done.
				Open Circuit Voltage	100V	
				Discharge energy level	(CF) 100 104	
				Discharge current	3A to 8.5A	
				Discharge Duration (*te*) (nominal)	100ns 180ns	
				Recharge frequency	(nominal) 150kHz	
6[21]	Applied Surface Science 276 (2013) 731– 743	reaction-bonded silicon carbide	Additives in dielectric fluid	**Machining condition**		1:Micro EDM of RBSiC is employed to machine l ocal material migration phenomenon between tool electrode and work piece material.
				Work piece material	RB-SiC	
				Electrode	Tungsten	
				voltage	60-110V	
				Additive(µm)	CNF(diameter 0.15 ,length 6-8)	
				Concentration(g/L)	0.06-0.28	
				Machining time	3min	

7[20]	Procedia CIRP 6 (2013) 135 – 139	single crystal silicon carbide	copper foil	**Machining conditions**		1: Negative polarity is advantageous for SiC foil EDM with higher machining speeds and lower tool wear proportions when the pulse duration is short. 2: By raising the discharge current and employing thinner foil electrodes, the cutting speed of SiC foil EDM can be enhanced.
				Assisting Electrode	Copper foil	
				Width/ Thickness	10mm /50,80,100,200	
				Open voltage	120V	
				Servo voltage	90V	
				Discharge duration	3μs	
				Discharge interval	30 μs	
				Dielectric fluid	EDM oil	
8[22]	Journal Of Materials Processing Technology 2 0 8 (2 0 0 8) 245–250	Al2O3 ceramic	Red copper	**Machining conditions**		1: When we utilize a water-based emulsion as the working fluid, no hazardous gases are produced and the equipment is not rusted during ED milling.
				Pulse duration	50-500 μs	
				Pulse interval	350 μs	
				Peak current	25A	
				Dielectric	water-based emulsion	

4. Conclusions

Wear resistance, high compressive strength, chemical resistance, and abrasion resistance are among the mechanical features of advanced engineering ceramics. Modern industry might get benefit from improved engineered ceramics. EDM is a viable machining procedure if the ceramics, composites have a low electrical resistivity and are adequately conductive to induce sparking. EDM is used to machine hardened materials that are conductive in nature that's why conductive ceramics can be easily machined by this method. For non-conductive ceramics, a conductive layer, as an assisting electrodehas to use in the form of a conductive material layer on the ceramics. Thus if any how it is possible to generate a spark between tool and workpiece for carrying-out EDM, very hard materials such as Advanced ceramic, Insulated ceramics, ceramic- composite could be machined. By reviewing previous various works it can also be said that:

1. When EDM is used to machine conductive ceramics, the major material removal mechanism is melting, evaporation. Thermal fracture or thermal cracking are used to remove non-conductive ceramics.

2. While milling non-conductive materials with the EDM process and an auxiliary electrode, thermal spalling or thermal cracking is observed to be the predominant material removal mechanism. Thermal spalling is not seen in materials with a higher toughness.

3. The rate of material removal increases as the input power increases and decreases when the material's hardness increases.

4. When machining conductive materials, surface roughness was better than when machining insulating materials. The MRRs on the insulating materials significantly higher than those on the conducting materials.

5. When a water-based emulsion is employed as the machining fluid, no hazardous gas is produced, and the equipment is not corroded during ED milling.

6. Due to the lower viscosity of water, EDM in de - ionized water produces a thinner recast layer.

References

1. *Baghel, R., Mali, H.S. & Biswas, S.K. Parametric optimization and surface analysis of diamond grinding-assisted EDM of TiN-Al_2O_3 ceramic composite; The International Journal of Advanced Manufacturing Technology (2018), Vol. 100, Issue 5-8, pp. 1183-1192, IF 2.601.*
2. *Baghel, R., Mali, H.S. & Biswas, S.K.; Micro hole fabrication in TiN-Al2O3 ceramic composite by SiC powder assisted micro- EDM; Engineering ResearchExpress 2 (2020), 015028,*
3. *Lok, YK; and Lee, TC ," Processing of Advanced Ceramics Using the Wire- Cut EDM Process" Journal of Materials Processing Teclmology 63 (1997) 839-843*
4. *Baghel, R., Mali, H.S. & Biswas, S.K.; An Experimental study on the fabrication of microchannels in Titanium Nitride Alumina Composite Using electro-discharge Milling; International Journal of Modern Manufacturing Technologies (2018), vol. 10,Issue 10, pp. 24-29. ISSN 2067-3604*
5. *Ji, Renjie; Liu, Yonghong; Zhang, Yanzhen and Wang Fei "Machining performance of silicon carbide ceramic in end electric discharge milling" Int. Journal of Refractory Metals and Hard Materials 29 (2011) 117–122*
6. *Norliana Mohd Abbas, Darius G. Solomon, Md. Fuad Bahari ," A review on current research trends in electrical discharge machining (EDM)" International Journal of Machine Tools & Manufacture 47 (2007) 1214–1228*
7. *Baghel, R., Mali, H.S.; The Scope of Machining of Advanced Ceramics by Electro-Discharge Machining and Its Hybrid Variants, International journal of advanced materials manufacturing and characterization (2018), vol. 8, Issue 2, pp. 88-101*
8. *Baghel, R., Mali, H.S. & Biswas, S.K Study of Vibration Assisted Micro Electro-Discharge Milling Of Titanium Nitride- Aluminium Oxide Composite, All India Manufacturing Technology, Design, and Research conference (AIMTDR) 16-18 Dec. 2016*

at COEP Pune, pp. 276-279. ISBN: 978-93-86256-27-0.

9. *Luis, CJ; Puertas, I; Villa, G " Material removal rate and electrode wear study on the EDM of silicon carbide": Journal of Materials Processing Technology 164– 165 (2005) 889–896*
10. *Lauwers,B; Kruth, JP; Liu W; Eeraerts, W; Schacht, B; Bleys , P "Investigation of material removal mechanisms in EDM of composite ceramic materials", Journal of Materials Processing Technology 149 (2004) 347–352*
11. *Mitra, Souren; Sarkar, Soumya ; Paul, Goutam; Bhaduri, D; Biswas, Sampad " Pareto Optimization Of Electro Discharge Machining Of Titanium Nitride-Aluminium Oxide Composite Material Using Genetic Algorithm" , Advanced Materials Research Vols. 264-265 (2011) pp 985-990*
12. *Zhang , Chengmao" Effect of wire electrical discharge machining (WEDM) parameters on surface integrity of nano-composite ceramics" , Ceramics International40(2014)9657–9662*
13. *Sabur, Abdus; Yeakub Ali, Mohammad; Maleque, Md. Abdul; Khan ; Ahsan Ali "Investigation of material removal characteristics in EDM of nonconductive ZrO_2 ceramic" Procedia Engineering 56 (2013) 696 – 701*
14. *Mali, H.S., Baghel, R., Unune, Deepak R.; Experimental Investigation on Hybrid Micro-Electro Discharge Machining of Inconel 718 and ceramics materials, International Journal of Conceptions on Mechanical and Civil Engineering (2017), Vol. 5, Issue. 1, pp. 1-6*
15. *Baghel R., Mali H.S., Baghela V. (2019) Micro-tool Fabrication and Micro-ED Milling of Titanium Nitride Alumina Ceramic–Composite. In: Shunmugam M., Kanthababu M. (eds) Advances in Micro and Nano Manufacturing and Surface Engineering. Lecture Notes on Multidisciplinary Industrial Engineering. Springer, Singapore.*
16. *Liu, Y.H. ; Ji, R.J. ; Li, X.P. ; Yu, L.L. Zhang, H.F. ; Li , Q.Y. " Effect of machining fluid on the process performance of electric discharge milling of insulating Al_2O_3 ceramic" , International Journal of Machine Tools & Manufacture 48 (2008) 1030– 1035*
17. *Muttamara, Apiwat; Fukuzawa Yasushi;, Mohri , Naotake ;Tani , Takayuki "Probability of precision micro-machining of insulating Si3N4 ceramics by EDM" , Journal of Materials Processing Technology 140 (2003) 243–247*
18. *Schubert, H. Zeidler, M. Hahna, M. Hackert-Oschätzchen, J. Schneider "Micro- EDM milling of electrically nonconducting zirconia ceramics" , Procedia CIRP 6(2013) 297 – 302*
19. *Zhao, Y. ; Kunieda, M. ; Abe K.," Experimental investigations into EDM behaviors of single crystal silicon carbide" Procedia CIRP 6 (2013) 135 – 139*
20. *Liew , Pay Jun; Yan, Jiwang; Kuriyagawa, Tsunemoto "Experimental investigation on material migration phenomena in micro-EDM of reaction-bonded silicon carbide" Applied Surface Science 276 (2013) 731– 743*
21. *Liu, Y.H. ; Li, X.P. ; Ji, R.J. ; Yu, L.L. ; Zhang, H.F. ; Li, Q.Y. "Effect of technological parameter on the process performance for electric discharge milling of insulating Al_2O_3 ceramic", journal of materials processing technology 2 0 8 (2 0 0 8) 245–250*
22. *Hu , C.F.; Zhou , Y.C. ; Bao , Y.W." Material removal and surface damage in EDM of Ti_3SiC_2 ceramic" Ceramics International 34 (2008) 537–541*

IOP Conf. Series: Materials Science and Engineering 1224 (2022) 012004 doi:10.1088/1757-899X/1224/1/012004

The combined implementation of Green, Lean and Six Sigma approaches for achieving environmental sustainability – A Review

Arsalan Ahmad[1] and Abid Ali Khan[1]
Department of Mechanical Engineering,
Zakir Husain College of Engineering and Technology
Aligarh Muslim University, Aligarh, UP, India
abid.khan.me@amu.ac.in

Abstract. Sectors like manufacturing, automotive, transport and logistics, etc. have greatly contributed to resource depletion all around the world. The large consumption of resources results in the rise of many greenhouse gases, waste, and other environmental problems. Organizations are now under stress to modify the way they manage their procedures and operations in response to various government environmental restrictions and the rising customer demand for ecologically friendly services and commodities. Management system approaches like green, lean and six sigma can help in achieving sustainability goals. However, no single approach is effective in achieving all sustainability goals as there are inherent limitations of these approaches. This has led people in both industry and academia to focus on the combined application of these approaches. In this paper, various research articles and case studies focusing on the combined implementation of Green, Lean, and Six Sigma methodologies for environmental sustainability across various industries is systematically summarized and reviewed. Another aim of the article is to find which combination of these approaches is best for achieving sustainability.

1. Introduction

Today the humans are facing a great many challenges which include a raging pandemic, rising population, climate change, resource shortages, and environmental degradation. Environmental waste, environmental degradation, and the loss of resources are some environmental issues. These environmental challenges pose a great threat to the existence of all living beings, their effects can be seen in the form of changing climate, scarcity of water, etc. Governments and international agencies have taken cognizance of these issues. Various agreements such as Paris Agreement 2016 and regulations both on legal and economic fronts are being discussed and notified. UN has come up with its 17 Sustainable Development Goals (SDG's) to address these issues at various fronts. The public interest is also increasing on these issues. Now consumers show a positive attitude towards eco-friendly products, thereby increasing demand for such products [1][2]. Organizations in various sectors are adopting sustainable practices in various operational areas [3][4][5]. The use of management systems approaches like green, lean and six sigma can help to tackle this problem.

In the Green approach reduction of unfavourable ecological impacts of products and processes while improving environmental performance as an organizational strategy for businesses is done [6]. Efficiently building products for waste and pollution minimisation by parts reduction, reusing components, and rationalising the materials used [7][8][9]. Whereas, in the Lean approach elimination of eight kinds of waste which include transportation, unnecessary movement, overproduction, waiting, processing, excess inventory, defects waste, and unused creativity in all of an organization's functional areas and its supplier network is done [10][11][12]. It was first adapted at Toyota Production System [13]. It helps organizations improve productivity and effectiveness by the reduction of waste [18] and as a result, it also improves resource efficiency and reduces environmental impact [14][15]. Common lean tools and techniques include Just-In -Time (JIT), 5S, Value Stream Mapping (VSM), Kanban, Kaizen, etc [16][13]. Six Sigma is another approach that is used to achieve sustainability goals. Traditionally this comprehensive quality improvement method has focused on reducing defects and helping companies to attain and sustain success [17] [18]. It provides an operational framework in the company that explores and promotes problem solving [19]. Although the primary goal of this approach is not to reduce environmental waste, it is capable of improving environmental performance by lowering air emissions, energy consumption, and wastewater [20]. Technically, this methodology yields a failure rate of 3.4 parts per million [21]. This methodology was first introduced in the 1980s by Motorola and later refined by General Electric to improve performance and reduce costs [12]. After implementing this methodology both the companies were able to save huge sums of money [17]. This methodology has become increasingly popular over the last two decades and has been widely used in numerous organizations and different kinds of industries [22]. It uses two kinds of approaches: DMAIC (D-Define, M-Measure, A-Analyse, I-Improve, C-Control) and DMADV (D-Define M-Measure, A-Analyse, D-Design, V-Verify) [19]. The former is used for improving an already existing product or process and the latter is used for designing or implementing new products or processes, that can deliver a Six Sigma performance [13].

In recent years combining these approaches to achieve sustainability is becoming popular in research [23]. These combined approaches can help many manufacturing and service organizations to successfully achieve sustainability goals and at the same time make profits. A combination of these three methodologies i.e., Green, Lean and Six Sigma needs to be reviewed and summarised. Another objective of this article is to find which combination of these approaches is best for achieving environmental sustainability.

2. Methodology

In this literature review, Google Scholar (scholar.google.com), Taylor and Francis (tandfonline.com), ScienceDirect (sciencedirect.com), and IEEE Xplore (ieeexplore.ieee.org) databases were browsed using the search words “green lean six sigma”, “lean six sigma for environment sustainability”, “Eco-lean six sigma”, and “lean six sigma for sustainability”. The literature for last the 10 years was looked into and 20 journal articles were selected for review. Inclusion criteria were to choose those research articles which proposed any framework, model, or method to combine Green, Lean, and Six Sigma approaches. Out of these most articles were from the journals ‘Production Planning & Control’ and ‘Journal of Cleaner Production’. The percentage distribution of the articles across different industries is shown in Figure 1.

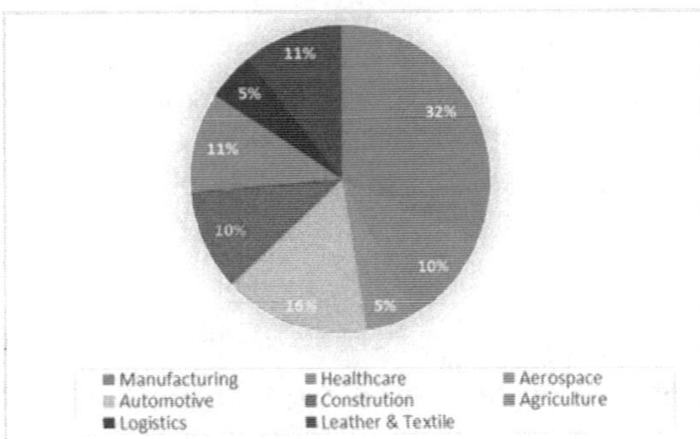

Figure 1: Percentage distribution of case studies sector wise.

3. Literature Review

3.1. Green-Lean Approach

As both lean and green approaches focus on eliminating waste, there exists a synergistic relationship between the two [24]. The basic difference between these two approaches has to do with the kind of waste they eliminate. In the lean approach elimination of non-value adding activities like unnecessary movement, waiting time, overproduction is done whereas in the green approach unnecessary usage of energy, natural resources, etc. is reduced [25]. Furthermore, not all lean tools are eco-friendly and can result in negative ecological effects [26]. Combining these approaches in form of models and methods can help in improving operational efficiency and environmental sustainability. A decent number of such models and frameworks have been proposed, which can be implemented in the manufacturing, automotive, agriculture, and logistics sectors.

An integrated model based on Gemba-Kaizen approach bolstered by continuous improvement approach was proposed [24]. A large aerospace manufacturing company based in the US implemented it to reduce electricity usage. It was found that there was a saving of $200,000 across six facilities over a 2-year period. It was also applied to an injection molding company and after its implementation, there was a consistent decrease in energy consumption and raw material, which resulted in a cost saving of 113,000 USD per year. A Green-Modified Value Stream Mapping (GMVSM) model was proposed for integration of Green and Lean approaches [27]. Carbon emission and carbon efficiencies were used as the evaluation indicators. The outcomes of this model were verified in a metal stamped parts producing unit and it was found that there was 39.2% and 107.4% improvement in carbon efficiency and total carbon emissions respectively. For implementing the Green-Lean approach at manufacturing cell level a 5 steps model which uses Kaizen to improve energy and mass flows was developed [28]. On an average there was a reduction in resource use between 30 to 50% after implementation. The entire cost of mass and energy flows per cell can be saved to 5-10%. Another 5-step methodology was developed in which Carbon-Value Efficiency is used as an indicator to evaluate Lean and Green performance was proposed [29]. On testing, this methodology in a metal stamped parts production plant it was found that Carbon-Value Efficiency and production lead improved by 36.3% and 64.7% respectively, and carbon footprint reduced by 29.9%. A novel tool and systematic methodology named Sustainable Transportation Value Stream Map (STVSM) for the transport and logistics sector was proposed [30]. It yielded in reduction of excess travelled distance by 57% and the fuel consumption by 13.5%. Thereby, reducing the release of CO_2, NOx, CO, and HC into the environment. Another systematic methodology and a novel tool named Green Integrated Value Stream Mapping (GIVSM) was proposed to achieve both environmental sustainability and operational efficiency [31]. This was applied in a packaging manufacturing SME in the U.K. to reduce carbon footprint in their packaging products. It was observed that GIVSM decreased the carbon footprint by 77% and also decreased the lead time by 63% hence increasing the operational efficiency of the company. For the agriculture sector, a framework was proposed that can be used in small and mid-size farms with the goal to enhance production and profit, and simultaneously achieving environmental sustainability [32]. Post implementing this framework, it was found that the search time for materials and tools reduced, workplace became safer and waste disposal and recycling routines improved. Also, due to improvement in work structure resource use efficiency increased.

3.2. Lean-Six Sigma Approach

Lean Six Sigma (LSS) refers to the combination of lean and six sigma approaches. It is a systematic approach that helps in quality improvement. The benefit of this approach includes reduction of defects, reworks, inventory levels, faster production, space requirement, transportation, and in waiting time [33]. It also results in increased production rate and employee motivation. LSS concepts can be effectively used to achieve environmental benefits. The manufacturing firm can profit in terms of defect reduction and variation reduction, which leads to lower resource consumption, if LSS and environmental sustainability ideas are properly aligned [34]. Achieving environmental sustainability is not the main objective of the LSS approach. Thus, in order to translate environmental opportunities into operational excellence, adequate leveraging of LSS ideas with environmental benefit is essential [35].

A framework based on DMAIC methodology for implementing LSS to reduce environmental impact in the automotive industry was proposed [34]. After successful implementation of this framework in an Indian automotive sector there was a reduction in net power consumption, net water consumption, net coolant consumption, and the use of raw material. Also, the total number of rejections of the product was reduced from 12 to 3 units and there was improvement in sigma level from 3.60 to 4.06. A generic framework also based on DMAIC was proposed and tested on case studies from manufacturing, healthcare, and construction [36]. Post implementation it resulted in several benefits like reduction in energy usage, pollution prevention, reduction in various kinds of wastes, etc.

3.3. Green-Lean-Six Sigma Approach

Any individual or combination of any two approaches is not able to deal with all the aspect of sustainability in an organization [37]. However, these limitations can be overcome by using a combination of all the three approaches viz., Green, Lean, and Six Sigma. The Green Lean Six Sigma (GLSS) approach can assist in developing a high-quality, low-cost, and environmentally friendly process or product [6][38]. GLSS approach reduces environmental impact with the help of the 3R's i.e., Reduce, Reuse, and Recycle [39]. It is a very promising approach which can help organizations in significantly reducing environmental impact of their products. However, only four proposed frameworks for implementing this approach were found.

A generic framework that methodically guides businesses to successfully incorporate and adopt integration of these approaches through a 5-stage and 16-steps process was proposed [4]. It was applied across different sizes industry namely (Agri-food, Textile, Tannery and Hotel), and it enabled companies to minimize their use of resources on average from 20% to 40 % and to reduce energy costs and mass streams by 7-12 %. Another generic GLSS framework that can be implemented in a methodological manner 8 steps was developed [40]. DMAIC approach was implemented in every lean cycle of this framework and organizational culture is the central tenant in this model which binds the GLSS initiatives. Green initiatives are implemented at every phase of LSS, and which results in decrease of energy consumptions, emissions, environmental costs, business waste, and will increase environmental revenues. This framework was successfully used for reducing graphite and dust pollution in an open cast mine. An integrated framework to enhance quality and improve environmental impacts of construction process was developed [17]. In this framework DMAIC methodology was used to evaluate a chosen process, to identify and measure the impact of the generated waste. Then with the help of Six Sigma tools possible reasons for the waste generation were found. For implementing GLSS approach in hospital supply chains a framework was developed [41]. This framework is based on IDEF model adjusted to healthcare setting. In this IDEF model input is the current supply chain system, the output is a leaner and greener supply chain, controls are the hospital value proposition which includes lean- six sigma philosophy and green values, and mechanism are the stakeholders and tools like 5S, VSM etc. This model helps in reducing several kinds of wastes like inventory waste, paper waste, etc. which will ultimately lead to efficient energy consumption and reduction carbon emissions.

4. Conclusion and Future Scope

Since many organizations in sectors like manufacturing, automotive, transport and logistics, etc are responsible for environmental degradation, they are under great pressure due to environmental regulations and the rise in customer demand for environmentally sustainable services and goods [24]. Using combinations of approaches like green, lean and six sigma is one best way that these organizations can achieve sustainability goals at the same time and continue to make profits. In this paper, various research articles and case studies that provided a framework for implementing Green-Lean, Lean-Six-Sigma, and Green-Lean Six Sigma across various industries for achieving environmental sustainability were reviewed. LSS being a data-driven and structured approach to eliminate sources of variation and waste surely is beneficial, but environmental waste reduction is not its main goal. The number of articles implementing LSS for environmental sustainability is scarce, and only 2 frameworks that were verified through a case study were found. Lean-Green approach is effective achieving in sustainability goals. All Green-Lean frameworks and models reviewed resulted in a significant reduction in the use of resource consumption which in turn translated into reduction of negative environmental impact. But this combined approach inherits limitations of the individual approaches [3]. Therefore, it is recommended to integrate Six-Sigma into Green Lean approach to overcome these limitations [3]. This makes the Green-Lean-Six Sigma (GLSS), the most promising approach as it leverages the benefits of all the individual approaches and overcomes the limitation of the individual approaches. The four proposed GLSS frameworks significantly reduces the impact on environment, but there is a dearth of such frameworks. Therefore, there is need of further research in developing these types of models. The COVID-19 pandemic brought a drastic change on how various business activities are undertaken. Disruptions in the supply chain have put the existing businesses into restructuring their operations into more sustainable fashion. On demand side there is a more awareness among consumers towards more sustainable products and services. Governments across the globe are coming up with policies to address the issues of Global Warming. There is no better time to further the existing research in GLSS approach.

GLSS approach can make a profound impact across various industries. In the Manufacturing sector with emergence of Industry 4.0 researchers need to explore ways to reduce environmental waste, eliminate non-value-added activities and reduce defects to cater to the environmentally cautious consumer and in scenario where supply chain is hit globally. The automotive sector which is witnessing a momentum shift with increasing demand for electric vehicles, challenging the traditional petrol and diesel engines has helped, in India alone, a lot of EV companies are setting up. GLSS can play a major role at various steps of the product life cycle. It can be paramount in Vehicle Scrappage policy set up by Government of India 2021, which also aims at reducing prices by recycling of components in old vehicles and reducing pollution. Future research is needed for GLSS in this domain. In a world where e-commerce is at its peak, logistics is the backbone of these businesses. The increase in logistic activity in the consumer segment has also added to the carbon footprint. With the rise in e-commerce businesses, future research is needed in setting up a GLSS based network. The healthcare industry is a beneficiary of LSS, this methodology prevents medical errors. Future research is needed with application of GLSS, to address the surge in telemedicine and digitalisation in various aspects of healthcare. Globally, agriculture is a major source of pollution and over exploitation of natural resources adds to the situation. From farm to fork, a lot of carbon footprint is added. Future research is needed for the inclusion of a GLSS based approach to address the issue. The leather and textile industry are key polluters globally. The whole supply chain polluting and at most stages follows a traditional methodology. These are also water guzzlers and polluters. Future research is needed to implement the GLSS methodology in the traditional supply chain. Construction is a vital industry especially in emerging markets where the government and private players are heavily invested in the construction industry. As governments and consumers are demanding sustainability in this industry, GLSS needs to be explored in future research especially at the material procurement and execution phases where a lot of excess waste is created.

5. References

[1] Arvola A, Vassallo M, Dean M, Lampila P, Saba A, Lähteenmäki L, Shepherd, R 2008 Predicting intentions to purchase organic food: The role ofaffective and moral attitudes in the Theory of Planned Behaviour, *Appetite,***50(2)** 443-454.

[2] Schmeltz L 2012 Consumer-oriented CSR communication: focusing on ability or morality?," *Corporate Communications: An Int. J.* **17 (1)** 29-49.

[3] Garza-Reyes, J.A 2015 Green lean and the need for Six Sigma," *Int. J. of Lean Six Sigma* **6(3)**, 226-248.

[4] Cherrafi A, Elfezazi S, Govindan K, Garza-Reyes, J.A Benhida K, and Mokhlis A 2017 A framework for the integration of Green and Lean Six Sigma for superior sustainability performance. *Int. J. of Production Research,* **55(15)** 4481-4515.

[5] Erdil N.O, Aktas C.B and Arani OM 2018 Embedding sustainability in lean six sigma efforts. *J. of Cleaner Production,* **198** 520-529.

[6] Sagnak M and Kazancoglu Y 2016 Integration of green lean approach with six sigma: an application for flue gas emissions. *J. of Cleaner Production,* **127** 112-118.

[7] Sharma S and Vredenburg H 1998 Proactive corporate environmental strategy and the development of competitively valuable organizational capabilities. *Strategic management J.* **19(8)** 729-753.

[8] González-Benito J and González-Benito Ó 2008 Operations management practices linked to the adoption of ISO 14001: An empirical analysis of Spanish manufacturers. *Int. J. of Production Economics,* **113(1)** 60-73.

[9] Schoenherr T 2012 The role of environmental management in sustainable business development: a multi-country investigation. *Int. J. of Production Economics,* **140(1)** 116-128.

[10] Chauhan G and Singh T.P 2012 Measuring Parameters of Lean Manufacturing Realization," *Measuring Business Excellence,* **16(3)** 57–71, 2012.

[11] Kanigolla D, Cudney E.A and Corns S.M 2013 Employing project-based learning in Six Sigma education. *J. for Quality and Participation,* **36(1)** 34-38.

[12] Cudney E.A, Venuthurumilli S.S.J, Materla T and Antony J 2020 Systematic review of Lean and Six Sigma approaches in higher education. *Total Quality Management & Business Excellence,* **31(3-4)** 231-244.

[13] Amitkumar D.M and Gajanan S.P 2019 "A methodical literature review on application of Lean & Six Sigma in various industries," *Australian J. of Mechanical Engineering.*

[14] King A.A and Lenox M.J 2001 Lean and green? An empirical examination of the relationship between lean production and environmental performance. *Production and operations management,* **10(3)** 244-256.

[15] Aguado S, Alvarez R and Domingo R 2013 Model of efficient and sustainable improvements in a lean production system through processes of environmental innovation. *J. of Cleaner Production,* **47** 141-148.

[16] Chiarini A 2017 Environmental policies for evaluating suppliers' performance based on GRI indicators. *Business Strategy and the Environment,* **26(1)** 98-111.

[17] Banawi A and Bilec M.M 2014 A framework to improve construction processes: Integrating Lean, Green and Six Sigma. *Int. J. of Construction Management,* **14(1)** 45-55.

[18] Sperl T, Ptacek R and Trewn J 2013 Practical lean Six Sigma for healthcare: Using the A3 and lean thinking to improve operational performance in hospitals, clinics, and physician group practices. MCS Media.

[19] Parast M.M 2011 The effect of Six Sigma projects on innovation and firm performance. *Int. J. of Project Management,* **29(1)** 45-55.

[20] Farrukh A, Mathrani S and Taskin N 2020 Investigating the Theoretical Constructs of a Green Lean Six Sigma Approach towards Environmental Sustainability: A Systematic Literature Review and Future Directions. *Sustainability,* **12(19)** 8247.

[21] Banuelas R, Antony J and Brace M 2005 An application of Six Sigma to reduce waste. *Quality and Reliability Engineering Int.* **21(6)** 553-570.

[22] "Improve the Extrusion Process in Tire Production Using Six Sigma Methodology," in *Procedia*

Manufacturing, 2017.

[23] Chiarini A 2017 Environmental policies for evaluating suppliers' performance based on GRI indicators. *Business Strategy and the Environment*, **26(1)** 98-111.

[24] Cherrafi A, Elfezazi S, Hurley B, Garza-Reyes J.A, Kumar V, Anosike A and Batista L 2019 Green and Lean: a Gemba–Kaizen model for sustainability enhancement. *Production Planning & Control,* **30(5-6)** 385-399.

[25] Duarte S and Cruz-Machado V 2013 Modelling lean and green: A review from business models. *International Journal Of Lean Six Sigma,* **4(3)** 228–250.

[26] Sartal A, Martinez-Senra A.I and Cruz-Machado V 2018 Are all lean principles equally eco-friendly? A panel data study. *J. of Cleaner Production,* **177** 362-370.

[27] Zhu X.Y, Zhang H and Jiang Z.G 2020 Application of green-modified value stream mapping to integrate and implement lean and green practices: A case study. *Int. J. of Computer Integrated Manufacturing,* **33(7)** 716-731.

[28] Pampanelli A.B, Found P and Bernardes A.M 2014 A Lean & Green Model for a production cell. *J. of Cleaner Production*, **85** 19-30.

[29] Ng R, Low J.S.C and Song B 2015 Integrating and implementing Lean and Green practices based on proposition of Carbon-Value Efficiency metric. *J. of Cleaner Production*, **95** 242-255.

[30] Garza-Reyes J.A, Villarreal B, Kumar V and Molina Ruiz P 2016 Lean and green in the transport and logistics sector–a case study of simultaneous deployment. *Production Planning & Control,* **27(15)** 1221-1232.

[31] Choudhary S, Nayak R, Dora M, Mishra N and Ghadge A 2019 An integrated lean and green approach for improving sustainability performance: a case study of a packaging manufacturing SME in the UK. *Production planning & control,* **30(5-6)** 353-368.

[32] Barth H and Melin M 2018 A Green Lean approach to global competition and climate change in the agricultural sector–A Swedish case study. *J. of cleaner production*, **204** 183-192.

[33] Albliwi S, Antony J, Lim S.A.H and Van der Wiele T 2014 Critical failure factors of Lean Six Sigma: a systematic literature review. *Int. J. of Quality & Reliability Management,***31(9)** 1012-1030

[34] Ben Ruben R, Vinodh S and Asokan P 2017 Implementation of Lean Six Sigma framework with environmental considerations in an Indian automotive component manufacturing firm: a case study. *Production Planning & Control,* **28(15)** 1193-1211.

[35] Cherrafi, A., S. Elfezazi, A. Chiarini, A. Mokhlis, and K. Benhida 2016 The Integration of Lean Manufacturing, Six Sigma and Sustainability: A literature review and future research directions for developing a specific model. *J. of Cleaner Production,* **139** 828-846.

[36] Erdil N.O, Aktas C.B and Arani O.M 2018 Embedding sustainability in lean six sigma efforts. *J. of Cleaner Production,* **198** 520-529.

[37] Pandey H, Garg D and Luthra S 2018 Identification and ranking of enablers of GreenLean Six Sigma implementation using AHP. *Int. J. of Productivity and Quality Management,* **23(2)** 187-217.

[38] Kumar S, Luthra S, Govindan K, Kumar N and Haleem A 2016 Barriers in green lean six sigma product development process: an ISM approach. *Production Planning & Control,* **27(7-8)** 604-620.

[39] Sreedharan V and Raju R 2016 A systematic literature review of Lean Six Sigma indifferent industries. *Int. J. Lean Six Sigma* **7 (4)** 430–466.

[40] Sony M and Naik S 2020 Green Lean Six Sigma implementation framework: a case of reducing

graphite and dust pollution. *Int. J. of Sustainable Engineering,* **13(3)** 184-193.

[41] Zhu Q, Johnson S and Sarkis J 2018 Lean six sigma and environmental sustainability: a hospital perspective. In Supply Chain Forum: *An Int. J. Taylor & Francis* **19(1)** 25-41.

3D printed Polyether ether ketone (PEEK), Polyamide (PA) and its evaluation of mechanical properties and its uses in healthcare applications.

Sandeep Shetty, Nandish B T, Vivek Amin, *Jayaprakash K, Shahira, Stanly Selva Kumar G S, Faizan Khan, Pooja Harish

Department of Orthodontics, Yenepoya (Deemed to be University), Mangalore, Karnataka, India.

***Corresponding author:** Dr. Jayaprakash K, jayanperla@gmail.com

Abstract. Additive manufacturing (AM) is evolving continuously which enables 3D printing a given idea into a functional prototype, the complex structures can be 3D printed with versatility of changing the density depending on the in-fills programmed during printing thus it can be light in weight with less material wastage which cannot be achieved by other methods. Polyether ether ketone (PEEK) is widely used in implantable devices such as spinal, dental and orthopedic implants due to its superior biocompatibility and nearly matching properties with bone. Polyamide (PA) is another biomaterial widely used as sutures, catheters, and dentures due to its biocompatibility. In the present study, the specimen of PEEK and polyamide were 3D printed and the mechanical properties evaluated and compared for healthcare applications. **Methodology:** In this work, a comparison of mechanical strength such as compressive strength (CS), tensile strength (TS), flexure strength (FS), and impact strength (IS) of 3D printed PEEK samples by fused deposition modelling (FDM) and PA specimens by selective laser sintering (SLS), respectively. Samples for mechanical strength were prepared according to ASTM and ISO specifications. **Result:** The strength of the Polyether ether ketone was higher than that of Polyamide. **Conclusions:** The results of this experiment provide a comprehensive understanding of the mechanical strength of PEEK and PA, that can be used as a valuable guide for their healthcare applications.

Keywords: 3D printing, SLP method, mechanical strength, healthcare

1. Introduction

Additive manufacturing (AM), is a fastest evolving technology that has been evolving continuously and have grown exponentially since a decade. The basic principle of this technology is to fabricate objects from Stereolithography (STL) format from 3D design data, extruding the molten polymer from the printer layer by layer, also enabling the printing of lightweight and complex structures [1]. 3D printing is used to accurately fabricate complex structures in a wide range of materials such as polymers, metals, rubbers, or ceramics. It reduces the wastage of materials compared to the CAD-CAM technique [2,3,4]. 3D printing creates objects based on digital models created by computer aided design software which then fed into to printer for the manufacturing [3,4].

Polyamide 12 (PA12) has a broad temperature range and is able to lessen the crystallization of printed parts during the printing processes hence it is widely used polymer for powder bed fusion (PBF) processes because [3]. Polyamide (PA) is a biomaterial widely used as suture materials and catheters

due to its biocompatibility. It has a tunable mechanical properties and desirable chemical stability making it an ideal material of choice for sutures. PA can be drawn easily into fibres, and molded parts, and is also used in orthopedic applications [5]. One of the recent advances in medical and dental biomaterials is PEEK polymer known for its superior biocompatibility, high strength, light in weight and superior tissue acceptability, these are some of the properties that drawn an attention towards its use in orthopedic, spinal and dental implants [6]. One of the major advantages of this biomaterial is the radiolucent feature of the PEEK, which aid in the diagnosis which otherwise by using metal implant causes artifact in the imaging.
This study aimed to compare and evaluate the mechanical properties between Polyamide and PEEK for various healthcare applications.

2. Materials and methods:

The specimens for the mechanical test were designed digitally by using Autodesk Fusion 360. The specimens for mechanical tests of PEEK have been prepared at the Prototyping Lab with a 3D-Printer - INTAMPSYS - FUNMAT HT by using unreinforced PEEK (450G - Victrex Plc., Lancashire, UK) (Fig. 1). PA specimens were prepared with a 3D-Printer -Sinterit Lisa Pro and PA12 smooth v2 (Fig. 1). The details of printer parameters used during were given in table 1. The compressive, tensile, and flexure tests were performed using an electromechanical universal testing machine (UTM) model- Zwick/Roell Z020 with a 20kN load cell. The setup of a specimen and test fixture for mechanical tests was shown in Fig. 2-A, B, C. The impact strength was tested using Zwick/Roell HIT 50P impact tester by the Izod method with a pendulum weight = 5.5J (Fig 2-D). Specimens were tested in triplicate for the mechanical test. The geometrical details of digitally 3D designed test specimens for mechanical tests such as compression, tension, flexure, and impact test (Unit: mm) were shown in Fig. 3. Graphical representations of Stress-strain behavior of two tested materials for various mechanical properties are shown in Fig. 4.
Samples for compressive strength were prepared according to EN ISO 604. The cylindrical specimens a diameter-height ratio (x/L) greater or equal to 0.4. Consequently, a cylindrical specimen with a diameter of x=10mm and a height L=25mm was prepared. The tests were performed with UTM at a crosshead speed of 5 mm/min. until failure of the specimen [7].
For tensile strength evaluation, specimen was prepared according to ASTM D368 and designed with a dimension L = 135 and, Width = 10 and thickness = 4mm. The samples were loaded up to material failure at a displacement rate of 50mm/min. until failure of the specimen [8].
Flexural strength by 3-point bending test was performed according to ASTM D7264 at a loading rate of 2mm/min until failure of the specimen. The load applied at the time of fracture was noted. The flexural strength of the specimen was calculated using a standard equation, S= 3FL/2bd2
Where, F= Fracture load, L= Distance between two supports, set at 25mm., b= Width of the specimen, and d= Thickness of the specimen [9,10,11].

To determine the impact strength of the PEEK and PA by the amount of energy absorbed during the fracture, an un-notched impact test was performed according to ASTM D4182. A rectangular specimen of dimension 64 mm lengths, 13mm depth, and 3 mm thickness was made according to ASTM D4182 using the 3D printing technique.
The specimens were clamped into the pendulum impact test fixture with a thin edge facing the striking edge of the pendulum. The pendulum was released and allowed to strike through the specimen. The impact strength was calculated by dividing the impact energy in J by the thickness of the specimen (m) [12].

Results

The results of this study are shown in table 2 and fig. 5. Analysis of the results was conducted using SPSS software (V-22). Tests were performed at a confidence level of 95% and with a significant P-value of ($P \leq 0.05$). These statistical tests included inferential statistical analysis of variation (ANOVA). A one-way ANOVA (Post-hoc, Tukey test) showed statistically significant differences ($P < .001$) in mechanical properties evaluated between PEEK and PA. The PEEK showed higher mechanical properties such as tensile, compressive, flexural, and impact strength than PA.

IOP Conf. Series: Materials Science and Engineering 1224 (2022) 012005 doi:10.1088/1757-899X/1224/1/012005

Discussion

The ability to process powders by laser sintering (e.g., ease of deposition and spreading over the powder bed) and the quality of laser-sintered parts (e.g., surface roughness, density/porosity) depend on the powder morphology, size, and size distribution. Particles that are too large negatively influence surface finish and part density and powder particles with too small a diameter exhibit poor bulk flow at high temperatures, presumably due to the higher interparticle friction found in extremely fine powders.

The optimum particle size in processing ability is generally around 45–90 lm for laser sintering of macro-sized parts. Spherical with regular morphology is the ideal powder for laser sintering. Powders tend to arrange themselves more efficiently with spherical morphology that increases the density of parts. In contrast, powders with irregular morphology cannot achieve this form of efficient arrangement resulting in low-density parts.

Moreover, spherical morphology will help powder flow and ensures that a flat and thin powder layer is deposited during the sintering process. The PA12 powder used in this study was PA12 smooth v2. The mechanical parameters, resistance to heat and flexibility, are a ideal for functional prototypes. PA 12 by Laser printing does not require supportive structures and hence it enables printing the most complicated designs [13].

In this study, the PEEK demonstrated superior mechanical properties such as higher tensile, compressive, flexural, and impact strength and lower plastic deformation in the elastic region than PA. On the other hand, the PA polymer's flexibility and highly flexible nature, as shown in the stress-strain curve (fig. 4), could potentially improve the retention of the partial denture, which are the most desirable properties required for partial denture in dentistry.

Polyamide has a tunable mechanical properties and desirable chemical stability making it an ideal material of choice for sutures. PA can be drawn easily into fibres, and molded parts, and is also used in orthopedic applications [14]. PEEK materials are currently used in dentistry for temporary clinical abutments, temporary bruxism devices, long-term implant fixtures into bone, implant-supported prosthetic substructures, crowns and bridges, removable dentures [15]. PEEK is widely used in medical implants due to its approximate properties with bone, excellent biocompatibility, mechanical properties similar to those of human cortical bones and radiolucency which prevent imaging artifacts [16,17].

The following conclusions may be made within the limitations imposed by the methods of the current study:

- PEEK showed higher mechanical properties than PA.
- PEEK materials are currently used in dentistry for temporary clinical abutments, temporary bruxism devices, long term implant fixtures into bone, implant-supported prosthetic substructures, crowns, and bridges, removable dentures, and also used as medical implants due to its approximate property with bone
- Polyamide can be drawn easily into fibres, and molded parts, and is also used in orthopedic applications. Due to its biocompatible nature, has a tunable mechanical properties and desirable chemical stability making it an ideal material of choice for sutures, etc.
- Even though two materials tested in this study show different mechanical properties, each material has advantages and diverse healthcare applications.

Conflict of Interests

There is no conflict of interest regarding the publication of this paper.

FIGURES, TABLES, AND LEGENDS

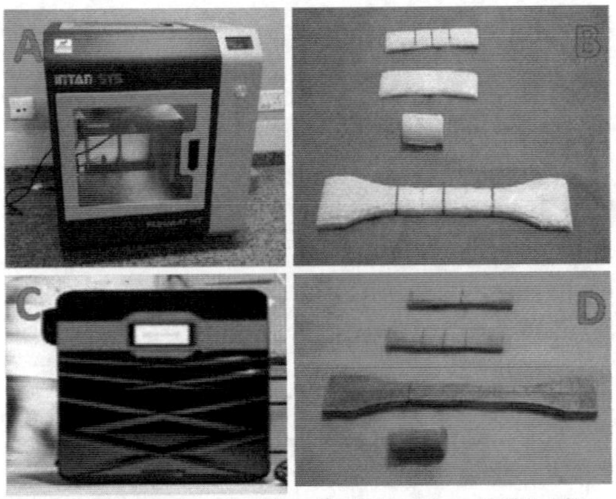

Fig. 1 3D Printer used for PEEK (A) and PA (C), printed specimens PEEK (B), and PA (D)

Fig. 2 Experimental set-up for compression (A), tensile (B), flexural (C) and impact (D) test

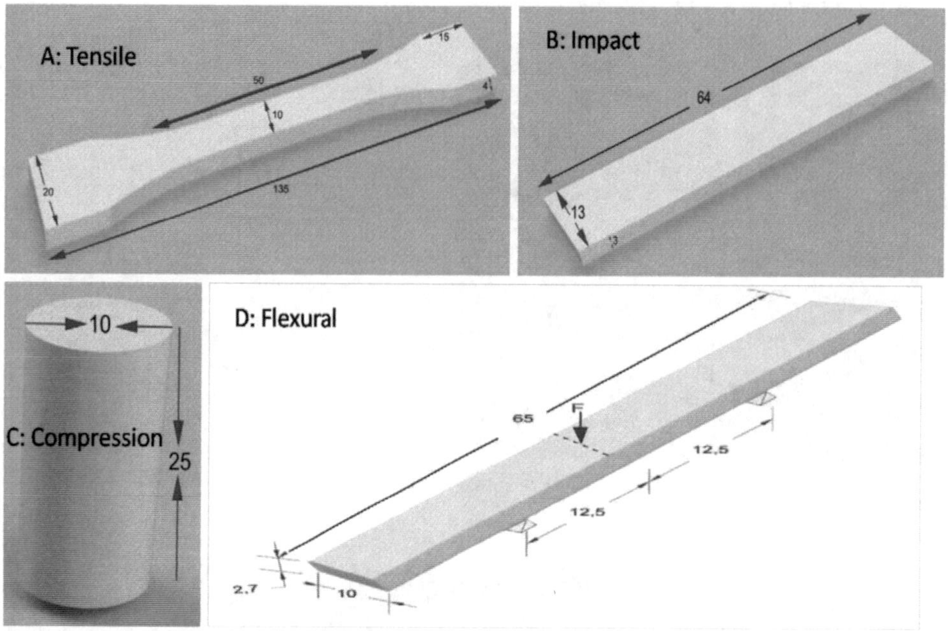

Fig. 3 Geometrical details of tested specimens for tensile (A), impact (B), compression(C), and flexural test (Unit: mm)

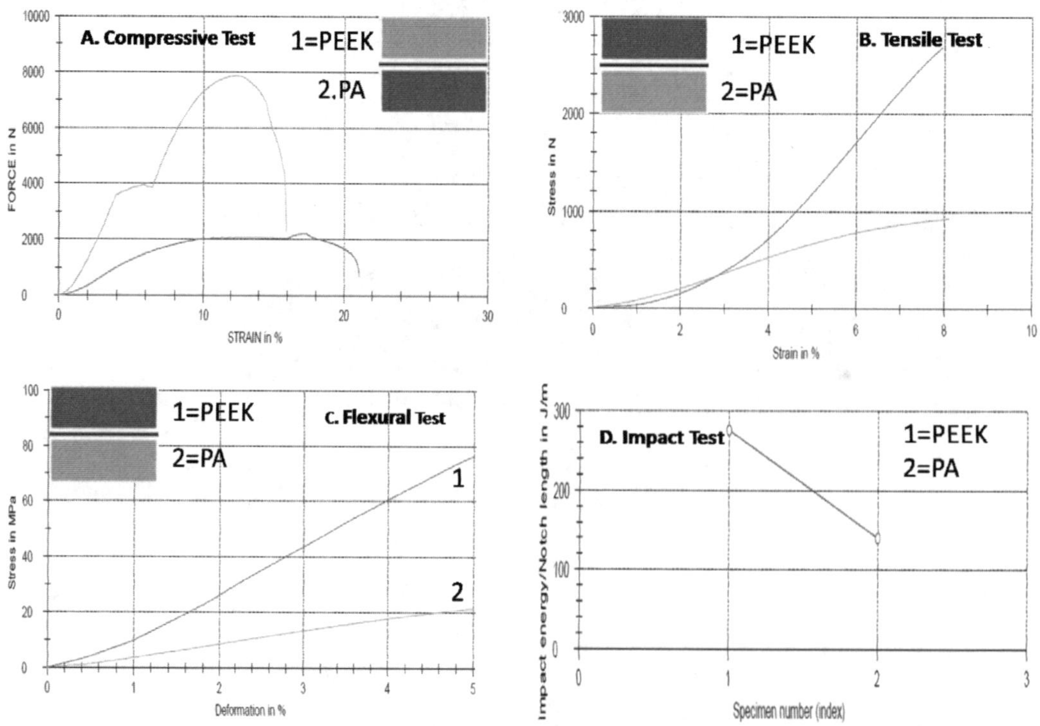

Fig. 4 Machine diagram shows the Stress-Strain behavior of PEEK and PA specimen during compression (A), tensile (B), flexural(C) and impact test (D)

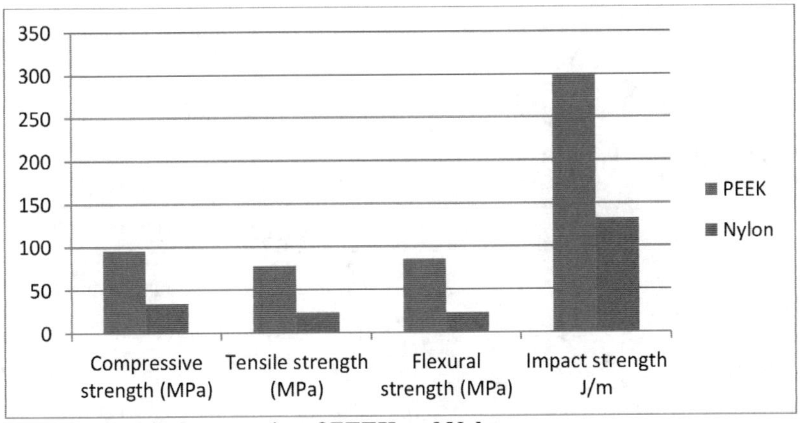

Fig. 5 Mechanical properties of PEEK and Nylon

Table 1: Printing parameters details for PEEK and PA

Printing parameters for PEEK	**Printing parameter for PA**
Printing temperature: 380-410°C	Build volume: 150×200 ×260[mm]
Build plate temperature: 130-160°C	Laser system: IR laser diode 5[W]
Printing speed: 15-40mm/min	Wavelength =808[nm]
Color: Natural	Min wall thickness: from 0.4[mm]
Raft top layer thickness: 0.3mm	Min detail size: from 0.1mm
Raft top layer width: 0.4mm	Heating system: heated-piston, cylinder, feed bed, print bed.
Raft middle layer thickness: 0.15mm	Max temp in chamber 200[°C]/392°F]
Raft base layer thickness: 0.36mm	
Raft base layer width: 0.8mm	
Raft top printing/scan speed: 25mm/s	
Raft middle printing speed: 18.75mm/s	
Raft base printing speed: 18.75mm/s	
Ambient temperature: 24°C	
Nozzle temperature: 400°C	
Nozzle diameter: 1.75mm	
Bed temperature: 145°C	

Table 2: Mechanical properties of PEEK and Nylon with mean values and standard deviation (SD)

Test materials	**Compressive strength (MPa) Mean(SD)**	**Tensile strength (MPa)**	**Flexural strength (MPa)**	**Impact strength J/m**
PEEK	95.8(SD=3.3)	77.5(SD=5.6)	85.5(SD=8.4)	301(SD=22.2)
Nylon	34.5(SD=4.6)	23.5(SD=4.56)	23(SD=4.09)	132.6(SD=6.2)

REFERENCES:

1. Martynkova GS, Sliva A, Krotosova G, Barbaszova KC, Studentova S, et al. Polymers 2021, 13, 810. https://doi.org/10.3390/polym13050810.

2. Bhat VS, Nandish BT, K Jayaprakash., Science of Dental Materials with Clinical Applications, 3rd ed., CBS Publishers and Distributors, P:230 (2019).

3. Cai C, Tey WS, Chen J, Jhu W, Liu X et al. Comparative study on 3D printing of polyamide 12 by selective laser sintering and multi-jet fusion Journal of Materials Processing Tech. 2021; 288 :116882.

4. Ngo, T.D., Kashani, A., Imbalzano, G., Nguyen, K.T., Hui, D., 2018. Additive manufacturing (3D printing): A review of materials, methods, applications, and challenges. Compos. Part B-Eng. 143, 172–196.

5. Shakiba M, Ghomi ER, Khosravi F, Jouybar S, Bigham A et al. Nylon—A material introduction and overview for biomedical applications. Polym Adv Technol. 2021;32:1–16).

6. McNamara A, Turner RM. Potential of polyetheretherketone (PEEK) and carbon-fibre-reinforced PEEK in medical applications. J Mater Sci Lett; 1987;6:188-90.

7. Percoco G, Lavecchia F, Galantucci LM. Compressive properties of FDM rapid prototypes treated with a low-cost chemical finishing. Res. J. Appl. Sci. Eng. Technol 2012(19):3838-42.

8. Liu Z, Lei Q, Xing S. Mechanical characteristics of wood, ceramic, metal and carbon fiber-based PLA composites fabricated by FDM. J Mater Res and Technol. 2019:8(5):3741-51.

9. Caminero MA, Rodríguez GP, Muñoz V. Effect of stacking sequence on Charpy impact and flexural damage behavior of composite laminates. Composite Structures. 2016;136:345-57.

10. D7264/D7264M.ASTM standard test method for flexural properties of polymer matrix composite materials. Annual book of ASTM standards 2007;7:1–1.

11. Jayaprakash K, Nandish BT, Rijesh M, Nayak J, Bhat SM et al. Fabrication of Hair and Copper Fiber Reinforced Polymethyl Methacrylate (PMMA) Composites and Evaluation of Their Mechanical Properties, Thermal Conductivity and Color Stability for Dental Applications, Trends Biomater Artif Organs., 30, 8-12 (2016).

12. Lay M, Thajudin NLN, Hamid ZAA, Rusli A, Abdullah MK, et al. Comparison of physical and mechanical properties of PLA, ABS, and nylon 6 fabricated using fused deposition modeling and injection molding. Composites Part B: 2019;176:107341.

13. Bai J, Goodridge RD, Hague RJM, Song M. Improving the Mechanical Properties of Laser-Sintered Polyamide 12 through Incorporation of Carbon Nanotubes. Polym. Eng. Sci. 2013. DOI 10.1002/pen.23459.

14. Ghomi ER, Khosravi F, Jouybar S, Bigham A, Mina Zare et al. Nylon—A material introduction and overview for biomedical applications. Review. Polym Adv Technol. 2021;1–16. DOI: 10.1002/pat.5372.

15. Siewert B, Plaza-Castro M, Sereno N, Jarman-Smith M. PEEK Biomaterials Handbook P: 333-42. https://doi.org/10.1016/B978-0-12-812524-3.00020-X.

16. Horák Z, Pokorný D, Fulín P, Slouf M, Jahoda D, Sosna A. Polyetheretherketone (PEEK). Part I: Prospects for use in orthopaedics and traumatology. Acta Chir Orthop 17. Traumatol Cech 2010;77 (6):463-9.

17. Guo Y, Chen S, Wang J, Lu B. Medical applications of polyether ether ketone. Transl Surg 2018;3:12-6.

Acknowledgment:

This work was supported by the BIRAC (Project number- BIRAC/KIIT0555/BIG-14/19) and Yenepoya Research Centre (Yenepoya University). The authors would like to thank Mr. Asim Syed Sheeraz, Incubator Manager, Yenepoya Technology Incubator, and Dheeraj Poojary, Design Engineer, Yenepoya Technology Incubator, for assisting in preparing specimens for the mechanical test.

3D PRINTING AND ITS ROLE DURING THE COVID PANDEMIC

Mohammed Zainul Arefeen[1], Hussam Bin Mehare[1], Mohammad Hassan Murtaza[1], & Mustafa Shamsi[2]

[1]Department of Mechanical Engineering, Zakir Husain College of Engineering and Technology, Aligarh Muslim University, Aligarh, India

[2]Department of Mechanical Engineering, Faculty of Engineering and Technology, Jamia Millia Islamia, New Delhi, India

hussambm9909@gmail.com

Abstract. The current COVID-19 pandemic took the entire world by surprise, making us realize how far we had been lagging behind in the development of new technologies which makes the entire fabrication process more efficient and faster. During the pandemic, worldwide there was a shortage of basic medical equipments PPE kits, masks, syringes, ventilator pipes & masks, etc. All we needed at that time was a faster and more cost-efficient fabrication technology which would have kept pace with the growing demands of these basic equipments as the situation went from bad to worse. 3D-Printing, we believe is the future of the manufacturing and fabrication sector as it's faster compared to our current technology and much more efficient when seen from an economic and an environmental point of view. In this paper, we have talked about how the COVID-19 pandemic wreaked havoc worldwide on their healthcare systems with thousands of people dying every day due to the lack of basic medical equipments. All these challenges could have been met head-on if the abilities of 3D Printing were exploited on a larger scale. We have tried to offer a brief insight into how 3D Printing can change our lives for the better and make it a lot easier once 3D Printing reaches its full potential.

1. INTRODUCTION:

1.1 COVID-19

The COVID-19 or the coronavirus disease is a malady that causes utmost respiratory disorders in the human body. This virus got the name "Corona" from the spikes, shaped like a crown on the surface of the virus. This disease was first spotted on the 8th of December 2019 in Wuhan, China, and was named "2019 Novel Coronavirus"[1-2]. The Coronavirus pandemic offered enormous difficulties to the public health, scientific and medical sectors just like the two other coronavirus breakouts SARS (2002-2004) and MERS (Middle East respiratory syndrome, 2012 – present) [1]. Due to the global spread of SARS-CoV-2, WHO declared a pandemic on March 11, 2020. This virus has had a great impact throughout the world; more than 200 million individuals have been infected by this, killing around 5 million people as of October 2021 [3]. Figure 1 gives a visual representation of the confirmed COVID-19 cases and deaths all around the world[4].

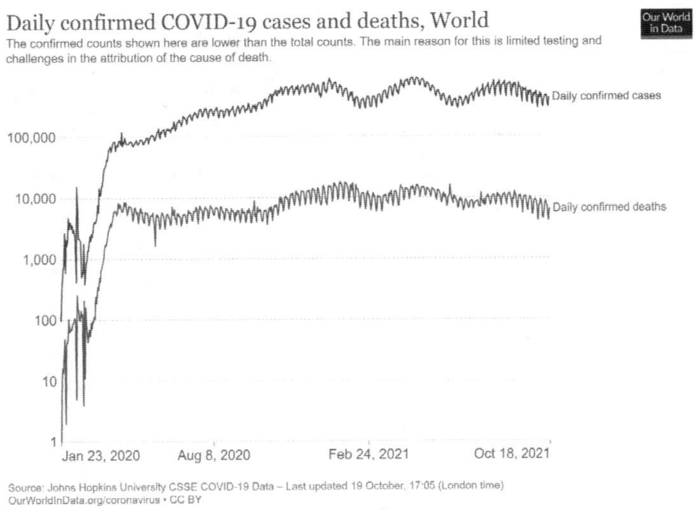

Figure-1

SARS-CoV-2 can spread from human to human through the medium of air threatening every single human on the planet. The rate of transmission of this virus is way too high than in previous outbreaks [5]. Due to these complications, it's endorsed that people must wear masks, surgical or non–surgical; they should also wear face shields, gloves, goggles, gowns, and N95 masks at public places where -- maintaining physical distance is arduous. Doing this will prevent physical contact between people who are infected, with or without symptoms; which will help in slowing the communication of this virus from person to person. This virus led to economic and social unhappiness. The healthcare systems worldwide were under tremendous pressure due to extreme shortages of beds in the hospitals; personal protection equipment (PPE) and also medical equipment because of the increasing number of patients in the hospitals globally [6].

This pandemic is a "wicked dilemma", comprising many competing issues that appear to contradict one another [7], like saving lives of people are prioritized second, and conserving livelihood is prioritized first. Although, crucial actions to limit this pandemic have begun; through creative and inventive majors, which led to the astounding methods of producing some useful essentials and one of the methods is 3D printing. Amid COVID-19, 3D printing has proven to be helpful through its ability to perform competitive tasks during emergency situations. It replaces a few conventional techniques that are implemented during public health emergencies, by acting more productively; reducing the time that was utilized in drafting, manufacturing, packaging, and transporting the product [8].

1.2 3D-Printing

3D printing is a process of manufacturing three-dimensional objects using materials viz. plastic, metal, carbon, etc. These objects are generated with the help of supporting materials -- that are applied layer by layer on a base. 3D printing for medical applications used to be aspiring. However, through hard work, dedication, and investing money and time 3D printing was made physical. Nowadays, medical and pharmaceutical firms exploit 3D printing technology in developing better-specialized medications, by manufacturing medical implants quickly and changing the way doctors and surgeons use to perform surgeries. As 3D printing or additive manufacturing has provided an easy and rapid way of prototyping 3D structures from different types of materials, it has attained huge attention. This approach of 3D printing is prominent for its way of flexibility, waste reduction, mass customization, constructability, and architectural manufacturing [9-11].

3D printing provides a flexible way of manufacturing, by designing on software like CAD; and straight away manufacturing products like face masks, ventilator components, and even nasal swabs for testing covid, rather than waiting for the components and medical items to be produced in a mass amount by third party manufacturers, who will dispatch the product - to the hospitals. Outsourcing these digital designs of 3D printed models for COVID-19 has improved the efficiency of the products.

2. **Benefits of 3D Printing in Medical Applications**

3D printing has been beneficial in the healthcare field offering additional applications every year saving and improving the lives of patients in a way that was never imagined. It has been useful in a broad range including healthcare settings. 3D printing is not only possible for cardiothoracic surgery, cardiology, gastroenterology, neurosurgery, oral ophthalmology, otolaryngology but also possible in other healthcare fields due to the benefits which this technology provides vastly. Applications of 3D printing in the medical field and healthcare are as follows.

The substantial benefit which 3D printing provides in healthcare and medical applications is that 3D printing provides feasibility to produce custom-made products and equipment. For instance, 3D printed custom prosthetics and implants provide amazing value for both the patient as well as the physicians. Also, made-to-order fixtures and jigs can be produced by 3D printing to use in operating rooms. In terms of required time for surgery and recovery of the patient, custom-made fixtures, implants, and surgical tools can possess a positive impact produced by 3D printing.

A key benefit of 3D printing is that it provides the production of items cheaply compared to traditional methods. Although these methods of manufacturing things traditionally are less expensive when production is at a large scale. 3D printing however is becoming more competitive for manufacturing products at small-scale production.

The manufacturing cost can be reduced in 3D printing by reducing the usage of resources that are not necessary. For instance, a pharmaceutical tablet of 10mg weight can be fabricated by customizing according to the requirement as 5 tablets of 1mg. Some medications can also be easily printed in the form of dosage, which will be more cost-effective for the patients to get delivered.

2.1. Personal Protective Equipment - PPE

Compared to other byzantine devices like valves and ventilators, these PPE (Personal protective equipment) were produced/printed in abundance due to their low tolerance requirement geometrically, relative simplicity, and lower risk of classification within the FDA during the pandemic [12].

During the COVID-19 pandemic, PPE kits were in shortage and this was mainly needed by the doctors to keep their bodies away from COVID exposure while they treat their patients. Complete PPE kits usually consist of gloves, gowns, shoe covers, head covers, respirators, eye protection, and goggles.

2.2. Face Shields

Face shields were considered a class A medical device by FDA, as they constitute most of the work among PPE [12]. These face shields were considered as an alternative by WHO to the medical mask during shortages. They protect the user's face from direct contamination through sprays and splashes, as they are clear plastic sheets with a head-worn frame to protect the face of the user [13-14].

The medical staff faced critical challenges while interacting with the patients, one of the major problems was a shortage of face shields. Extra variations on personal protective equipment to be included, Equipment designed by Prusa provided in a protective face shield. These types of devices provide a reusable printable headpiece in which a separate transparent plastic sheet can be attached to assemble a face shield that can protect the mouth, nose, and eyes of a person. Using Fused Deposition Modeling [FDM] face shield can be printed as it provides an improved and simple geometrical printable material. This method of using 3D printers to create face shields has completely changed the era of 3D printers [15]. Figure-2 below shows 3D Printed Face Sheilds[16].

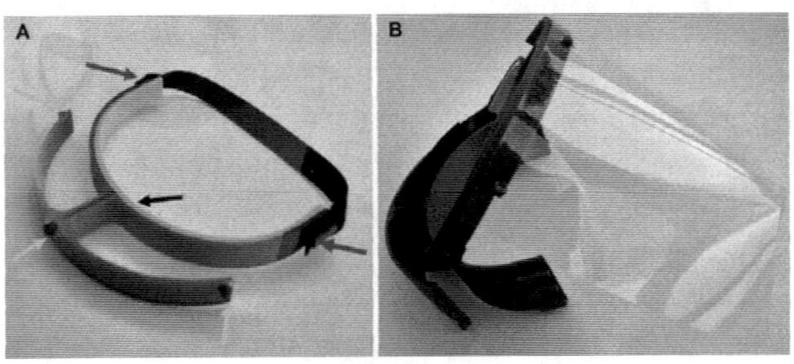

Figure-2

2.3. Face Masks

On account of the shortage of masks globally, designers are trying to achieve this demand in a different way. Adapting scuba and snorkel masks using them in a different manner, great conversion has been planned[17-18] In this conversion, a filter can be held to a face mask rapidly being smaller in size, by using a 3D printer at a very low cost. But it requires access to filter paper and mask by the person who is using it, to fit it inside the holder. Secondly, these masks are reusable just like the face shields, which could be easily used again by disinfecting them. To combat the COVID-19 pandemic, retrofitting equipment like face masks to PPE kits is one of the best ways to comprehend designers' creativity and also benefit society globally [19]. Figure-3 shows a 3D Printed N95 mask[20].

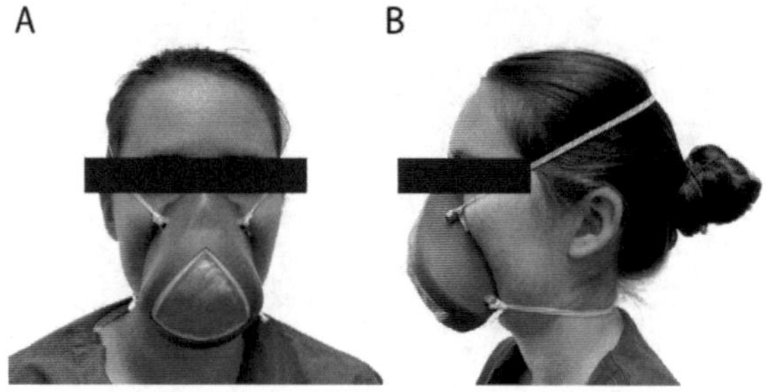

Figure-3

2.4. *3D Printed Medical Accessories*

Mechanical Ventilators are one of the most essential components supporting the patients that might be struggling in breathing. Severe respiratory complications experienced by patients are treated by getting incubated in IMV (Invasive Mechanical Ventilation) procedure [19]. During the pandemic, emergency valves like ventilator valves being the key components are 3D printed which are used to assist the breathing of the patients. The patients are supplied with oxygen through these ventilator valves, which are attached to the patient's mask. [20]

3. **Conclusion**

COVID-19 had caused almost every country to go into lockdown at some point of the pandemic, this move had really rocked the world supply chains of important medical parts and equipment such as masks, face shields, PPE kits, etc. as there were difficulties in catering to the ever-growing demands caused by the pandemic. It will not be unfair to assume the fact that many lives could have been saved if adequate supplies were present at their disposal. The gap between supply and demand was exploited and it was made clear to the world the shortcomings of the current manufacturing techniques which take quite some time to produce bulk orders, especially for critically important medical parts and equipments. 3D Printing can provide consumers with mass customization and offer on-demand manufacturing and these characteristics of 3D Printing make it a better fabrication technique when the world is suffering from a pandemic. Also, it's almost a completely computerized process that does not need human interaction at each step of the process makes it another good choice to be used during the pandemic. This review paper has talked about a few of the parts which were developed using 3D Printing during the current pandemic. But it also must be kept into consideration that the current 3D Printing still is in its dormant stage and it will take a few more years to be able to reach its capacity and this technology might be the solution to improvise the world supply chains

4. References

[1] Fauci, Anthony S., H. Clifford Lane, and Robert R. Redfield. "Covid-19—navigating the uncharted." (2020): 1268-1269.
[2] Ishack, Stephanie, and Shari R. Lipner. "Applications of 3D printing technology to address COVID-19–related supply shortages." The American journal of medicine 133.7 (2020): 771.
[3] World Health Organization. "World Health Organisation, Coronavirus disease 2019 (COVID-19): Situation Report 46." Geneva: WHO (2020).
[4] COVID-19 Data Repository by the Center for Systems Science and Engineering (CSSE) at Johns Hopkins University
[5] Ather, Amber, et al. "Coronavirus disease 19 (COVID-19): implications for clinical dental care." Journal of endodontics 46.5 (2020): 584-595.
[6] Longhitano, Guilherme Arthur, et al. "The role of 3D printing during COVID-19 pandemic: a review." Progress in Additive Manufacturing 6.1 (2021): 19-37.
[7] Cohen, Alison K., and Johnathan R. Cromwell. "How to respond to the COVID-19 pandemic with more creativity and innovation." Population Health Management 24.2 (2021): 153-155.
[8] Choong, Yu Ying Clarrisa, et al. "The global rise of 3D printing during the COVID-19 pandemic." Nature Reviews Materials 5.9 (2020): 637-639.
[9] Browne, Michelle P., Edurne Redondo, and Martin Pumera. "3D printing for electrochemical energy applications." *Chemical reviews* 120.5 (2020): 2783-2810.
[10] Kumar, Kandambath Padinjareveetil Akshay, and Martin Pumera. "3D-Printing to Mitigate COVID-19 Pandemic." Advanced Functional Materials 31.22 (2021): 2100450.
[11] Attaran, Mohsen. "3D printing role in filling the critical gap in the medical supply chain during COVID-19 pandemic." American Journal of Industrial and Business Management 10.05 (2020): 988.
[12] Novak, James I., and Jennifer Loy. "A quantitative analysis of 3D printed face shields and masks during COVID-19." Emerald Open Research 2 (2020).
[13] World Health Organization. Advice on the use of masks in the context of COVID-19: interim guidance, 5 June 2020. No. WHO/2019-nCoV/IPC_Masks/2020.4. World Health Organization, 2020.
[14] Kalyaev, Vladimir, et al. "Fast mass-production of medical safety shields under COVID-19 quarantine: Optimizing the use of university fabrication facilities and volunteer labor." International journal of environmental research and public health 17.10 (2020): 3418.
[15] Lam, Simon Ching. "Sourcing Personal Protective Equipment During the COVID-19 Pandemic: Challenging the principle of 'reasonably practicable by the flooding of counterfeit and fake face masks during the COVID-19 pandemic." JAMA-Journal of the American Medical Association (2020): 9-10.
[16] Amin, D.; Nguyen, N.; Roser, S. M.; Abramowicz, S. 3D Printing of Face Shields During COVID-19 Pandemic: A Technical Note. *J. Oral Maxillofac. Surg.* **2020**, *78* (8), 1275– 1278, DOI: 10.1016/j.joms.2020.04.040
[17] Radfar, Payar, et al. "The role of 3D printing in the fight against COVID-19 outbreak." Journal of 3D printing in medicine 5.1 (2021): 51-60.
[18] Radfar, Payar, et al. "The role of 3D printing in the fight against COVID-19 outbreak." Journal of 3D printing in medicine 5.1 (2021): 51-60.

[19] Carter, Chris, et al. "COVID-19 disease: Invasive ventilation." Clinics in Integrated Care 1 (2020): 100004.
[20] Imbrie-Moore, A.M.; Park, M.H.; Zhu, Y.; Paulsen, M.J.; Wang, H.; Woo, Y.J. Quadrupling the N95 Supply during the COVID-19 Crisis with an Innovative 3D-Printed Mask Adaptor. Healthcare 2020, 8, 225. https://doi.org/10.3390/healthcare803022

Fabrication and Automation of Drilling Machine by Using Arduino

[1]Mohammad Nasir Khan, [1]Mohd Shadman Ansari, [1]Md Irfan Ansari, [1]Saifuddin, [2]Mohd Mukhtar Alam

[1]Department of Mechanical Engineering, Aligarh Muslim University, Aligarh, Uttar Pradesh, India
[2]Department of Mechanical Engineering, Vivekananda Global University, Jaipur, Rajasthan, India

Mohammad Nasir Khan (nasir4k@gmail.com)

Abstract. For precision workpiece manufacturing, the system should have good dimensional accuracy and surface finish. In applications such as drilling, punching, marking, boring, tapping, etc., the workpiece is first positioned, and then the tool executes its action while the moving axis remains stationary. In the traditional method of such applications, manufacturers use very expensive CNC machines to program the cycle and perform the same work. Large manufacturers can afford such expensive machines, but for the small machinery manufacturing industry, we must consider low-cost solutions that can provide high-quality output. In this study, we tried to propose a low-cost design that can be used to achieve functions similar to CNC. By applying this machine in industry, multiple generations can be obtained in a short time. It is very difficult to estimate the drilling depth when manually drilling with a traditional drilling machine, and the work will usually fail due to over-drilling. In many cases, it is difficult to measure the depth after the drilling is completed; especially the depth of the fine hole cannot be measured. Therefore, an automatic drilling machine that performs the drilling function according to the generated drilling depth and transmitted to the control circuit is indispensable; therefore, undertaking the study, it exposed the technology of the dedicated drilling machine. The automatic drilling machine designed here is very useful for the mechanical workshop. The machine is built with power feed technology and is designed to drill the job to a certain specified depth.

Keywords: Fabrication, automation, drill machine, manufacturing

1. Introduction

The drilling process is an important process in industry because it is used to drill holes in metal, wood, and other materials. The drilling machine uses a drilling tool with a cutting edge at its tip. The cutting tool is fixed in the drilling machine through the taper of the chuck (sleeve), and rotates at a constant speed and is fed into the workpiece. Drilling requires manpower, the depth of the hole cannot be estimated correctly, the work may be destroyed by human error, and holes of different sizes cannot be drilled without replacing the drill bit. Repeatedly performing multiple tasks will consume a lot of time, these are disadvantages. In order to overcome all these problems, this automatic drilling machine is designed to automatically drill holes on the job according to the drilling depth data programmed through the keyboard. The main concept of this machine is to repeatedly drill holes in specific jobs at different depths, keeping the sequence. Since the machine contains a drilling motor, the motion control is accurate.

Mardane et al. [1] designed and manufactured an automatic attachment for positioning the drilling machine body according to the CAD model. A special accessory is provided to convert any drilling machine into an automatic drilling machine. With this attachment, the exact position of the working hole placed on the bed through its CAD model is achieved. In addition, Niranjan et al. [2] proposes a method to control mechanical movement by programming a microcontroller that will interface with the drilling depth feed motor. Seth et al. [3] carrying out such an arrangement on a radial drilling machine can reduce the time it takes to mark the center of the hole on the workpiece, and can avoid the need to use a microcontroller to make expensive and difficult-to-make fixtures. Alwis et al. [4], introduced the design of the PCB drilling machine, which eliminates the need to manually input the drilling coordinates and automatically detects the drilling in the circuit image. In addition, the machine uses a path planning algorithm that can estimate the effective travel path for the drill bit travel time. Jodh et al. [5] determined the design parameter considerations and configuration options, and they suggested that the gantry configuration is most suitable for better rigidity, precision and easy-to-program operation. Hong et al. [6] proposed a new multi-objective optimization method based on ant colony optimization (ACO). By expressing the efficiency and accuracy of riveting as a function of multi-degree-of-freedom point coordinates, a multi-objective optimization model for automatic drilling and riveting sequence planning is established. The time cost and accuracy of automatic drilling and riveting are described by point coordinates.

Automation is one of them, and many researchers have done a lot of research on it. The implementation of the automation system essentially eliminates the human aspect in the manufacturing process and brings many benefits. Automation will help move the drill to the desired drilling point, rather than the operator taking the time to move the drill to that point with sufficient accuracy. Automated production usually achieves higher consistency and accuracy in processing. In order to gain a competitive advantage in today's manufacturing market, a simplified manufacturing process approach is needed that focuses on increasing throughput while reducing or eliminating operator intervention. The most effective way to accomplish this task is through machine automation. Automation is the use of computers to control industrial machinery instead of manual operations. This is a step beyond mechanization. In mechanization, operators are provided with machines to help them complete their work.

2. Idea and Motivation

Manual drilling machines are the simplest and most common type of drilling machines in use today. These are light machines that are manually fed by the operator using the feed handle. In this way, the operator can "feel" the action of the cutting tool when cutting the workpiece. These drilling machines can be installed on a table or floor. They are driven by an electric motor that turns a drive belt on a motor pulley connected to the main shaft pulley. Manual feeders are essentially high-speed machines used in small workplaces that require holes of 1/2 inch or smaller. Usually, by loosening the locking bolt, the head can move up and down on the column. This allows the drilling machine to drill work at different heights. However, hand-delivered drilling rigs also have certain limitations. One of the most important issues is that a technician must always be present during operation on the machine. The operator's job is to provide manual feed to the drilling machine through the feed handle. There is also a torque limit when drilling larger holes. And because of the limited torque, the machine cannot handle harder materials.

All these limitations can be solved by using power feed drilling machines. Power feed drills are usually larger and heavier than manual feed. They can automatically feed the cutting tool into the workpiece at a pre-set cutting depth per spindle revolution, usually one thousandth of an inch per revolution. These machines are used in maintenance workshops for medium-sized work, or work with large drilling rigs that require electricity. Drill bits or cutting tools with diameters greater than 1/2 inch require power feed capability because they require more force to cut than provided by hand pressure. The speeds available on power feed machines can vary from about 50 RPM to about 1,800 RPM. The slower speed allows special operations such as counter bore, counter sinking, and reaming. These machines usually range in size from 17 inches to 22 inches with centre drilling capacity and are usually installed on the ground. They can handle drill bits up to 2 inches in diameter, which are installed in tapered Morse sockets. Larger workplaces usually use T-bolts and clamps to clamp directly on the workbench or base, while small workplaces are fixed with vice. There is a depth limiting device on the drill bit close to the spindle to help drill the hole to a precise depth. The use of power feed can solve many problems related to manual feed drilling machines. The main improvement is that there is a skilled operator who can "feel" the movement of the cutting tool when cutting the workpiece is no longer needed, because using the data acquisition system, the power feed drill can easily perform operations on a variety of different materials . Secondly, due to the introduction of the electric motor used to provide the feed, the problem of the large drilling force requirement that the manual feed drill cannot provide due to the limitation of the torque generated by the manpower is solved. In contrast, the electric motor can generate a huge amount of torque, enough to easily drill large-diameter holes.

3. Methodology

Through a brief study on the feed automation of the drilling machine, we have concluded that the electric motor used to power the drilling machine can be effectively controlled by using ARDUINO. After choosing the microcontroller, we must choose the motor according to our requirements. Our requirement is to control the vertical movement of the drilling rig. For this we need a motor with angular accuracy, speed control and torque required for operation. For these tasks, we can use servo motors as well as stepper motors.

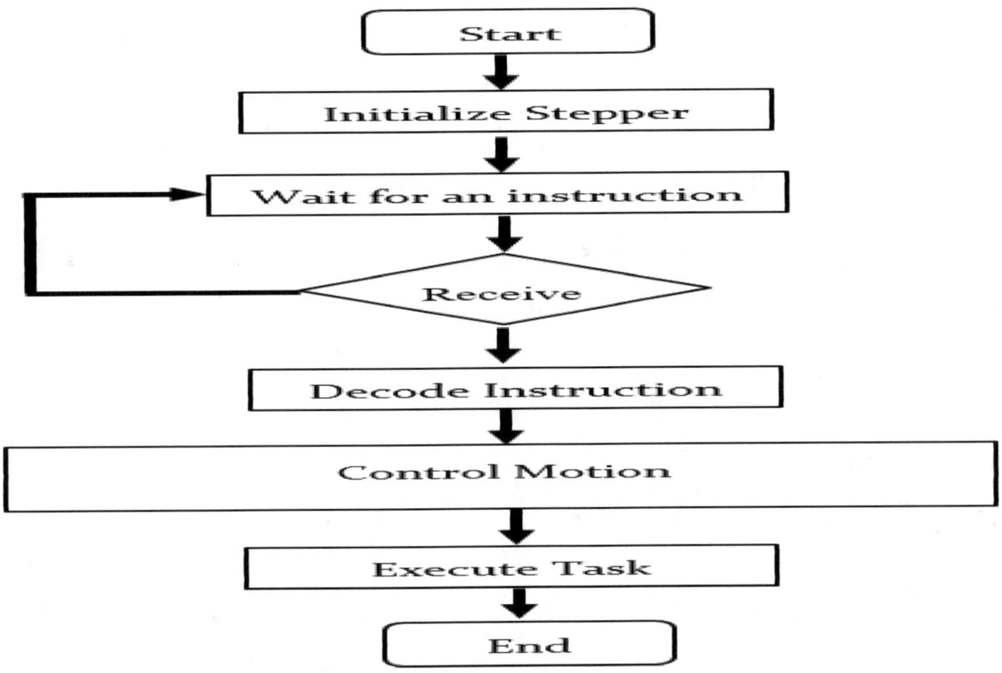

Flow Chart of the System

3.1. Selection and Fabrication

There are different types of microcontrollers available, among which we have chosen Arduino as our microcontroller. It was chosen because of its low cost and convenient operation. Choose wood as the base material of the production device. The model was designed using Solid Works software version 20.0, and then manufactured as shown in Figure 1. The material of choice for our model manufacturing is plywood. We chose this material because it provides enough strength for our model and it is easily available in the market at a reasonable price.

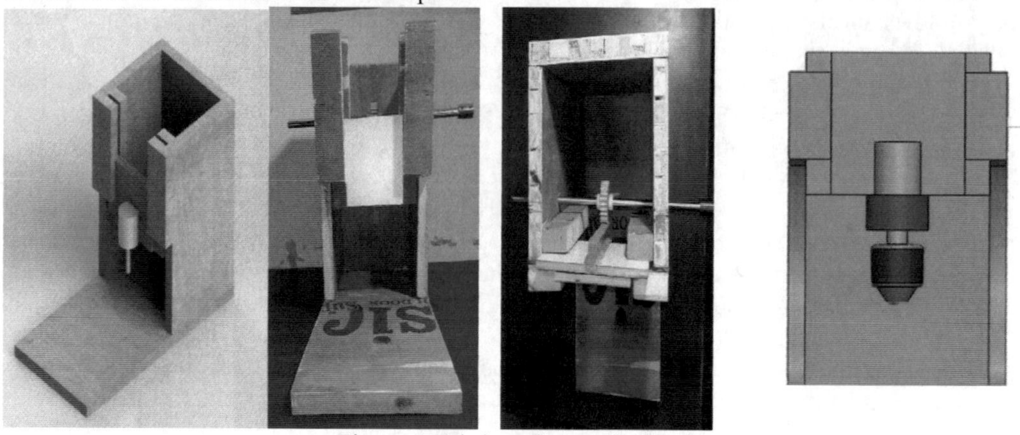

Figure 1 Fabricated experimental model

3.2. Programming for Angular Control

```
// defines pins numbers
const int stepPin = 5;
const int dirPin = 2;
const int enPin = 8;
// Button
const int b1 = A0;
const int b2 = A1;
const int b3 = A2;
const int b4 = A3;
int currentAngle = 0;
int angle = 0;
float stepPerAngle = 1.8; // full step = 1.8
int numstep;
void setup() {
Serial.begin(9600);
// Sets the two pins as Outputs
pinMode(stepPin,OUTPUT);
pinMode(dirPin,OUTPUT);
pinMode(enPin,OUTPUT);
digitalWrite(enPin,LOW);
digitalWrite(dirPin,HIGH);
pinMode(b1, INPUT);
pinMode(b2, INPUT);
pinMode(b3, INPUT);
pinMode(b4, INPUT);
    }
void loop() {
int n;
```

```
if( digitalRead(b1) == HIGH){
angle = 0;
  }
else if( digitalRead(b2) == HIGH){
angle = 45;
  }
else if( digitalRead(b3) == HIGH){
angle = 225;
  }
else if( digitalRead(b4) == HIGH){
angle = 270;
  }
if( currentAngle != angle ){
if( currentAngle < angle){
digitalWrite(dirPin,HIGH);
n = angle - currentAngle;
numstep = n / stepPerAngle;
  }
else if( currentAngle > angle){
digitalWrite(dirPin,LOW);
n = currentAngle - angle;
if( angle == 0){
n =currentAngle;
  }
numstep = n / stepPerAngle;
  }
for(int x = 0; x < numstep; x++) {
  igitalWrite(stepPin,HIGH);
  delayMicroseconds(1000);
  digitalWrite(stepPin,LOW);
  delayMicroseconds(1000);
  }
  currentAngle = angle;
  }
  delay(500);
}
```

4. Results and Discussion

We can see that all the production based industries wanted low production cost and high work rate which is possible through the utilization of multi-function operating machine which will less power as well as less time, since this machine provides working at different center it really reduced the time consumption up to appreciable limit. This project is a combined effort and the goal was to produce a cost effective drilling machine which would help the small scale industry. It would help to drill holes easily at any desired depth. Using small machine tools to fabricate small scale parts can provide both flexibility and efficiency in manufacturing approaches and reduce capital cost, which is beneficial for small business. Arduino based drilling machine is designed and implemented under very limited budget. From this paper, we can get a machine which has high accuracy and low cost as compared to large CNC machine.

4.1. Practical Implementation

a. Can be used very efficiently in small scale Industries.

IOP Conf. Series: Materials Science and Engineering 1224 (2022) 012007 doi:10.1088/1757-899X/1224/1/012007

b. Affordable and can be manufactured very easily.
c. Reduces labour cost as single computer can be used to control many drilling machines.
d. Increases product quality because human errors are eliminated.

4.2. Industrial Application

a. It is primarily utilized for holes of precise depths and various dimensions, on a variety of surfaces.
b. It is featured of performing a variety of tasks like tapping, spot facing, reaming, counter sinking and counter boring etc.
c. Precise drilling is always required for a vast number of manufacturing processes that are related to different types of metals.
d. It is also considered as one of the multi-purpose equipment's; as it also has the skill to execute a number of functions.
e. It normally utilizes one or the other type of drill bit to drill very precise holes in a specific material. Besides this, it can also be utilized for bonding two pieces of materials together with the help of some kind of fastener.
f. When the tool is utilized for placing fasteners; it simultaneously drills holes that are extremely precise, through the work piece while installing the fastener.
g. Accessible in a variety of specifications, and can also be supplied as per the customized needs of the clients in the national as well as international markets. As a result, there is a huge demand throughout the world

5. Conclusion

The method proposed in this study needs to be further improved for accurate machine in drilling operation. On large size drilling, for example in the case of PCB drilling operations, the proposed model may not be possible and provides the best solution. In addition, the joint effort and goal was to develop a cost-effective drilling machine that would help small-scale industry. The Arduino based drilling machine is designed and implemented on a very limited budget.

Conflict of interest

None.

References

[1] R. Mardane, U. Gulhane and A. Sahuis, "Design and fabrication of automated attachment for positioning bed of drilling machine with respect to cad model," *IJARIIE,* vol. 2, no. 4, pp. 95-100, 2016.

[2] G. Niranjan, A. Chandini and P. Mamatha, "Automated Drilling Machine with Depth Controllability," *International Journal of Science and Engineering Applications,* vol. 2, no. 4, pp. 90-93, 2013.

[3] K. Barad, M. Balsara, G. Patel and S. Sheth, "Automation of conventional radial drilling machine," in *nnovations in Mechatronics Engineering*, 2009.

[4] P. Alwis, A. Premarathna, Y. Fonseka, S. Samarasinghe and J. Wijayakulasooriya, "Automated printed circuit board (PCB) drilling machine with efficient path planning," in *SAITM Research Symposium on Engineering Advancements*, 2014.

[5] G. Jodh, P. Sirsat, N. Kakde and S. Lutade, "Design of low Cost CNC Drilling Machine," *International Journal of Engineering Research and General Science,* vol. 2, no. 2, pp. 189-196, 2014.

[6] X. Hong, L. Yuan, Z. Kaifu, Y. Jianfeng, L. Zhenxing and S. Jianbin, "Multi-objective Optimization Method for Automatic Drilling and Riveting Sequence Planning," *Chinese Journal of Aeronautics,* vol. 23, no. 6, pp. 734-742, 2010.

Study and Design of Arduino Based CNC Laser Cutting Machine

Mohammad Nasir Khan, Ambuj Maheshwari, Himanshu Verma

Department of Mechanical Engineering, Aligarh Muslim University, Aligarh, Uttar Pradesh, India

Mohammad Nasir Khan (nasir4k@gmail.com)

Abstract. For precise job manufacturing, system should possess good dimensional accuracy and surface finish. In applications like drilling, punching, marking, boring, tapping work piece is positioned firstly and then cutting tool performs its action while motion axes are kept stationary. In traditional approach for such applications, manufacturer uses highly expensive CNC machine for programming job cycle and executing the same. Big scale manufacturer can afford such highly expensive machines but for small scale mechanical job manufacturing industry we have to think of low-cost solution that can deliver high class output. In this project we are trying to propose low-cost design which gives similar functionality as that of high-cost CNC machine. By applying this machine in industry, it can acquire the multi generation brief time. In order to perform some operation which requires continuous cutting over a pre-defined curve, CNC machines are used as the traditional machines in the industry. Installation of a CNC machine requires a high financial investment and a skilled labour to perform the different operations on the machine. Apart from that the maintenance cost of such machine are also high. In small scale industries where the operations are performed on small job pieces, the installation of such high-cost CNC machine is very difficult; hence this project work is taken up in which a low-cost CNC laser cutting machine is designed using the Arduino. This CNC laser cutting machine will be quite useful for small scale industries where the job size is not very large. Also, the machine does not require a highly skilled labour to perform operations.

Keywords: CNC Laser cutting machine, Arduino, Proteus, design, modelling.

1. Introduction

A number of techniques have been developed by the researchers for optimizing multiple characteristics of the products. Kumar et.al. [1], design and fabricated a portable laser cutting and engraving machine and provides the fundamental goal to design and fabricate the laser cutting and engraving machine which is conveniently controlled by the Arduino CNC. It was made affordable than that of CO_2 laser cutters which is one of the precise and advantageous forms of cutting available today. CO_2 lasers are commonly pumped by passing a current through the gas mix or using radio frequency energy. They used the ELEKSMAKER Software to convert the graphical image into a G code language. Laser Diode Modules of 500mw power with 450nm wavelength are utilized as a part of an assortment of uses that require little sizes not withstanding low power utilization with long working life-time. All design outlines were done on Solid works software. Stepper motor used in the model controlled by a microcontroller called Arduino CNC Shield Driver. The key point was proficient that was to produce a laboratory modeled Laser cutting and engraving machine. The result seems for the better quality cutting and engraving. Reza et.al. [2], described the design and fabrication of a microcontroller-based plotter and focused on the sketching aspect of the industry. The complex designs from the different industrial workshops can be obtained by the use of such types of machines. In this project three motors were used to obtain the range of motion across three translational axes X, Y, and Z. The Arduino IDE was used for the programming of the microcontroller while CAMotics software was used in order to generate the G-codes.

Madekar et.al. [3], fabricated the automatic mini CNC machine for PCB drawing and drilling operation. The main intention of this study was to design and drill PCBs at a low-cost CNC system. In order to achieve this low-cost goal, the machine incorporates features of a computer with the ATMEGA 328 controller in an Arduino. For the connection of the computer and the hardware, FTDI (Future Tech Devices International) module is used which is a serial to USB converter. All the modeling was initially done on the Solid works interface and later on fabrication was done. In addition, Firangi et.al. [4], developed the model of a CNC plotter machine which is able to draw a circuit layout on PCB or any other solid surface using the electronic component. In this project, Arduino Uno is used with the atmega328p microcontroller and is used as the control device. A servo motor is used along the Z-axis for positioning the pen which moves up and down. The stepper motors used in this project are bi-polar stepper motors whereas the servo motor used is the most common S90G servo motor particularly used in the low torque robotics applications. In another study by Joshi et.al. [5], a mini pen plotter machine that is capable of drawing difficult designs and write on paper or writable surfaces using G codes was fabricate. This small machine was portable and easily assembled as per requirement. For accurate functioning, they used two Stepper motors for precise movement in the X and Y axis respectively, and one servo motor for lifting up the pen in the Z direction. The author used drivers to supply the G codes in Sequence to the stepper motors and servo motor. The author used the Arduino UNO microcontroller board which controls the position of motors with help of a program. CNC shield mounted on Arduino which distributes the current in the command of the Arduino and converts the command of G codes in digital pulse by Stepper motor. The machine worked on input as G codes of design and converted it via the use of software further transferring to Arduino, CNC Shield, Stepper motors, Servo Motor.

Jayachandraiah et.al. [6], provided the idea to develop the low-cost Router system which is capable of 3 axes simultaneously interpolation. The low cost prototyping is achieved by incorporating the features of the standard PC interface with microcontroller base CNC system in an Arduino based embedded system. With a limited budget, the author concludes that small machine tools to fabricate small parts can provide flexibility and efficiency in the manufacturing approach and reduce the capital cost, which is beneficial for small business owners. In a review, Badoniya [7] examined the recent experimental investigations in laser cutting of various engineering materials concerned with cut quality analysis. The objective of this study was to identify the most common process parameters analyzed, cut quality characteristics. The results indicated that the quality of cut depends upon many control factors or parameters such as laser beam, material, assist gas and processing parameters. Therefore, the purpose of present work is to develop a methodology that would allow us to design and simulate the portable laser cutting machine.

2. Objective

The present investigation was carried out to study and design the portable CNC laser cutting machine using Arduino and also to simulate the machine on a virtual platform and make it usable for cutting purpose for small scale industries.

2.1. Initial Design

The structure of laser cutting machine has been designed and modelling in Siemens NX 12.0 software with desired dimensions (Figure 1). All parts of the machine will be achieved before implementation of the hardware of the actual laser cutting machine. This design will be further modified by some iteration and simulation. Some iterations will be done in order to achieve a final design to make it more stable and realistic.

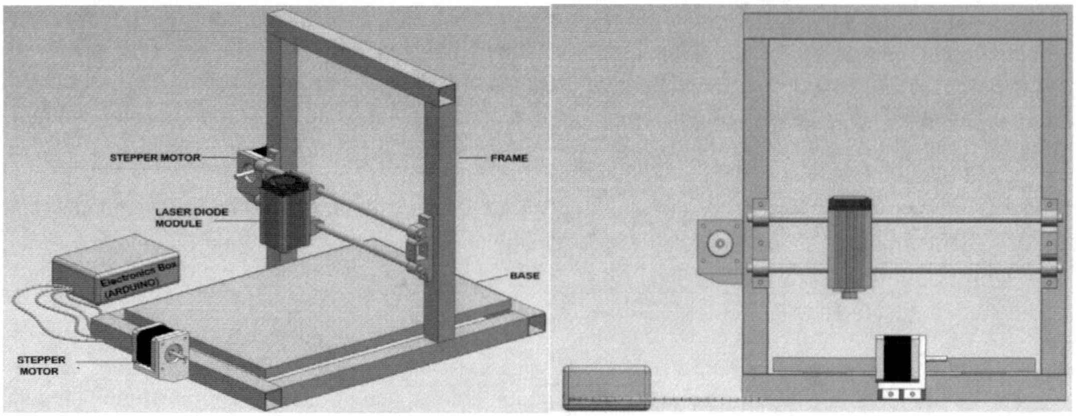

Figure 1 Isometric and Front view of the Assembly

2.2. Final Design

After some iteration and motion simulation, the final design is achieved based on the interaction between the different components of the assembly (Figure 2). The final positioning of the different components like motors, lead screws and mounting plates is decided by considering the minimum tolerance and desired stability.

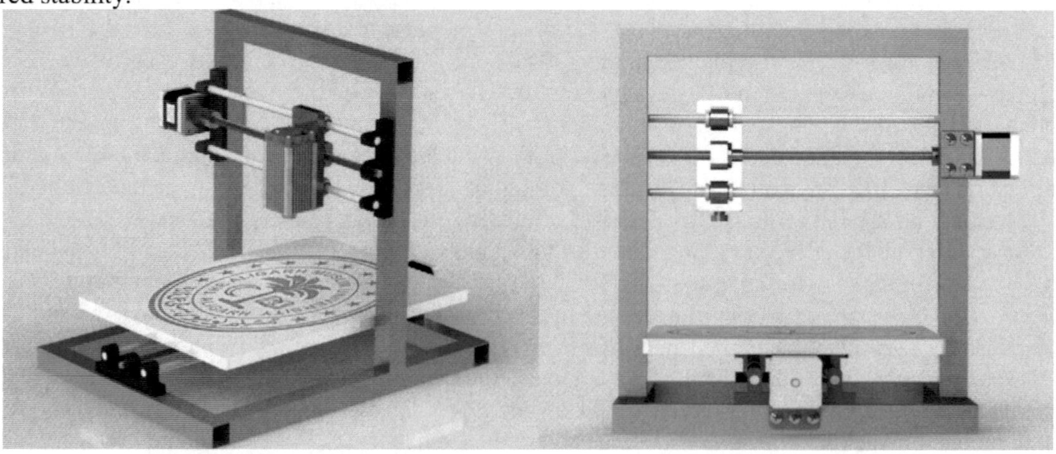

Figure 2 Isometric and Front view of the final rendered Assembly

3. Results and Discussion

3.1. Implementation of the process

Step 1 -Firstly programming of Arduino has been done.

Step 2 - Now in the Inks cape software, the design imported or the shapes to be cut sketched on the interface.

Step 3 - In the third step the design traced by the feature called bitmap tracing and after this, objects converted to paths.

Step 4 - Now the G-code file generated.

Step 5 - The G-code file opened by using GRBL controller/ Universal G-code sender.

Step 6 - The home of the machine set and the file sent to the hardware to start the cutting operations.

Step 7 – The electronic simulation is done on the Proteus software in order to validate the Arduino and stepper motor circuit.
Step 8 - In the last, motion study of the final CAD model is done to check any type of clearances between the machine parts on Solid works.

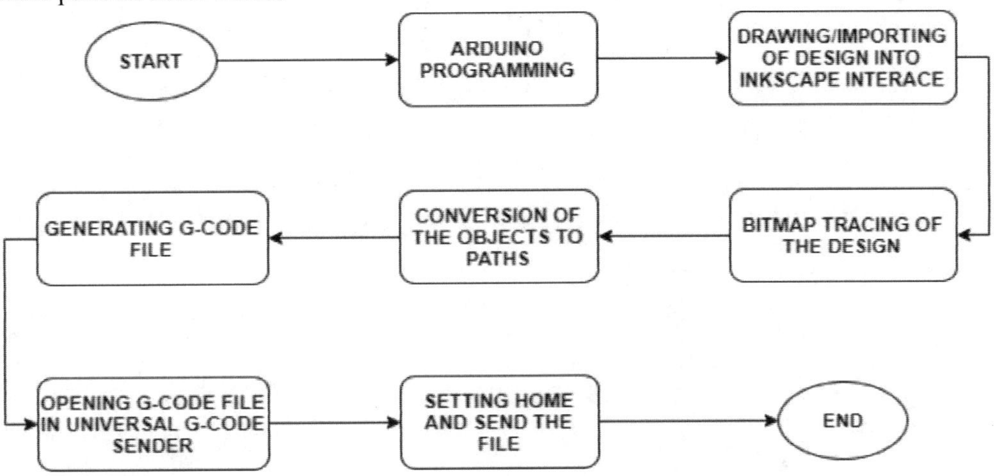

Flowchart representing the implementation of the process

3.2. Arduino Coding
For controlling the machine, Arduino UNO is used as the microcontroller. For the processing of the G-code and sending the signals to the individual driver is done by an Arduino coding using grbl library. In case of controlling the machine axis using different buttons, Arduino code was developed in the platform Arduino IDE.

3.3. Simulation of Arduino based CNC by Proteus and GRBL firmware.
In order to simulate the Arduino, stepper motor and CNC code, Proteus software is used with GRBL firmware. The Proteus Design Suite is a proprietary software tool suite developed by Labcenter Electronics used primarily for electronic design automation, schematic capture and PCB design. Its simplicity and user friendly design made it popular among electronic design engineers and technicians to create schematics and electronic prints for manufacturing printed circuit boards. Arduino can't operate NC by its own. Some firmware is used for doing that job. GRBL is one of popular and open source firmware. For virtual simulation, there is a need of virtual port driver. Virtual Serial Port Emulator named software application is used that replicates physical COM ports and are treated in the same way as a real port. G-code file is needed for the simulation that was generated previously from the Inkscape software. The main focus was to run the stepper motors through Arduino. After the code was executed. The designed circuit was simulated in Proteus software. Simpler layout of the circuit is shown in figure 3.
In order to connect the GRBL firmware with the Proteus software, there are several components used in the Proteus software like ARDUINO UNO R3, LEDBIBY, LOGICSTATE, MOTO-BISTEPPER, DRV8825,COMPIN, VIRTUAL TERMINAL. The virtual COM pin from the Virtual Serial Port Emulator pinned in Proteus to work as a real physical port. The Hex file of GRBL to Arduino is introduce in Proteus software.

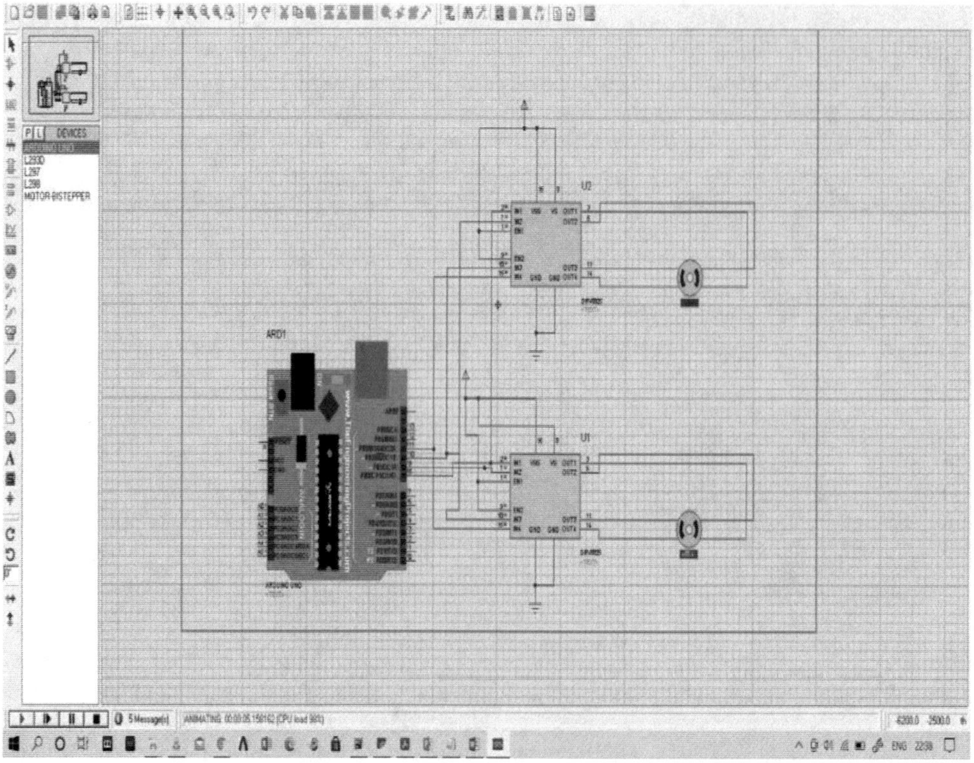

Figure 3 Simpler layout of stepper motor simulated circuit

3.4. Design Motion study

The design motion study has been carried out on the solid works software. Motion studies are graphical simulations of motion for assembly models (Figure 4). They simulate and animate the motion of a prescribed model. The main objective of the motion simulation is to see how components visibly interact with each other.

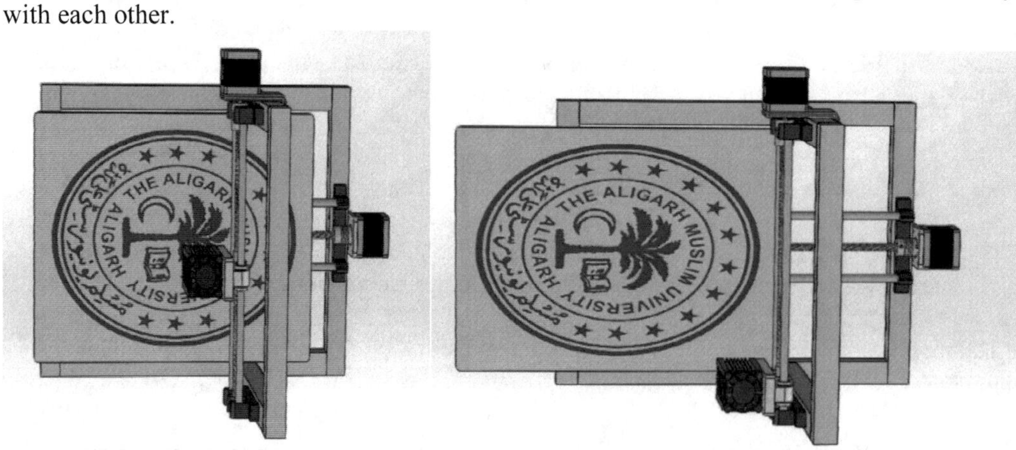

Figure 4 Motion study Frames of the final rendered Assembly

4. Conclusion

We can see that all the production-based industries wanted low production cost and high work rate which is possible through the utilization of multi-function operating machines which will consume less power as well as less time, since this machine provides working at different environments it really reduced the time consumption up to appreciable limit. This project is a combined effort and the goal of this project is to design a cost-effective CNC laser cutting machine which would help the small-scale industry. It would help laser cutting easily at certain desired depth. Using small machine tools to fabricate small scale parts can provide both flexibility and efficiency in manufacturing approaches and reduce capital cost, which is beneficial for small business. Arduino based CNC laser cutting machine is designed and implemented under very limited budget. From this paper, we can get a machine which has high accuracy and low cost as compared to large traditional CNC machines.

5. Practical Implementation

1. Can be used very efficiently in small scale Industries.
2. Affordable and can be manufactured very easily.
3. Reduces labour cost as single computer can be used to control many machines.
4. Increases product quality because human errors are eliminated.

6. Limitations

1. Cannot be used in large scale manufacturing system because of Arduino.
2. There will be the maximum size limit of the job because of the limited work area.

Conflict of interest

None.

References

[1] P. J. Kumar, A. S. S. Tarun, M. Gowtham, P. T. Rao, and G. Yashwanth, "Design and fabrication of portable laser cutting and engraving machine," Int. J. Eng. Technol., vol. 7, no. 1.1 Special Issue 1, pp. 570–573, 2018, doi: 10.14419/ijet.v7i1.1.10170.
[2] M. M. Hasan, M. R. Khan, A. T. Noman, H. Rashid, N. Ahrned, and S. T. Reza, "Design and Implementation of a Microcontroller Based Low Cost Computer Numerical Control (CNC) Plotter using Motor Driver Controller," 2nd Int. Conf. Electr. Comput. Commun. Eng. ECCE 2019, pp. 7–9, 2019, doi: 10.1109/ECACE.2019.8679123.
[3] J. M. Kajal, R. N. Kranti, R. P. Pooja, and S. M. Vikas, "Automatic mini CNC machine for PCB drawing and drilling," Int. J. Recent Innov. Trends Comput. Commun., vol. 3, no. 12, pp. 2041–2042, 2016.
[4] S. Firangi, "An Automated CNC Machine," vol. 5, no. 3, pp. 1145–1151, 2019.
[5] H. Joshi and N. Kulkarni, "A Review on Pen plotter," pp. 44–46, 2018.
[6] B. Jayachandraiah, O. V. Krishna, P. A. Khan, and R. A. Reddy, "Fabrication of Low Cost 3-Axis CNC Router," Int. J. Eng. Sci. Invent., vol. 3, no. 6, pp. 1–10, 2014.
[7] P. Badoniya, "CO_2 Laser Cutting of Different Materials – A Review," Int. Res. J. Eng. Technol., vol. 05, no. 06, pp. 1–12, 2018.

Evaluation of Surface Grinding Response Parameters of EN31 Alloy Steel

Vishal Sharma, Ganesh Kantak, Mohd Mukhtar Alam, Kunal Yadav

Department of Mechanical Engineering, Vivekananda Global University, Jaipur, Rajasthan, India

vishal_sharma@vgu.ac.in

Abstract. The purpose of this study is to understand the influence of processing parameters on surface roughness and material removal rate of EN31 Alloy Steel. Three input parameters are considered in this study, namely, the speed of the grinding wheel, number of passes and the depth of cut. The L16 orthogonal array design is used to analyse these process parameters. The results show that cutting depth and grinding wheel speed are the most effective parameters affecting material removal rate and cutting depth, and the number of passes is the most effective parameter affecting surface roughness. In addition, result showed that, the maximum MRR is 0.79051 g/sec and the surface roughness is 0.14000 µm. Therefore, it can be concluded that compared with other materials, cutting depth and wheel speed are less effective parameters, but compared with EN31 alloy steel, cutting depth and grinding wheel speed are the most effective parameters.

Keywords: Response Parameters, EN31 Alloy Steel, material removal rate, cutting depth, and surface roughness

1. Introduction

Surface grinding is one of the most commonly used processing techniques in the industry at present, and surface grinding is the best way to produce a good finish on a flat surface. George et al. [1], through Taguchi optimization technology, studied the grinding parameter of the minimum surface roughness. They experimented with different materials EN31, EN24 and die steel. Analysis of variance is used to optimize response parameters. They found that when the depth of cut changes, the surface roughness decreases from 10 microns to 20 microns. At workpiece speed (120) and cutting depth (20), the minimum surface roughness obtained is 0.47. However, Tao et al. [2] performed the investigation on the same material, but the selected input parameters were different, such as feed rate, abrasive size, grinding depth, and material removal rate. Surface integrity is preferred as the response parameter. They observed that the hardness is greatly affected by the depth of cut, and the material removal rate is most affected by the size of the abrasive. Padda et al. [3], considered the input parameters, such as wheel speed, depth of cut, and wheel particle size. In the course of the experiment, they noticed that stress will be generated when the grinding wheel is in contact with the workpiece, and observed that the speed of the grinding wheel is the most important factor in surface grinding. Lal et al. [4], taken the depth of cut, workpiece hardness and wheel type as input parameters, and observe that the output parameters are affected by the interface temperature of the tool and workpiece. In another study by Trung et al. [5], analyzed the input parameters of SUJ2 material, such as grinding wheel speed, cutting depth and working speed. After analysis, they found that as the workpiece speed and volume fraction increase, the surface roughness value increases.

Demirl et al. [6] studied ASISI 1050 steel and regarded grinding force, grain size and depth of cut as input parameters. When using an 80 grain size grinding wheel to increase the depth of cut from 0.04 mm to .06 mm, the change in grinding force increased by 50%, and burns and cracks appeared on the machined surface. However, Vinay et al. [7] check AISI D3 steel on a surface grinder. The input parameters are drying and pool conditions, feed rate, depth of cut, and dressing depth. RSM is used to find the relative influence of input parameters on surface roughness. They pointed out that the surface finish obtained under dry grinding conditions is better than wet grinding conditions. Jejurkar et al. [8], performed an experiments on AISI 321 stainless steel on a surface grinder, considering such input parameters as working speed, feed rate and depth of cut. Taguchi method and analysis of variance were used to check the values from input parameters. They pointed out that speed is the parameter that has the most influence on MRR, but high MRR can be achieved at wheel speed (1150rpm), cutting speed (16m/min) and depth of cut (10.6 μm). In a study by Puerto et al. [9], the finishing operations of hard materials were checked to improve the minimum surface roughness and precision tolerances of F5229 steel. The input parameters are considered as grinding wheel dressing, grinding conditions, grinding force and radial wear of the grinding wheel. They concluded that under aggressive grinding conditions, rough grinding wheel topography will be produced. However, in an investigation by Patel et al. [10], the surface roughness of EN8 steel was checked. The preferred input parameters are wheel speed, depth of cut, wheel material, surface grinding and working speed, and different grinding wheels for cylindrical grinding. This method applies Taguchi's method, analysis of variance, and signal-to-noise ratio, and then they conclude that the material of the grinding wheel and the grade of the grinding wheel are the most important factors in the surface and cylindrical grinding process. Therefore, the main purpose of this article is to predict the grinding behavior and achieve the best operating process parameters. This study uses the following input process parameters: number of passes, depth of cut, and wheel speed.

2. Material and Method

EN31 steel alloy is used as the workpiece material. EN31 alloy steel is widely used in various industrial applications such as ball bearings and roller bearings, spinning tools, ball roller punches and dies. 16 flat steel bars of EN31 steel were selected as workpieces with a length of 100 mm, a width of 20 mm, and a thickness of 15 mm. As shown in Figure 1. The experiments were performed at Bhartiya Skill Development University, Jaipur. The range of input parameters for the selected machine shown in Table 1.

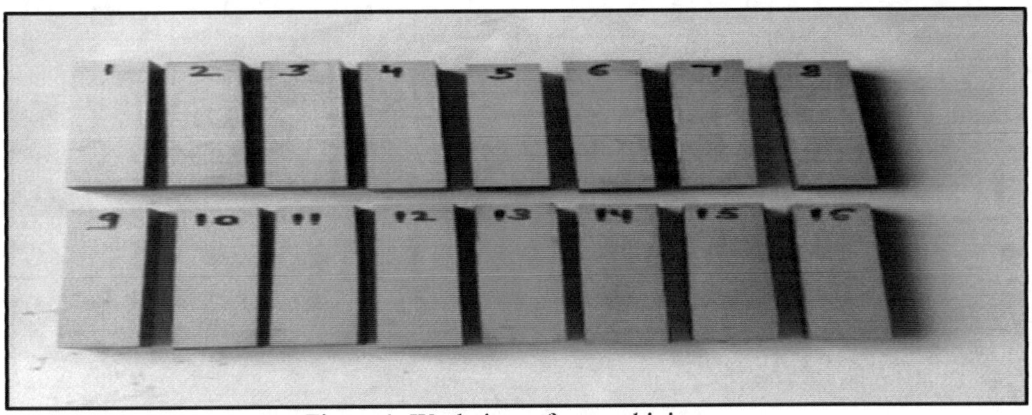

Figure 1. Workpiece after machining

Table 1: Input Parameters

Sr. No.	Input Parameters	Units	Range
1	No. of Passes	Nil	1-4
2	Depth of Cut	mm	0.01mm-1mm
3	Wheel Speed	Mm/Min	375-150mm/Min

2.1. Design of Experiments

Design of experiments is known as the technique of defining and investigating all possible conditions in an experimental involving multiple factors. This technique is also referred to as factorial design and design of experiments concepts have been in use since Sir Ronald A fishers work in agricultural experimentation. Successfully to apply designed of experiments to determine optimum treatments of land for agriculture to achieve maximum yield by Ronald A. fisher. In manufacturing process are consisting the controls factors, response factors and several noise factors. The following considerations for application of the technique

- Definition of quality
- Standardized DOE
- Loss function
- Signal to noise (S/N) ratio
- Robust design strategy

For present work L16 orthogonal array is used. Properties of L16 orthogonal array are as follow:
No. of experiments = 16 No. of levels= 4 No. of factors = 3
Here we selected L16 OA means level 4 in Taguchi design Experiment because when we select L9 the nos. of workpiece are 9 means level 3 and when we select L16 the nos. of workpieces are 16 so with the level for output result would be more optimized but with the level 3 the nos. of workpiece is less than the level 4 so output result comparatively not more optimized with the level 4 so here we selected L16 OA in this experimental work. Most of the author selected the range of the parameters near to this range and depth of cut is selected maximum up to 1mm because above the 1mm the burning of metal started. The input parameter range and DOE are shown in Table 2 and 3.

Table 2: Input parameters range

Control factors	Units	Levels			
		I	**II**	**III**	**IV**
No. of passes	Nil	1	2	3	4
Depth of Cut	mm	0.02	0.04	0.06	0.08
Wheel speed	Mm/min	375	750	1125	1500

Table 3: Design of experiment

Run order	No. of passes	Depth of Cut	Wheel speed
1	1	0.02	375
2	1	0.04	750
3	1	0.06	1125
4	1	0.08	1500
5	2	0.02	750
6	2	0.04	375
7	2	0.06	1500
8	2	0.08	1125
9	3	0.02	1125
10	3	0.04	1500
11	3	0.06	375
12	3	0.08	750
13	4	0.02	1500
14	4	0.04	1125
15	4	0.06	750
16	4	0.08	375

2.2. Measurement of Surface Roughness
Surface roughness is measured with the help of SURTRONIC S-128 surface roughness tester for each experiment as shown in Table 4.

2.3. Signal to Noise ratio
The logarithmic function used to optimize the process or product design, minimizing the changeability is called signal-to-noise ratio. To analysis the performance and optimized the process parameters Taguchi used S/N ratio and provide the optimum results. The three categories of S/N ratio generally used and the selection of one type of S/N ratio in among the three type of S/N ratio are depending on the nature of problem. The three types of S/N ratio are Nominal is better, smaller is better, larger is better. Smaller is better (S/N ratio) for surface roughness and larger is better (S/N ratio) for maximum material removal rate.

(a) Nominal is better

$$\frac{S}{N} = 10\log\frac{\mu^2}{\sigma^2} \tag{1}$$

(b) Smaller is better

$$\frac{S}{N} = -10\log\frac{1}{n}\{\Sigma_{i=1}^{n} y_i^2\} \qquad (2)$$

(c) Larger is better

$$\frac{S}{N} = -10\log\frac{1}{n}\{\Sigma_{i=1}^{n} \frac{1}{y_i^2}\} \qquad (3)$$

Where

μ^2= desired square of mean

σ^2 = Variance of observation of replicated response variable.

Table 4 MRR and surface roughness value

No. of experiments	No. of Passes (N)	Depth of cut (D)	Wheel speed (V)	Material Removal rate (MRR)	Surface roughness (Ra)
1	1	0.02	375	0.41667	0.22667
2	1	0.04	750	0.78740	0.21000
3	1	0.06	1125	0.41841	0.15333
4	1	0.08	1500	0.79051	0.24667
5	2	0.02	750	0.11148	0.22667
6	2	0.04	375	0.13514	0.18667
7	2	0.06	1500	0.16667	0.14667
8	2	0.08	1125	0.33223	0.16667
9	3	0.02	1125	0.07937	0.21000
10	3	0.04	1500	0.10672	0.14000
11	3	0.06	375	0.20471	0.19667
12	3	0.08	750	0.10549	0.23667
13	4	0.02	1500	0.27397	0.16667
14	4	0.04	1125	0.08224	0.25667
15	4	0.06	750	0.07645	0.34333
16	4	0.08	375	0.07692	0.24000

3. Results and Discussion

The higher the signal to noise ratio in more favourable is the effect of the input variable on the output. The graph shows that the optimum value levels for best surface roughness (minimum) and MRR maximum. Table 5 shows the response table for signal to noise ratios (smaller is better) in which wheel speed is ranked 1 and Table 6 shows the response table for signal to noise ratios (larger is better) in which nos. of passes is ranked 1. In addition, the optimal value for material removal rate (MRR) is 0.79051 gram/ sec obtained and the optimum machining parameters are number of passes 1, depth of cut of 0.08 mm and 1500 rpm of wheel speed. When compared with the other materials depth of cut and wheel speed is less effective parameters but with the EN31 alloy steel depth of cut and wheel speed are most effective parameters. However with the increase in no of passes, decrease the depth of cut and increase the wheel speed surface roughness increases and material removal rate also decreases. However the minimum surface roughness obtain was 0.47 at workpiece speed (120), depth of cut (20) on ASISI 1050 steel but with the EN31alloy steel maximum MRR 0.79051 gram/ sec is obtain at depth of cut 0.08mm, no of passes 1 and wheel speed 1500rpm and minimum surface roughness surface roughness 0.14000 µm is obtain at 0.06mm depth of cut, no of passes 3 and wheel speed 1500rpm. Moreover, main effect plot of S/N ratio for surface roughness and MRR are shown in Figure 2 and 3 respectively.

Table 5 Response Table for Signal to Noise Ratios (Smaller is Better)

Level	No. of passes	Depth of cut	Wheel speed
1	13.72	13.73	13.50
2	14.93	14.26	12.06
3	14.32	14.09	14.30
4	12.26	13.16	15.37
Delta	2.66	1.10	3.31
Rank	2	3	1

Table 6 Response Table for Signal to Noise Ratios (Larger is Better)

Level	No. of passes	Depth of cut	Wheel speed
1	-4.823	-14.978	-15.261
2	-15.394	-15.149	-15.750
3	-18.689	-14.810	--15.211
4	-19.389	-13.357	-12.071
Delta	14.566	1.791	3.679
Rank	1	3	2

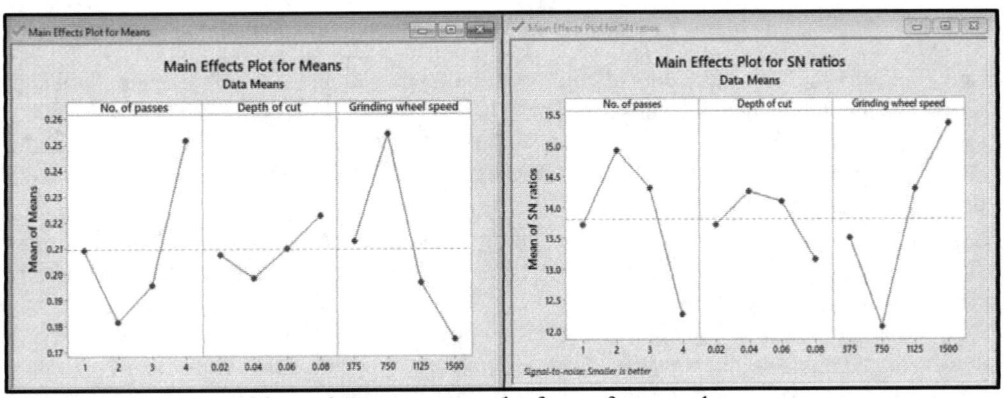

Figure 2 Response graphs for surface roughness

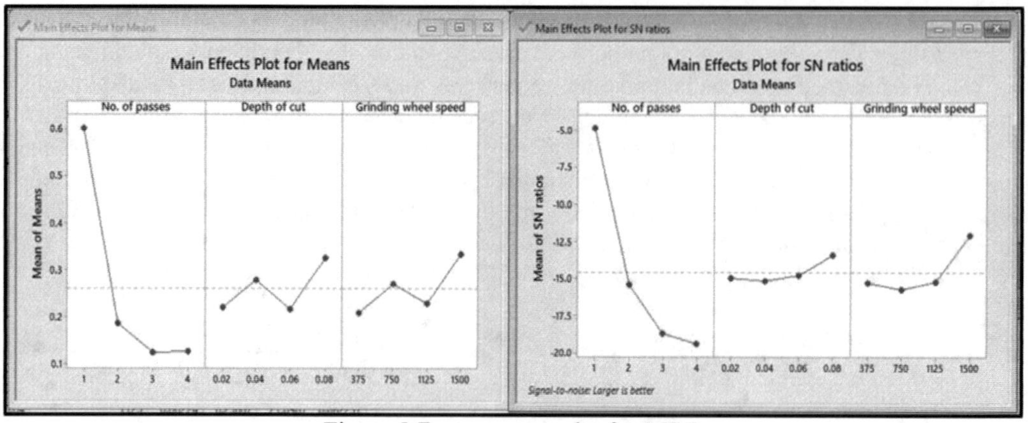

Figure 3 Response graphs for MRR

4. Conclusion

For EN31 steel following conclusion are analysed as follows:

For Material Removal Rate (MRR), depth of cut was the most effective parameter for EN31 alloy steel work material followed by grinding wheel speed, number of passes and depth of cut. So, to achieved the maximum material removal rate for EN31 alloy steel at high depth of cut 0.08 mm, moderate no. of passes 1 and high wheel speed 1500 rpm.

For Minimum surface roughness (Ra), no of passes was the most effective parameter for EN31 alloy steel work material followed by grinding wheel speed, number of passes and depth of cut. So, to achieved the minimum surface roughness for EN31 alloy steel moderate depth of cut 0.06 mm, No. of passes 3 and high wheel speed 1500 rpm.

Conflict of interest
None.

References

[1] L. P. George, K. V. Job and I. M. Chandran, "Study of Surface Roughness and its Prediction in Cylindrical Grinding Process Based on Taguchi Method of Optimization," *International Journal of Scientific and Research,* vol. 3, no. 5, pp. 1-5, 2013.
[2] Z. Tao, S. Yaoyaoa and S. Laaksoa, "Investigation of the Effect of Grinding Parameters on Surface Quality in Grinding of TC4 Titanium Alloy," *International Conference on Flexible Automation and Intelligent Manufacturing,* no. 11, pp. 2131-2138, 2017.
[3] P. A. Singh, K. Satish and M. Aishna, "Effect of Varying Surface Grinding Parameters on the Surface Roughness of Stainless Steel," *International Journal of Engineering Research and General Science,* vol. 3, no. 6, pp. 314-319, 2015.
[4] A. V.Lal and P.Dinesh, "Influence of Machining Parameters on the Surface Finish During Surface Grinding," *International Journal of Innovative Research in Science, Engineering and Technology,* vol. 6, no. 6, pp. 11900-11907, 2017.
[5] D. D. Trung, N. V. Thien and H. T. Dung, "Predictive Surface Roughness of Workpiece in Surface Grinding," *Roughness American Journal of Materials Research,* vol. 4, no. 6, pp. 37-41, 2018.
[6] H. Demirl, A. Gullu, C. Ibrahim and S. Ulvi, "An Investigation into the Influences of Grain Size and Grinding Parameters on Surface Roughness and Grinding Forces when Grinding," *Strojniski Vestnik Journal of Mechanical Engineering,* pp. 447-454, 2010.

[7] P. V. Vinay and C. S. Rao, "Experimental Analysis and Modelling of Grinding AISI D3 Steel," *International Journal of Recent Advances in Mechanical Engineering,* vol. 4, no. 1, pp. 47-60, 2015.
[8] A. S. Jejukar and V. L.Kadlag, "Optimization of Surface Grinding Process Parameter for AISI 321 by using Taguchi Method," *IJARIIE,* vol. 2, no. 4, pp. 108-185, 2016.
[9] P. Puertoa, R. Fernandeza, J. Madariagab and J. Arana, "Evolution of Surface Roughness in Grinding and its Relationship with the Dressing Parameters," *Engineering Society International Conference,* pp. 174-182, 2013.
[10] D. K. Patel, D. Goyal and B. Pabla, "Optimization of Parameters in Cylindrical and Surface Grinding for Improved Surface Finish," *Department of Mechanical Engineering National Institute of Technical Teachers Training and Research,* pp. 1-11, 2018.

Optimization of Surface Roughness and Material Removal Rate in Turning of AISI D3 Steel with Coated Carbide Inserts

Surya Prakash Anand, Amit Kumar Saraf
Department of Mechanical Engineering, Jagannath University, Jaipur, India

suryagupta357@gmail.com

Abstract. The aim of this paper is to optimize the metal removal rate and surface roughness through turning on AISI D3 Steel with coated carbide inserts. As per the current market scenario, for the metal cutting industries the most important challenge in the turning process is to satisfy both requirements quality as well as high productivity in less time in order to remain competitive. Among all the different cutting processes, the turning process is one of the most applied and fundamental metal removal operations in the real manufacturing environment. This work examines the impact of machining factors like speed, depth of cut & feed on roughness and MRR of AISI D3 steel which have a hardness value of 60 HRC with coated carbide inserts. At high speed and feed rate optimum MRR was achieved and opposite to that at low feed and depth of cut optimum surface roughness was obtained. Taguchi L9 (3)3 OA has been applied for experimental design. ANOVA analysis is performed just after DOE to identify the significant factor which influences MRR and surface roughness. Among the three machining factors, depth of cut and speed has more influencing effect on roughness and the feed influences MRR. Effort was made to find optimum solution to minimize the roughness and maximize the MRR by Taguchi and ANOVA technique.

Keywords: ANOVA, MRR, AISI D3 Steel, Surface Roughness, manufacturing.

1. Introduction

In the current scenario, all machining industry has its main focus in achieving high quality surface finish, dimensional accuracy, minimum tool wear, high-production rate and improving performance of product with less effort. Turning plays a very important role in the machining process, however with the turning other processes may also require several tools and for finishing several setup is required [1,2].

In this research, there are three processing parameters used which are cutting speed, depth of cut & feed. The roughness is dependent on these three parameters and tool geometry as well. In this work Taguchi L9 is used as an orthogonal array for experimental design followed by ANOVA analysis is performed on D3 steel to find out the significant factor on MRR and surface roughness.

Turning operation: Turning process is widely used for metal cutting in manufacturing companies. The primary function of turning is to remove metal from the outside diameter of the cylindrical shape work piece. Turning is used for reducing work piece diameter to get the desired dimensions and produces good surface finishing of metal [7].

Cutting speed: The cutting speed refers to what speed the tool cuts the work piece. Speed of cutting varies from metal to metal. All different materials have their optimum cutting speed. It may be defined as the speed at which cutting tools move with respect to a workpiece. It is mainly expressed in m/min. Cutting speed represented as:

$V = \pi DN/1000$

Hence, D=initial work piece diameter (mm), V=cutting speed (m/min), N=speed of spindle (rpm)

Depth of cut: Depth of cut mainly known to be layer thickness which is get removed by machining on work piece or it is the distance between the new surface obtained to original workpiece surface. It is expressed by mm. Depth of cut mathematically given as:

$d = (D1-D2)/2$

Where, D1 is work piece initial diameter (mm), D2 is work piece final diameter (mm)

Feed: It is basically stated that the distance of the tool moves a single or per revolution of the work piece. For better finish a finer feed is required. Feed is direct proportional to speed of spindle & is express by mm/rev [8,9].

It is given as: $F_m = f \times N$

Hence, Fm= speed of feed per min. (mm/min), F = feed (mm/rev), N = spindle speed (rpm)

2. Literature search methodology

Basically the term turning process has its own reputation in the metal cutting industry. It is mainly used in workshop practice for the purpose of performing conventional machine tools as well as CNC and NC machine tools, manufacturing system and machining centres. Turning process basically involves the lathe machine in its operation and its main function is to produce cylindrical and conical parts. Hence it is important to improve the tool life, reduce cutting force, increase surface finish & feed force and reduce temperature of machining zone in turning operation by optimization study. For minimizing the friction & wear, improving tool life & surface finish we mainly take use of cutting fluids [10,11,12]. Result signifies that the cutting speed has most influences on surface roughness and on MRR the feed rate gives more impact than other factors. At last for knowing the influence of input variables on MRR and surface roughness the normal probability plot, main effects plot and contour plot is to be found [13]. Grinding is a slow process with a low amount of material removal rate so it needs some replacement to increase MRR. Therefore the hard turning process is the new solution to that and it comes in early ninety [3]. There are 3 machining parameters like speed, feed & depth of cut but there are many other factors that may affect the work material. It was found out the impacting of these three factors on MRR & surface roughness. From this we get to know which parameter has more or less impact on roughness and MRR [4, 5].Calculation method & DOE is used to examine the quality of surface. After that for achieving the main objective the DOE method is further executed with appropriate values and data. DOE table has been created to use Taguchi method and ANOVA method is used for finding the parameter which affects the roughness and MRR [6, 14, 15].

3. Objective & Approach:

- In turning process, the main objective is to examine the impact of processing factors on MRR and roughness through using cutting tool on steel AISI D3 when performing these experiments.
- To examine result of optimal values for MRR and roughness then the operation is executed in D3 steel for CI level at 95%.
- To create an experiment setup with different input parameters and then execute the process to find the response.
- To create DOE, Taguchi L9 orthogonal array which use to examine effect of processing variables.
- ANOVA technique was used for analysing contribution of percentage & to get the significant factor from machining factors.

4. Experimental Detail

Work Materials: In this study, the use of D3 steel as a workpiece material in shape of cylinder & has its dimension i.e. diameter of 30 mm and length of 80 mm. Since the metal have very large application in metal cutting industry. It possesses different properties like wear resistance, high strength and good hardenability. Typical applications for D3 steel: Punching tools, drawing & deep drawing tools, shear blades, dies drawing, blanking, shear blades, cold die punches, trimming dies, etc.

4.1 Experiment procedure

In this section, we are performing the experiments basically to find out the main effects of different processing parameter on the value of (Ra &Rz) & material removal rate (MRR). Then the experiments with the three machining parameters were performed to three different levels, which is mention in Table 1. Taguchi L9 (33) OA in Minitab 18 software has 8 degree of freedom. Before executing the Taguchi process, turning operation was performed on different 9 work pieces in wet condition on a CNC lathe machine, it has 350 rpm spindle speed & have 16 KW max power. Before going to process main work, first removed the rust layer from 9 work pieces by cutting insert mainly for minimize the effects on experiment values. Then after performing turning operation on different work piece, following parameters were found out.

Table 1. OA L9 of Taguchi's experimental design and experiment result

Run No.	**v**	**f**	**d**	**Ra (μm)**	**Rz (μm)**	**MRR (gm/sec)**
1	200	0.05	0.20	**0.86**	**7.12**	**0.12613**
2	200	0.09	0.30	**0.98**	**8.14**	**0.21510**
3	200	0.13	0.40	**1.25**	**10.09**	**0.27686**
4	230	0.05	0.30	**1.15**	**10.01**	**0.15690**
5	230	0.09	0.40	**0.90**	**8.01**	**0.24616**
6	230	0.13	0.20	**0.87**	**7.20**	**0.22107**
7	260	0.05	0.40	**0.81**	**6.90**	**0.19672**
8	260	0.09	0.20	**0.74**	**6.02**	**0.25674**
9	260	0.13	0.30	**0.80**	**6.81**	**0.32981**

After completing the experiment, there are developed experiment data which were used for distinguishing the effects of processing parameters cutting speed (v), feed (f) and depth of cut (d) on our output variables, that is roughness (Ra &Rz) & MRR. All machining process completed, to get the results of roughness & MRR which are shown in tabulated form in Table 5. At run no. 8, finally achieved the minimum value of surface roughness (Ra &Rz) at v = 260 m/min, d = 0.20 mm & f = 0.09 mm/rev and at run no. 3, It is get the highest value of surface roughness Ra & Rz at d = 0.40 mm, v = 200 m/min & f = 0.13 mm/rev. At last the highest value of MRR was achieved at run no. 9 when d = 0.30 mm, f = 0.13 mm/rev & v = 260 m/min.

IOP Conf. Series: Materials Science and Engineering 1224 (2022) 012010 doi:10.1088/1757-899X/1224/1/012010

5. Results and Discussion

Results of ANOVA: By using the MINITAB software, ANOVA method gives final experimental and evaluates result. ANOVA result for MRR & surface roughness (Ra &Rz) are tabulated below (in Table 2, Table 3 & Table 4). At confidence interval (CI) 95% following analysis was done at α = 0.050 as acceptance level. Below tabulated table give the idea for percentage contribution of main parameters and that shows the impact value on result obtained.

Table 2 ANOVA for Ra

Source	DOF	Seq SS	Contribution (%)	Adj SS	Adj MS	F-value	P-value
v	2	0.134274	68.04%	0.134274	0.067137	24.09	0.040
f	2	0.010877	5.51%	0.010877	0.005438	1.95	0.339
d	2	0.046614	23.62%	0.046614	0.023307	8.36	0.0107
Error	2	0.005574	2.82%	0.005574	0.002787		
Total	8	0.197339	100.00%				
S = 0.0527904			**R-sq = 99.38%**		**R-sq(adj) = 88.70%**		

Table 3. ANOVA for Rz

Source	DOF	Seq SS	Contribution (%)	Adj SS	Adj MS	F-value	P-value
v	2	10.7710	58.71%	10.7710	5.38549	366.11	0.003
f	2	1.2066	6.58%	1.2066	0.60330	41.01	0.024
d	2	6.3382	34.55%	6.3382	3.16912	215.44	0.005
Error	2	0.0294	0.16%	0.0294	0.01471		
Total	8	18.3453	100.00%				
S = 0.121285			**R-sq = 99.84%**		**R-sq (adj) = 99.36%**		

After applying ANOVA, we get the table Ra, Rz& MRR. In Table 2, from this table we can see that which for Ra, our 1st main factor cutting speed mainly have 68.04% of contribution followed by depth of cut which have 23.62 % of contribution and last is the feed with 5.51% contribution.

In Table 3, in this table we see Rz is mainly affected by cutting speed which has a contribution of 58.71%, second factor is depth of cut with 34.55% contribution and the third parameter is feed with minimum 6.58% of contribution. In Table 4, the first most important factor which affects the MRR was feed with 71.19% of contribution and the 2nd factor that effect the MRR is our cutting speed which has mainly 17.15% of contribution & the last one factor was our depth of cut with 11.01% of contribution Last contribution we see for roughness (Ra & Rz) & MRR is error. For MRR, Ra & Rz, the error contribution is 2.02%, 2.82% and 0.65% respectively. From this we can say that error contribution has negligible effect, then it may be guided that important parameter is not forgotten and no involvement of measurement error.

Interpretation of Plots: We have use the MINITAB 18 software for different interpretation plots, with the help of it we show the probability graph &main effect plot for Ra, Rz & MRR in Fig.1 and Fig. 2.

From these graphs our all 3 machining parameter i.e. depth of cut, feed & cutting speed and will show deviation of response.

Table 4. ANOVA for MRR

Source	DOF	Seq SS	Contribution (%)	Adj SS	Adj MS	F-value	P-value
v	2	0.149324	17.15%	0.149324	0.074662	26.47	0.036
f	2	0.619868	71.19%	0.619868	0.309934	109.89	0.009
d	2	0.095894	11.01%	0.095894	0.047947	17.00	0.056
Error	2	0.005641	0.65%	0.005641	0.002820		
Total	8	0.870727	100.00%				
S = 0.0531082			R-sq = 99.35%		R-sq(adj) = 97.41%		

Table 5. Ra & Rz Mean Value

Level	Criteria Ra (μm)			Criteria Rz (μm)		
	$\overline{V}$	$\overline{f}$	$\overline{d}$	$\overline{V}$	$\overline{f}$	$\overline{d}$
1	1.0300	0.9400	**0.8233**	8.450	8.010	**6.780**
2	0.9733	0.8733	0.9767	8.407	7.390	8.320
3	**0.7833**	0.9733	0.9867	**6.577**	8.033	8.333
Delta	0.2467	0.1000	0.1633	1.873	0.643	1.553
Rank	1	3	2	1	3	2

Table 6. MRR Mean Value

Level	$\bar{V}$	$\bar{f}$	$\bar{d}$
1	0.2060	0.1599	0.2013
2	0.2080	0.2393	0.2339
3	**0.2611**	**0.2759**	0.2399
Delta	0.0551	0.1160	0.0386
Rank	2	1	3

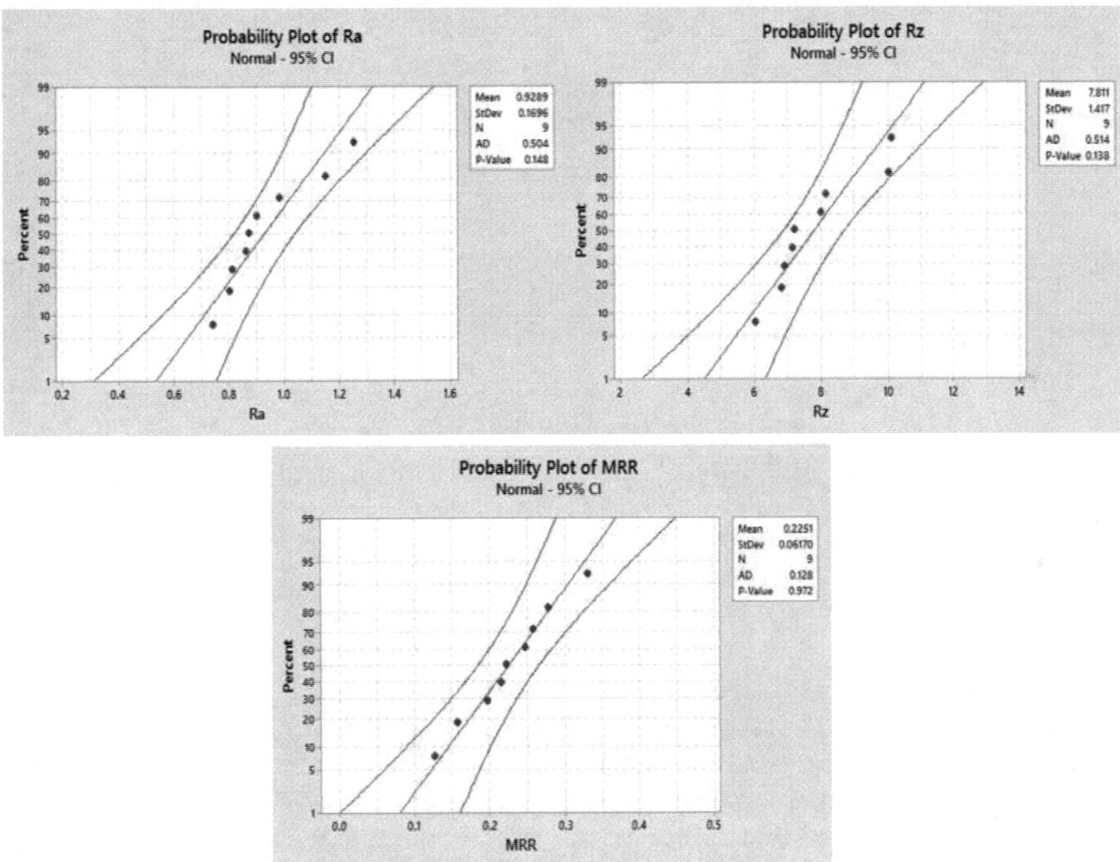

Fig. 1 Probability Plot for Ra, Rz & MRR

We made the inspection for diagnostic plot of model to test its statistical validity. Normal probability plot is plotted in between the residuals and predicted response for the surface roughness (Ra & Rz) & MRR which is shown in figure 1. The residuals are following the straight line closely. From the figures we made a conclusion that these values are in range of 95% CI level.

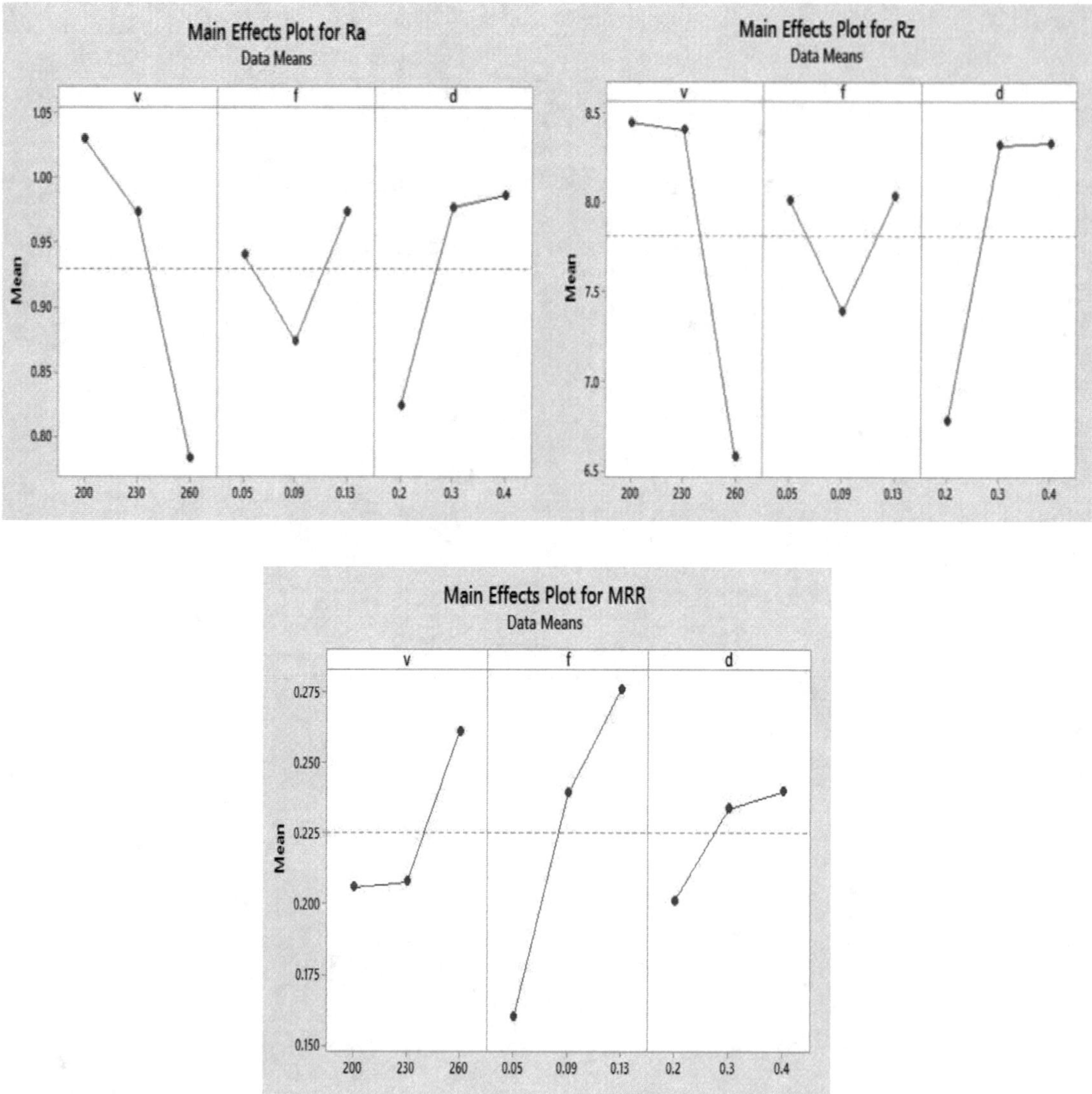

Fig. 2 Main Effects plot for Ra, Rz & MRR

For Ra & Rz, from the graph, the outcome result shows that if we increase depth of cut then roughness value get increases. Whereas when we see the graph for cutting speed, if we increase the value of speed then roughness value get decreases. In feed graph, when we increase feed rate then roughness value first decreases and then start increasing. From main effects graph, the optimum condition for roughness is achieved at feed is 0.09 mm/rev, depth of cut is 0.20 mm & cutting speed is 260 m/min.
For MRR, It can say that by increasing the value of cutting speed, feed and depth of cut then the value of MRR is also increases. From the graph, it visible to us that we obtain highest value of MRR for all machining parameters when feed is 0.13 mm/rev, depth of cut is 0.40 mm & cutting speed is 260 m/min.

Optimal Design:

From these above figures of main effects plot of MRR & surface roughness (Ra & Rz), we can see speed and depth of cut are most influencing variables on Ra & Rz and cutting speed & feed are most influencing variables for MRR. We achieve a smaller values of surface roughness at level 3 when cutting speed 260 m/min and at level 1 when depth of cut 0.20 mm and for MRR, we achieve a higher value at level 3 when cutting speed 260 m/min & at level 3 when feed 0.13 mm/rev.

From Table 5, **for roughness (Ra)** our approximate mean value solved as:

$\mu_{Ra} = \overline{V}_3 + \overline{d}_1 - \overline{T}_{Ra}$ (from Table 2 & 6, $\overline{T}_{Ra} = 0.9289$)

$= (0.7833 + 0.8233) - 0.9289$

$\mu_{Ra} = 0.6777$

At C.I. level 95% roughness is given mathematically:

$$\text{C.I.} = \sqrt{\frac{F95\%;(1,DOFerror)\ X\ Verror}{\eta eff}}$$

where, $\eta_{eff} = \frac{N}{1+\text{DOF associated to that level}} = \frac{9}{1+2+2} = 1.8$

$F_{95\%(1,2)} = 18.51$ & $V_{error} = 0.002787$ (from Table 3)

Hence, $CI_{Ra} = \sqrt{\frac{18.51\ X\ 0.002787}{1.8}}$

$CI_{Ra} = 0.2271$

C.I. 95% approximate range for Ra = $[\mu_{Ra} - CI_{Ra}] \leq \mu_{Ra} \leq [\mu_{Ra} + CI_{Ra}]$

That is, $(0.6777 - 0.2271) \leq \mu_{Ra} \leq (0.6777 + 0.2271)$

$0.4506 \leq \mu_{Ra} \leq 0.9048$ µm.

For roughness (Rz), approximate mean value solved as:

$\mu_{Rz} = \overline{V}_3 + \overline{d}_1 - \overline{T}_{Rz}$ (from Table 2 & 6, $\overline{T}_{Rz} = 7.8111$)

$= 6.577 + 6.780 - 7.8111$

$\mu_{Rz} = 5.5459$

where, $\eta_{eff} = \frac{9}{1+2+2} = 1.8$

$F_{95\%(1,2)} = 18.51$ & $V_{error} = 0.01471$ (from Table 4)

Hence, $CI_{Rz} = \sqrt{\frac{18.51\ X\ 0.01471}{1.8}} = 0.3889$

CI 95% approximate limit for Rz = $[\mu_{Rz} - CI_{Rz}] \leq \mu_{Rz} \leq [\mu_{Rz} + CI_{Rz}]$

That is, $(5.5459 - 0.3889) \leq \mu_{Rz} \leq (5.5459 + 0.3889)$

$5.1570 \leq \mu_{Rz} \leq 5.9348$ µm.

For MRR, approximate mean value determined as:

$\boldsymbol{\mu_{MRR} = \bar{f}_3 + \bar{v}_3 - \bar{T}_{MRR}}$ (from Table 2 & 7, $\bar{T}_{MRR} = 0.2250$)
$= 0.2759 + 0.2611 - 0.2250$
$= 0.3120$

Where, $\eta_{eff} = \frac{9}{1+2+2} = 1.8$

$F_{95\%(1,2)} = 18.51$ & $V_{error} = 0.00282$ (from Table 5)

Hence, $CI_{MRR} = \sqrt{\frac{18.51 X 0.00282}{1.8}} = 0.1702$

CI 95% approximate limit for MRR = $[\ \mu_{MRR} - CI_{MRR}\ ] \leq \mu_{MRR} \leq [\mu_{MRR} + CI_{MRR}\ ]$

That is, $(\ 0.3120 - 0.1702\ ) \leq \mu_{MRR} \leq (\ 0.3120 + 0.1720\ )$

$0.1418 \leq \boldsymbol{\mu_{MRR}} \leq 0.4822$ gm/sec

6. Conclusion

In present work, test of turning has being performed on AISI D3 steel which have hardness of 60 HRC machined with coated carbide tool to find the effects of processing parameter on roughness (Ra & Rz)& MRR.

1. In MINITAB 18 software, we create DOE, Taguchi L_9 OA and ANOVA method for evaluating the influence of processing parameter on roughness and MRR. We find out the optimal values of maximum MRR and least roughness.
2. After this experiments we got a clear image in this cutting speed is the most influencing factor on Ra & Rz with % contribution of 68.04% & 58.71% respectively. Depth of cut was 2nd affecting variable on Ra & Rz that have contribution of % with 23.62% &34.55% respectively. Feed is last parameter which affects Ra & Rz which has the percentage contribution 5.51% & 6.58 % respectively.
3. For MRR, first influencing factor is feed with contribution of 71.19% following factor is speed with 17.15% & the last is depth of cut with % contribution of 11.01%.
4. In turning process, the value of roughness Ra is given in some limit of lower than 1.60 µm. This optimum calculation shows our value for Ra is in given range.
   We get the minimum roughness value for Ra & Rz at feed (0.09 mm/rev), depth of cut (0.20 mm) & cutting speed (260 m/min). For MRR, we get maximum value at feed (0.13 mm/rev), depth of cut (0.40 mm) & cutting speed (260 m/min).
5. The calculated optimum range for Ra & Rz value is $0.4506 \leq \boldsymbol{\mu_{Ra}} \leq 0.9048$ & $5.1570 \leq \boldsymbol{\mu_{Rz}} \leq 5.9348$ resp. & for MRR is $0.1418 \leq \boldsymbol{\mu_{MRR}} \leq 0.4822$ gm/sec.

Through this review, we try to reveal various confounding factors that have become barriers to the prevention of musculoskeletal diseases, and try to systematically solve these barriers. Various factors that have a significant impact on causing musculoskeletal diseases have been discovered, such as posture, strength, frequency, and working hours.

Conflict of Interest

None to report.

References

[1] M. Kumara Swamy, B.PadmaRaju and B.RaviTeja, "Modeling and Simulation of Turning Operation", *Journal of Mechanical and Civil Engineering*, vol. 3, Issue 6, pp. 19-26, 2012

[2] SaurabhSinghvi, M. S. Khidiya, S. Jindal and M.A. Saloda, "Experimental Investigation of Cutting Force in Turning Operation", *International Journal of Advance Engineering and Research Development,* vol. 3, Issue 3, 2016

[3] S.R. Das, D. Dhupal, and A. Kumar, "Study of surface roughness and flank wear in hard turning of AISI 4140 steel with coated ceramic inserts", *Journal of Mechanical Science and Technology*, vol. 29, No. 10, pp. 4329-4340, 2015

[4] D.I. Lalwani, N.K. Mehta, and P.K. Jain, "Experimental investigations of cutting parameters influence on cutting forces and surface roughness in finish hard turning of MDN250 steel", *Journal of Materials Processing Technology*, vol. 206, pp. 167-179, 2008

[5] Ravi K. Goyal, "*Production Engineering - II*", Ashirwad Publication, 2006

[6] Douglas C. Montgomery, "*Design and Analysis of Experiments*", Arizona state university, John Wiley & Sons, Inc., 2012

[7] Jitendra Sharma, Ajay Kumar Aggrawal, "Literature Review on Optimization of Surface Roughness during Turning Operation*", International Journal of Innovative Research in Science, Engineering and Technology,* vol. 6, Issue 2, February 2017

[8] A. Rohit, B. Naga Raju, M. Raja Roy, "Optimization of Tool Temperature and Surface Roughness in Wet and Dry Conditions During Turning of Mild Steel Using Response Surface Method", *International Journal of Engineering Research in Management &Technology*, vol. 4, Issue-5, 2015

[9] Mohd. Arif. I. Upletawala, Prof. TusharKatratwar, "A literature Review on various factors affecting Turning operation", *International Journal of Engineering Technology, Management and Applied Sciences*, vol. 4, Issue 4, April 2016

[10] Dr. Osama Khayal, "Literature Review on Turning operations", Researchgate, August 2019

[11] Ranganath M S, Vipin, "Effect of machining parameters on surface roughness with turning process – Literature Review", *International Journal of Advance Research and Innovation*, vol. 2, Issue 2, 2014

[12] M. Kaladhar, K. Venkata Subbaiah, Ch. Shrinivasa Rao and K.Narayana Rao, "Optimization of process parameters in turning of AISI 202 austenitic stainless steel", *ARPN Journal of Engineering and Applied Sciences*, vol. 5, No. 9, 2010

[13] Arshad Qureshi, Prof. Madhukar Sorte, Prof. S.N. Teli, "A literature review on optimization of cutting parameters for surface roughness in turning process", *International Journal of Engineering Research and Development,* vol. 11, Issue 5, May 2015

[14] A Joshi, AK Saraf, RK Goyal, "EDM machining of die steel EN8 and testing of surface roughness with varying parameters", https://doi.org/10.1016/j.matpr.2020.05.277 Materials Today: Proceedings Vol 28,Part 4, PP 2557-2560, May 2020.

[15] Singh, Ajay Pal and Samad, Abdul and Saraf, Amit Kumar, Enhancement of Surface Finish by Optimization Technique Employed for Al 6061 Considering Different Parameters Using RSM (October 12, 2019). Proceedings of International Conference on Advancements in Computing & Management (ICACM) 2019, SSRN: https://ssrn.com/abstract=3468726 or http://dx.doi.org/10.2139/ssrn.3468726.

Experimental Exploration on Performance Advancement of Solid Desiccant Dehumidifier

Zafar Alam[1]*, Taliv Hussain[2]

[1,2]Mechanical Engineering Department, Aligarh Muslim University, Aligarh, India.

* Author corresponding Email: zafaralam003@gmail.com

Abstract. The foremost idea of this current effort is to study the performance parameters of silica gel based desiccant dehumidifier and optimize its performance on the basis of Process air inlet velocities and air inlet temperatures. Desiccant dehumidifier considered in this work is coupled with the VCRS system. This Analysis is carried out for hot and humid conditions at four different air velocities (i.e. 2.8, 3.5, 4.5, 5.2 m/s) and different ambient temperatures (i.e. 32, 33, 34, 35°C). Regeneration rate and adsorption rate at different air flow rates and at different ambient temperatures is analysed. Regeneration is occurring through an electric heater fitted in the regeneration sector. Thermal factors, for instance temperature (dry bulb) and relative dampness are measured to enhance performance parameters like Adsorption rate, latent COP, Regeneration rate and Effectiveness of desiccant of wheel. Revamp values of this parameters are obtained at 3.5 m/s entering velocity of process air and 34°C ambient temperature which results in better performance of desiccant dehumidifier with less energy utilization for the conditioned space.

1. Introduction

Decreasing energy utilization and furthermore to dislodge petroleum derivatives with environmentally friendly power sources fuel, we need advancements energy productive and very much coordinated with environmentally friendly power assets. Dehumidification measure is one of the energy burning-through cycles with which we are managing in an unexpected way, going from drying items to giving wanted air quality to molded spaces. Three different ways to eliminate dampness from air are applied: warming and ventilation, buildup and adsorption. Adsorption dehumidification by strong desiccant wheels is an energy-productive approach to eliminate dampness from air particularly when waste energy is being utilized.

Golubovic et al. (2007) [1] assessed, exhibition of rotational device to be specific based on three extents, cleanse, cycle and recovery, the presentation of a desiccant wheel with a warmed viable cleanse point was contrasted and the exhibition of a similar wheel without a cleanse point, it was discovered that warmed compelling cleanse point had a general beneficial outcome on the exhibition of a rotational dehumidifier. Jeong et al. (2010) [2] improved COP by 94% when contrasted with ordinary vapour pressure fridge created and indicated that an ideal rotational speed existed which amplified the dehumidification execution. Goldsworthy and White (2011) [3] discovered the electrical coefficient of performance (COP_e) was more noteworthy than 20 once the temperature of recovery was 70°C by source/recovery wind stream proportion of 0.67 at encompassing conditions, consequently this framework could accomplish considerable energy saving and decreased ozone harming substance emanation. Jani et al. (2016) [4] utilized standard of strong desiccant cooling framework and its mechanical applications and progressions was given, also various arrangements of cooling desiccant cycles, ordinary in addition crossover cycles of cooling desiccant, various sorts of numerical replicas of turning desiccant wheel, execution assessment of framework of cooling desiccant, mechanical enhancement and bit of leeway it might proposed as far as cost and energy reserve funds were signified. Vivekh et al. (2020) [5] uncovered that the recovery cycle added to huge entropy age rates and by decreasing 10 °C the boiling water temperature. Also, by choosing to cover the warmth exchanger with a material of polymer composite rather than silica gel, the proficiency of second law of desiccant wheel is upgraded via 2.6. Hussain et al. [6] compare the performance of evaporatively cooled and steel wire mesh condenser and improve the heat rejection rate with less energy utilization.

The objective of this work is to find out the optimum values of these parameters for hot and humid conditions at Refrigeration and Air Conditioning Laboratory, Mechanical Engineering department, Aligarh Muslim University, Aligarh, India. The present work only focused on the hot and humid conditions that is why temperature range is taken into consideration is from 32°C to 35°C. The range of Process Air Velocities ranges from 2.8 m/s to 5.2 m/s is chosen for the present work because of the restrictions of the investigational arrangement.

2. Experimental Arrangement

The analyses have been performed to examine the exhibition of desiccant based half breed framework. Different trials at various wind stream rates and at various encompassing temperatures have been performed. The adsorption interaction has been completed following recovery. In this arrangement, measure air is prepared through the blower which further went through the silica based desiccant wheel framework, in the wake of going through the desiccant wheel moistness diminishes and temperature of interaction air expands which further disregards the evaporator loop of the crossover VCRS framework as demonstrated in figure 1.

Figure 1. Shows Actual Experiment setup involved in this study.

1) Blower 4) Desiccant Wheel
2) Compressor 5) Capillary tube
3) Condenser 6) Evaporator Coil

2.1. Assessing device and Instruments

Relative humidity and Dry bulb temperature of air are measured with help of thermo hygrometer. Air velocity is measured by anemometer.

Table 1. Assessing device

Equipment	Range	Accuracy
thermo	-10 to +60 °C	±0.5°C
Hygrometer	0 to 100 %RH	±2.5%

2.2. *Specification of Desiccant Wheel*

Table 2. Desiccant Specification

Factor	**Measurement**
desiccant Volume ratio, (Φ)	0.48
Route of flow pitch, 2b (m)	0.0032
silica gel warmth conductivity, kd (W/mK)	0.175
Height of flow passageway, 2a (m)	0.0018
Solid desiccant specific heat, cd (J/kgK)	921
channel Width of wall, δ (m)	0.00034
silica gel density of, ρd(kg/m3)	1129
Permeability, ε	0.4
matrix material density of, ρm (kg/m3)	625
Diameter of DW, D (m)	0.37
matrix materials Specific heat,cm (J/kgK)	1030
Extent of Wheel, Lw (m)	0.1
Radius of Pore, r (m)	11 × 10−10
The ratio of stream of air route to entire part of a passage, (Ar)	0.844

3. Methodology

Via the adsorption subdivision humid air flows. On the inlet of desiccant dehumidifier temperature of dry bulb and air relative dampness is measured via digital hygrometer. The wetness from the moist air is removed at adsorption part. On the exit of adsorption subdivision relative dampness and temperature (dry bulb) of air is taken by thermo hygrometer. On desiccant surface this dampness is adsorbed and it is condensed over. The temperature of the incoming air increases from heat released by condensation process.

By using electric heater which is fitted in the regeneration duct sensible heating of air is done. Air forced over the electric heater in the regeneration sector by a centrifugal fan which is connected in the regeneration duct. On recovery inlet relative humidity and dry bulb temperature is considered. Air velocity at regeneration entry is fixed throughout experiment i.e. 2.8 m/s. In the recovery part, to renew the desiccant wheel hot air is used, forced by another blower connected at the regeneration sector. The hot air loses its temperature by evaporating the water vapour from desiccant layer. At the outlet of regeneration sector measurement of relative humidity and dry bulb temperature have taken with the help thermo hygrometer.

3.1. *Performance Factors*

Various performance factors have been calculated and analyzed at different Air inlet velocities (i.e. 2.8 m/s, 3.5 m/s, 4.5 m/s & 5.2 m/s) at diverse temperatures (i.e. 32 °C, 33 °C, 34 °C & 35 °C) and Recovery Air inlet velocity is 2.8 m/s throughout entire experiment .

Adsorption rate or MRC characterises moisture mass flow rate detached by the desiccant [7].

$$\mathrm{MRC} = \rho_1 \times \dot{V}_P \times (\omega_1 - \omega_2) \tag{1}$$

Regeneration Rate signifies moisture mass flow rate of regained via dehumidifier at regeneration sector [7].

$$\text{Regeneration Rate} = \rho_1 \times \dot{V}_{reg} \times (\omega_3 - \omega_4) \tag{2}$$

Wheel effectiveness symbolizes the ratio among the actual to perfect dehumidification ability of DW [8].

$$\text{DW effectiveness} = \frac{\omega_1 - \omega_2}{\omega_1} \tag{3}$$

Latent COP signifies the ratio amid the power connected to the moisture removal from air and the power provided at recovery section [9].

$$\text{Latent COP} = \frac{\rho_1 \times \dot{V}_P \times \Delta h_{vs} \times (\omega_1 - \omega_2)}{\rho_1 \times \dot{V}_{reg} \times C_p \times (T_4 - T_1)} \tag{4}$$

Water latent heat of vaporization of $\boldsymbol{\Delta h_{vs}}$, assessed through the subsequent empirical cubic function [10].

$$\Delta h_{vs} = -0.614342 \times 10^{-4} \times T_1^3 + 0.0158927 \times 10^{-2} \times T_1^2 - 0.236418 \times 10 \times T_1 + 0.250079 \times 10^4 \tag{5}$$

3.2. Assumptions

- Constant specific heat, ideal gas model assumed.
- Adsorption section and the recovery section is separated into two segments equally.
- Mass diffusion & Heat conduction are ignored in humid air.
- No leakage among Dissimilar areas of desiccant wheel.
- Laterally via air passage Mass transmission and the heat coefficients between the desiccant wall and the air stream is constant.
- Entirely honeycombed passages of rotatory desiccant wheel are same.
- The adsorbent the capacity of hygroscopic of material matrix is insignificant.
- Along the channel air domain, the air inlet situations are uniform.
- Thermo physical properties of the dry air & properties of the dry desiccant material are constant.

4. Results

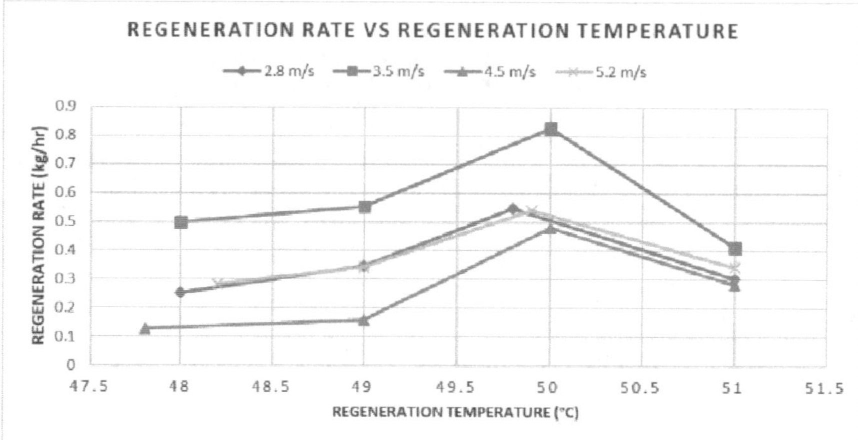

Figure 2. Shows the dissimilarity of regeneration rate with regeneration temperatures at different flow rates of air.

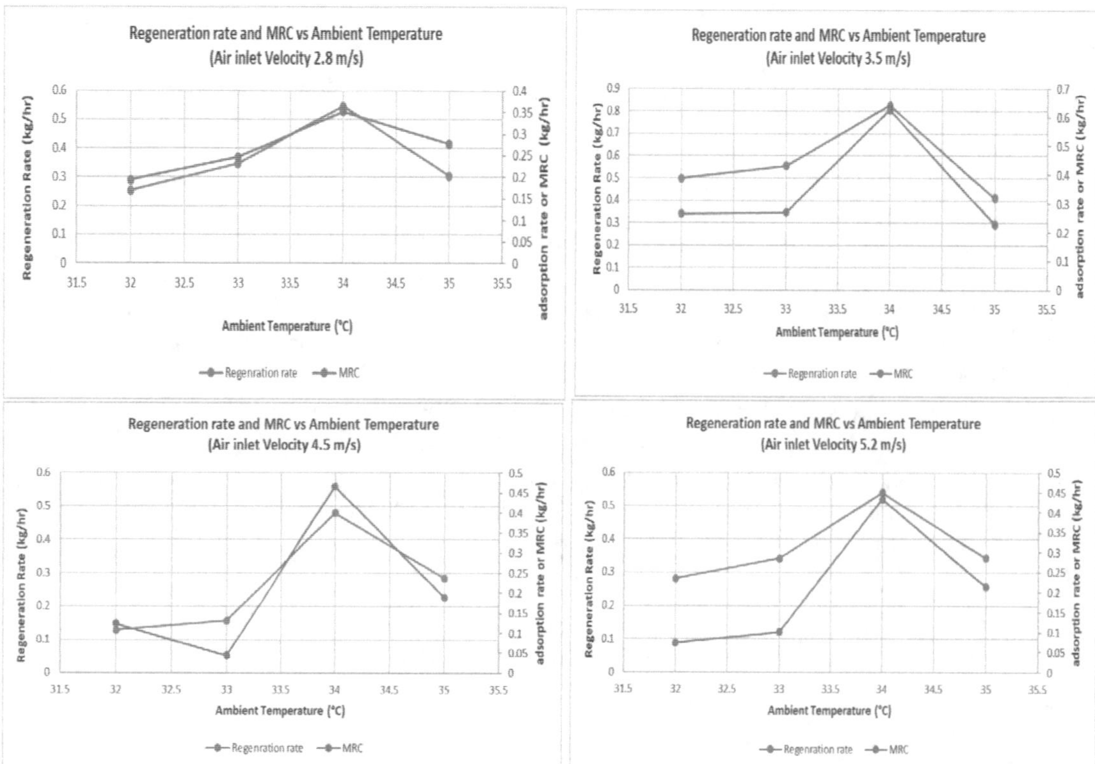

Figure 3. Illustrates the variation of regeneration rates and MRC with Ambient temperatures at 2.8, 3.5, 4.5 & 5.2 m/s air inlet velocity respectively.

IOP Conf. Series: Materials Science and Engineering 1224 (2022) 012011 doi:10.1088/1757-899X/1224/1/012011

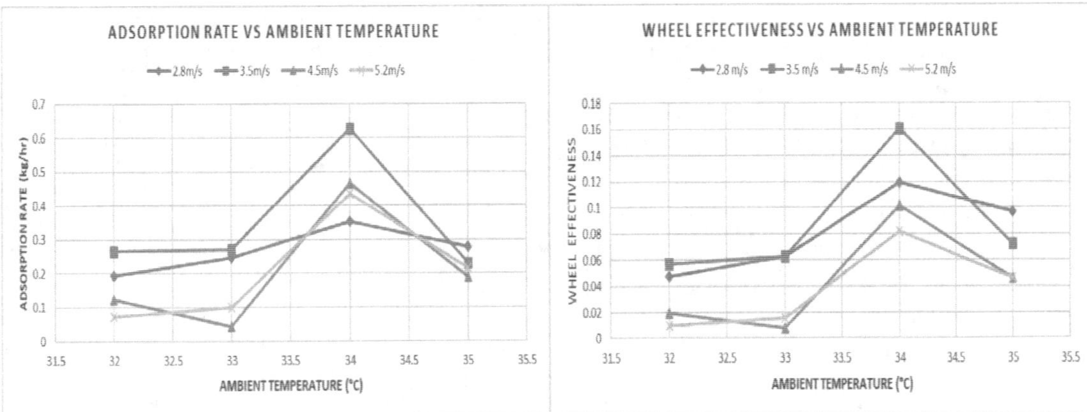

Figure 4. Signifies variation of Adsorption rate & Wheel Effectiveness with Ambient temperatures.

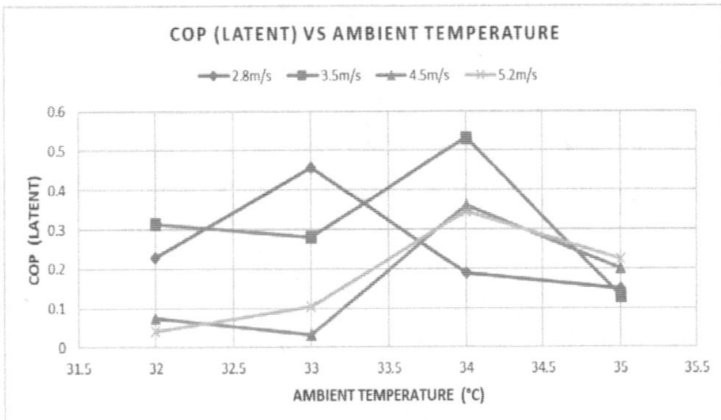

Figure 5. Expresses the dissimilarity of COP (latent) with different Ambient temperatures.

5. Error Analysis

The error investigation completed in present effort detailed by Kline and McClintock [11], depends on the root sum square strategy, the performance parameters described in this work (Wheel effectiveness, COP (Latent), Regeneration rate and Adsorption rate) are found by measurement from considered variables such as velocity of the air, relative humidity and temperature (DBT) all of these considered parameters is described via identified uncertainty values. Comparative uncertainty values obtained for the considered factors are, 8.2% for Wheel effectiveness, 11.5% for Adsorption rate, 12.9% for COP (Latent) and 12.5% for Regeneration rate.

6. Conclusion

In this work execution boundaries of desiccant wheel have been determined. At first, dry bulb temperature and relative moistness esteems have been estimated at the inlet and outlet of the interaction segment and recovery segment separately. Wind stream rates have been shifted (for example 2.8, 3.5, 4.5 and 5.2 m/s) at the cycle part of the desiccant wheel while on the recovery or recuperation area it is steady (for example 2.8 m/s). At various surrounding temperatures (for example 32, 33, 34 and 35°C) and at various wind current rates execution boundaries like Adsorption rate, dehumidification effectiveness, Latent COP and Regeneration rate have been broke down and discovered that ideal estimations of these exhibition boundary in this work is gotten at 3.5 m/s Process Air speed and 34°C surrounding temperature. Temperature of Regeneration is direct relating to the Regeneration rate and the adsorption rate equally. Rise in recovery temperature gives higher

Regeneration (recovery) and adsorption rate. Also, a critical working limit (i.e. Wind stream rate) which impacts the Regeneration and the adsorption speeds of desiccant dehumidifier. As rate of Air stream surges, both the Regeneration rate and the adsorption rate rises and the reverse way around. Also, it is assumed that the ascent in wind stream rate in territory of adsorption is more practical than rise in wind recovery rate for the better working of dehumidifier. To assemble the reasonability of wheel in mutual zones (Regeneration and adsorption), high recovery temperature and low flow rates of air in both the respective sections are required.

Nomenclature

MRC	Adsorption rate (kg/h)
h	specific enthalpy (kJ/kg)
c_p	specific heat of air (kJ/kg K)
COP	coefficient of performance (dimensionless)
$\dot{V}$	flow rate (volumetric) (m^3/h)
T	temperature (°C)
DW	wheel of desiccant
Greek signs:	
ρ	air density (kg/m^3)
ω	air humidity ratio (kg/kg)
Δh_{vs}	water heat of vaporization (latent) (kJ/kg)
Subscripts:	
P	process
reg	regeneration

References

[1] Golubovic, M.N., Hettiarachchi, H.D.M. and Worek, W.M., 2007, "Evaluation of rotary dehumidifier performance with and without heated purge", *International Communication in Heat and Mass Transfer*, Vol. 34, pp. 785-795.

[2] Jeong, J., Yamaguchi, S., Saito, K. and Kawai, S., 2010, "Performance analysis of four – partition desiccant wheel and hybrid dehumidification air conditioning system", *International Journal of Refrigeration*, Vol. 33, pp. 496-509.

[3] Goldsworthy, M. and White, S., 2011, "Optimisation of a desiccant cooling system design with indirect evaporative cooler", *International Journal of refrigeration*, Vol. 34, pp. 148-158.

[4] Jani, D. B., Mishra, M., & Sahoo, P. K. (2016). Solid desiccant air conditioning–A state of the art review. *Renewable and Sustainable Energy Reviews*, 60, 1451-1469.

[5] Vivekh, P., Bui, D. T., Islam, M. R., Zaw, K., & Chua, K. J. (2020). Experimental performance evaluation of desiccant coated heat exchangers from a combined first and second law of thermodynamics perspective. *Energy Conversion and Management*, 207, 112518.

[6] Hussain, T., Singh, A. K., Mittal, A., Verma, A., & Alam, Z. (2020). Performance Evaluation of Vapor Compression Refrigeration System by Varying Air Flow Rates in Air-Cooled and Evaporatively Cooled Condensers. *International Journal of Energy for a Clean Environment*, 21(1).

[7] Slayzak, S. J., & Ryan, J. P. (2000). Desiccant Dehumidification Wheel Test Guide. *National Renewable Energy Laboratory*. NERL/TP-550-26131.

[8] Mandegari, M. A., & Pahlavanzadeh, H. (2009). Introduction of a new definition for effectiveness of desiccant wheels. *Energy,* 34(6), 797-803.

[9] Ge, T. S., Ziegler, F., & Wang, R. Z. (2010). A mathematical model for predicting the

performance of a compound desiccant wheel (A model of compound desiccant wheel). *Applied Thermal Engineering*, 30(8-9), 1005-1015.

[10] Angrisani, G., Minichiello, F., Roselli, C., & Sasso, M. (2011). Experimental investigation to optimise a desiccant HVAC system coupled to a small size cogenerator. *Applied Thermal Engineering,* 31(4), 506-512.

[11] Kline, S. J. (1953). Describing uncertainty in single sample experiments. *Mech. Engineering*, 75, 3-8.

Effects of different air inlet temperatures and regeneration air inlet velocities on the performance of Hybrid Desiccant Air Conditioning System

Taliv Hussain

Mechanical Engineering Department, Aligarh Muslim University, Aligarh, India.

* Author corresponding Email: hussaintaliv@gmail.com

Abstract. A hybrid desiccant air conditioning unit is treated as an effective alternative to the conventional air conditioning system due to independent control of humidity, temperature, and eco-friendliness. The decrease in environmental impact and energy saving is more when the desiccant recovered by utilizing "free" thermal energy. This current work has experimented with a performance study of a hybrid desiccant air conditioning unit for typical humid and hot Indian (Aligarh) climatic conditions. Here, the effects of different air inlet temperatures at the process section, i.e. (28, 29.5, 31, 32.5, 34, 35.5 and 37°C) and regeneration section, i.e. (50-58.8°C) and different regeneration air inlet velocities (2.5 and 4.5 m/s) on the performance of hybrid desiccant air conditioning system are analyzed. Thus, optimization of performance parameters i.e. refrigerating effect, heat rejection rate, VCOP, ECOP, compressor work, dehumidification effectiveness, moisture removal capacity, DCOP, regeneration effectiveness and regeneration rate is identified for attaining maximum efficiency of hybrid desiccant air conditioning system under above operating states.

1. Introduction

People need to produce comfort conditions in the current world scenario, where the average temperature grows progressively. Vapour Compression Refrigeration System (VCRS) eliminates moisture from incoming air by cooling air beneath the dew point temperature through coolants, thus removing water vapour via condensation. In other words, moisture gets condensed and removed from the air below the dew point temperature. Thus, conventional vapour compression refrigeration systems concurrently cool and dehumidify incoming air. Whereas in Hybrid Desiccant Air Conditioning System comprises a rotor packed with desiccant substantial in which the incoming process air is dehumidified via desiccant substantial to compensate latent atmosphere loads. Here, the cooling coil has to handle the only sensible load in a desiccant-based air conditioning system. Thus, a desiccant-based air conditioning unit can be considered a possible alternative to a conventionally utilized refrigeration unit [1].

Some of the researchers worked in the field of hybrid desiccant air conditioning systems. Jia et al. (2006) [2] performed experiments on a desiccant-based air conditioning unit, incorporating a solid desiccant wheel and air conditioning system. It was noticed that when relative humidity and the process air temperature was sustained at 55% and 30°C, respectively, a hybrid desiccant vapour compression refrigeration system economizes 37.5% electricity power as that of a conventional system. Fatouh et al. (2009) [3] experimentally examined performance data of solid desiccant vapour compression refrigeration system. This unit was comprised of the packed bed desiccant incorporated thru an R407C conventional air conditioning unit. It was noticed that with increment in mass flow rate of renewed air from 7.4 to 10.2 kg/min, provides decrement in renewal period via 87.5% and augmentation in desorption rate through 16% after 10 minutes of recovering. Furthermore, with an increment in renewal temperature from 45 to 55°C, reactivation time decreases by 25%. Described outcomes publicized that a desiccant based air conditioning unit lowers the compressor electric power and number of electric units around 10.2%. Mittal and Khan (2010) [4] estimated the performance and energy were preserving the capability of desiccant air conditioning unit in India, which was consist of a silica gel bed, a split type air conditioner of 1.0 ton, air ducts, a blower, and fitted in a room with a volume of 86.4 m^3. During the test settings in the experimentation, suggested a 7 cm bed thickness with an extreme moisture removal capacity of 403 g/h. Optimal proportions of air ratios were: 10% of the outer air, 10% of return air and 80% of indoor air mixed with dry air leaving desiccant. The correspondent electricity saving was around 19%. Mandegari and Pahlavanzadeh (2010) [5] examined a desiccant wheel incorporating a vapour compression refrigeration system at different supplied-air humidities and temperatures. They had compared the performance of a desiccant air conditioning system to that of a conventional vapour compression refrigeration system on an experimental basis. They found that this technique of hybridization was much operative at high latent loads. Based on an economic study, they notified that this unit was high operative in nations where the cost of energy was high. Belguith et al. (2020) [6] examined a desiccant based air conditioning unit. Here, they discussed three arrangements of this refrigeration expertise for a test area of cooling capability of 1.8 kW underneath hot and dry weather conditions. It was noticed that arrangement which comprises of a desiccant wheel attached by a conventional vapour compression system, had the most satisfactory COP equivalents 2.48.

The prime aim of the current research paper is to determine various performance parameters of hybrid solid desiccant air conditioning system utilizing the waste heat from condenser for recovering at different air inlet temperatures at process section, i.e. (28, 29.5, 31, 32.5, 34, 35.5 and 37°C) and different air inlet temperatures at regeneration section, i.e. (50-58.8°C) and different regeneration air inlet velocities (2.5 and 4.5 m/s). Thus identifying the optimum conditions of performance parameters for effective working of hybrid desiccant air conditioning system under above operating states.

2. Methodology and design of the experiment

Air first passes through the air pre-treatment section, which consists of electric heaters and a humidifier. Electric heaters can heat incoming treated air to desired hotness and, after that, humidified through a humidifier. After that, the hot and humid air is blown over the adsorption sector of the rotary desiccant wheel. When processed air streams over the desiccant exterior in the process zone, its vapour pressure is higher than desiccant outward; a more humid air condition generates a significant difference. Thus, the process part takes water vapour from moist air. The water vapour is adsorbed on the exterior of the desiccant and is condensed above it. Therefore, adsorption occurs. Then the air passes over the cooling coil of the vapour compression refrigeration unit and then sensible cooled down to ideal temperature. In the regeneration section, when regeneration air (waste heat from the condenser) streams over a desiccant surface, it kicks the desiccant exterior. It increases its vapour pressure, enhancing the difference between the desiccant surface and the regeneration air. Thus, regeneration air evaporates moisture from a desiccant layer. Hence, regeneration takes place.

The experimental device (figure 2.1) consists of a pretreatment segment, a rotary desiccant wheel and a vapour compression refrigeration system. The air pretreatment segment consists of an electric heater and a humidifier. The experimental data has been collected for a desiccant based air conditioning unit in which the desiccant wheel is using waste heat from condenser for renewal at different air inlet temperatures at the process section, i.e. (28, 29.5, 31, 32.5, 34, 35.5 and 37°C) and different air inlet temperatures at regeneration section, i.e. (50-58.8°C) and different regeneration air inlet velocities (2.5 and 4.5 m/s). Here the air velocity at the process inlet is fixed, i.e. 1.5 m/s. The discharge pressure (P_{dis}) and suction pressure (P_{suc}) of refrigerant R-410a are measured with the help of Bourdon tube pressure gauges. The voltage (V) is measured through a voltmeter. The current (I) is measured with the help of an ammeter. A data logger measures the temperature of refrigerant R-410a at the different points of VCRs. The data logger also provides the different temperatures and relative humidity of air at entry and exit of process and regeneration segment of desiccant wheel. Specific humidity of air has been determined from relative humidity and dry bulb temperature through a Psychrometric calculator.

Figure 2.1 Photograph of an experimental setup for Desiccant Air Conditioning System utilizing waste heat of condenser for the regeneration of Desiccant wheel

2.1 Assessing device and Instruments

Dry bulb temperature and relative humidity and air are measured through Thermo-hygrometer. An anemometer measures air velocity. Table 1 shows the assessing device and instruments.

Table 1 Assessing device and Instruments

Equipment	Voltage	Sampling Rate	Range	Accuracy
Tweex 8 Channel Data Logger	Single Phase 220V Standard supply	5 seconds to 255 minutes	-10 to +85 °C 0 to 100 %RH	±0.5°C ±2.5%
Digital Anemometer	3.0 V DC	-	0 to 30 m/s	±5%

2.2 Performance Factors

Different performance parameters of hybrid solid desiccant air conditioning unit using the waste heat from condenser for regeneration are shown in equations (1-10) at different air inlet temperatures at the process section, i.e. (28, 29.5, 31, 32.5, 34, 35.5 and 37°C) and different air inlet temperatures at regeneration section, i.e. (50-58.8°C) and different regeneration air inlet velocities (2.5 and 4.5 m/s):

1. Refrigerating Effect = (h_{ref1} – h_{ref4}) (1)
2. Heat Rejection Rate = (h_{ref2} – h_{ref3}) (2)
3. Compressor Work = (h_{ref2} – h_{ref1}) (3)
4. VCOP of hybrid system [7]=Refrigerating Effect/ Compressor Work = (h_{ref1} – h_{ref4})/ (h_{ref2} – h_{ref1}) (4)
5. ECOP of hybrid system [8] = Total Cooling Effect/ Total Electric Power Consumption (5)

6. Dehumidification Effectiveness (ηdeh): symbolizes the ratio between actual and ideal dehumidification ability of desiccant wheel [9].

$$\eta_{deh} = \frac{\omega_1 - \omega_2}{\omega_1} \tag{6}$$

7. Moisture Removal Capacity (MRC): signifies mass flow rate of dampness detached by desiccant wheel [9].

$$MRC = \rho_1 \times \dot{V}_P \times (\omega_1 - \omega_2) \tag{7}$$

8. Dehumidification Coefficient of Performance (DCOP): signifies ratio among thermal power linked to air dehumidification and thermal energy provided for regeneration process [9].

$$DCOP = \frac{\rho_1 \times \dot{V}_P \times \Delta h_{vs} \times (\omega_1 - \omega_2)}{\rho_1 \times \dot{V}_{reg} \times C_P \times (T_4 - T_1)} = \frac{\dot{V}_P \times \Delta h_{vs} \times (\omega_1 - \omega_2)}{\dot{V}_{reg} \times (h_4 - h_1)} \tag{8}$$

The latent heat of vaporization of water Δhvs has been assessed using subsequent empirical cubic function [9].

$$\boldsymbol{\Delta h_{vs} = -0.614342 \times 10^{-4} \times T_1^3 + 0.0158927 \times 10^{-2} \times T_1^2 - 0.236418 \times 10 \times T_1 + 0.250079 \times 10^4}$$

9. Regeneration Effectiveness (ηreg): implies latent load held by a desiccant wheel about thermal regeneration power needed for adsorption procedure of desiccant wheel [10].

$$\eta_{reg} = \frac{\omega_3 - \omega_4}{\omega_4} \tag{9}$$

10. Regeneration Rate (RR) signifies the mass flow rate of dampness detached by a wheel from the process side to the regeneration side [10, 11].

$$RR = \rho_1 \times \dot{V}_R \times (\omega_3 - \omega_4) \tag{10}$$

3. Results and Discussion

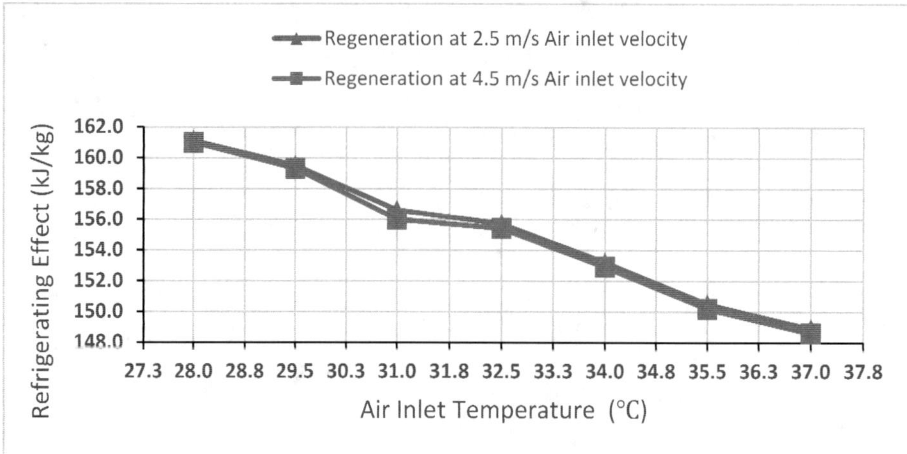

Figure 3.1 Refrigerating Effect variation with air inlet temperature at 2.5 m/s and 4.5 m/s velocity of air at regeneration inlet and fix 1.5 m/s velocity of air at process inlet

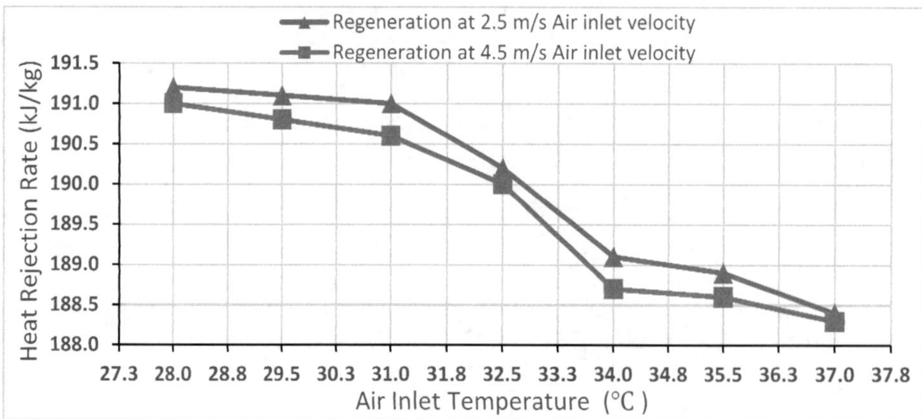

Figure 3.2 Heat Rejection Rate variation with air inlet temperature at 2.5 m/s and 4.5 m/s velocity of air at regeneration inlet and fix 1.5 m/s velocity of air at process inlet

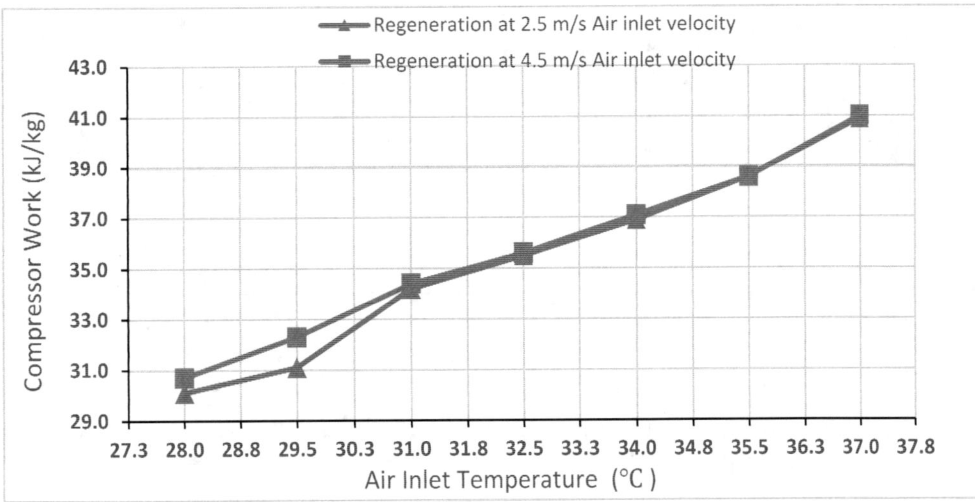

Figure 3.3 Compressor Work variation with air inlet temperature at 2.5 m/s and 4.5 m/s velocity of air at regeneration inlet and fix 1.5 m/s velocity of air at process inlet

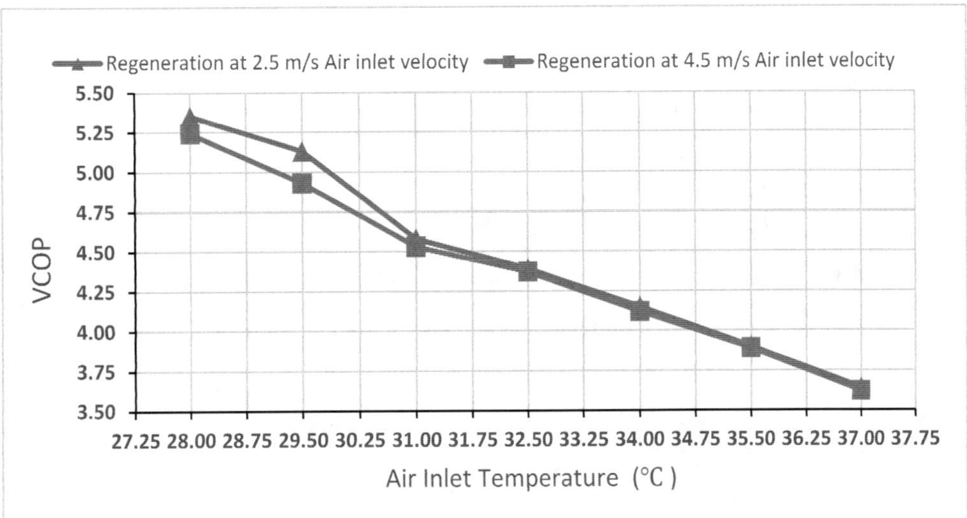

Figure 3.4 VCOP variation with air inlet temperature at 2.5 m/s and 4.5 m/s velocity of air at regeneration inlet and fix 1.5 m/s velocity of air at process inlet

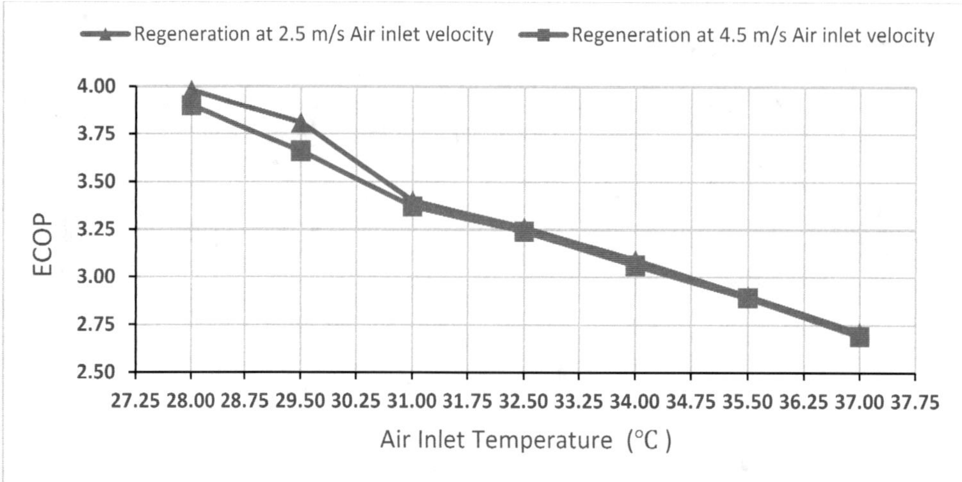

Figure 3.5 ECOP variation with air inlet temperature at 2.5 m/s and 4.5 m/s velocity of air at regeneration inlet and fix 1.5 m/s velocity of air at process inlet

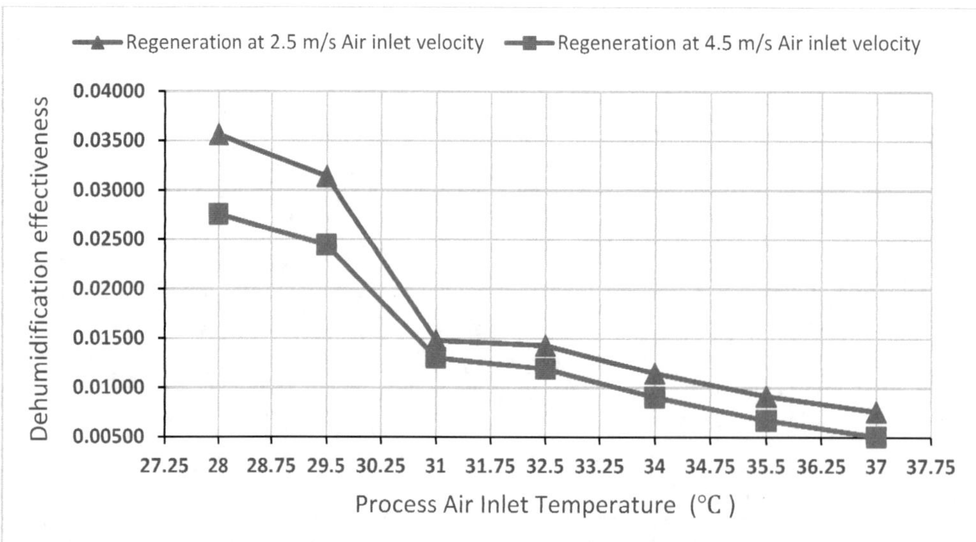

Figure 3.6 Dehumidification Effectiveness variation with air inlet temperature at 2.5 m/s and 4.5 m/s velocity of air at regeneration inlet and fix 1.5 m/s velocity of air at process inlet

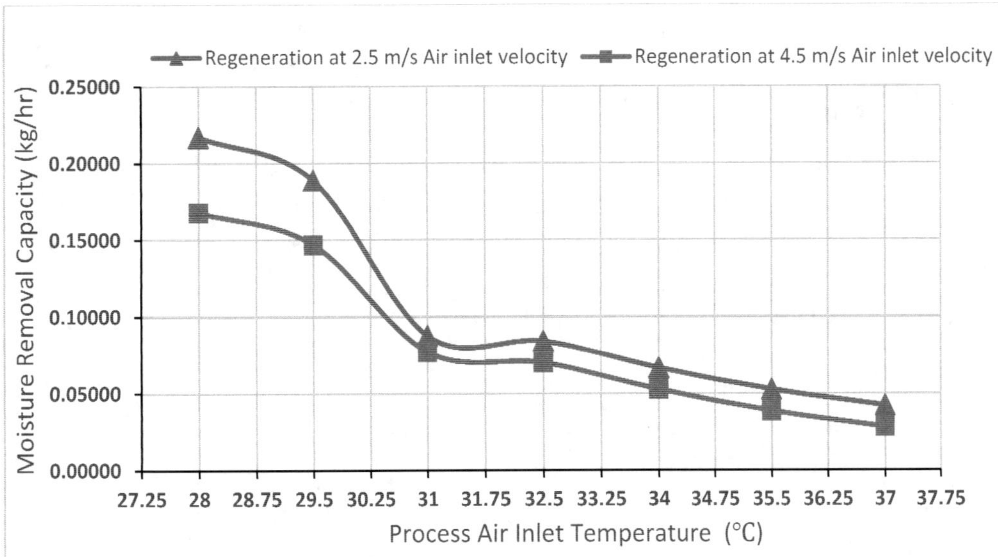

Figure 3.7 Moisture Removal Capacity variation with air inlet temperature at 2.5 m/s and 4.5 m/s velocity of air at regeneration inlet and fix 1.5 m/s velocity of air at process inlet

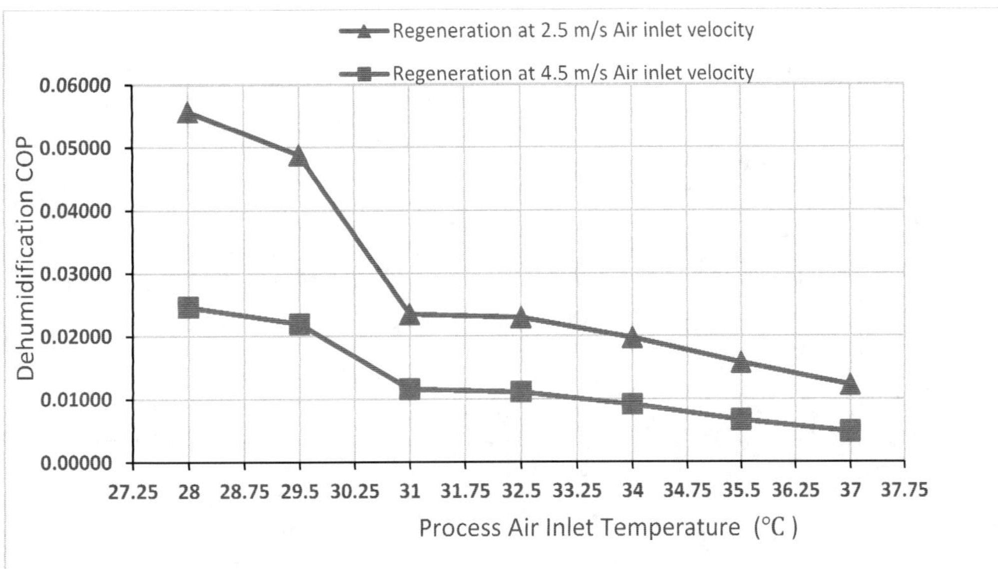

Figure 3.8 Dehumidification COP variation with air inlet temperature at 2.5 m/s and 4.5 m/s velocity of air at regeneration inlet and fix 1.5 m/s velocity of air at process inlet

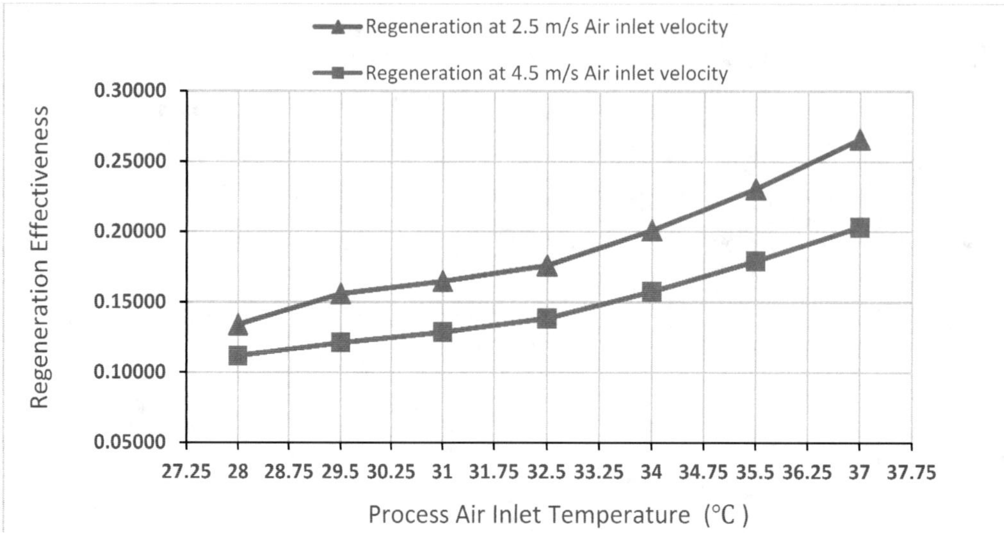

Figure 3.9 Regeneration Effectiveness variation with air inlet temperature at 2.5 m/s and 4.5 m/s velocity of air at regeneration inlet and fix 1.5 m/s velocity of air at process inlet

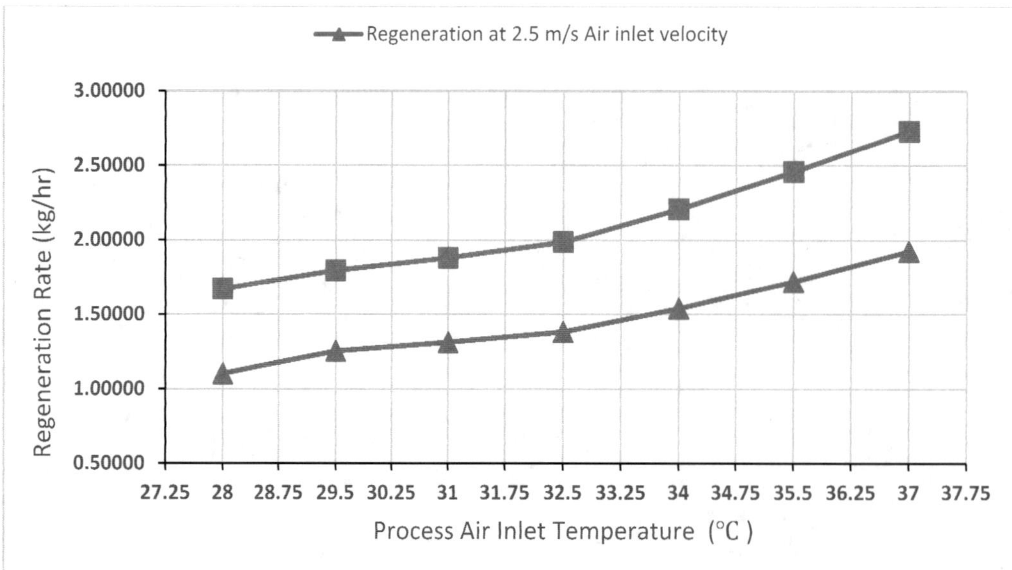

Figure 3.10 Regeneration Rate (kg/hr) variation with air inlet temperature at 2.5 m/s and 4.5 m/s velocity of air at regeneration inlet and fix 1.5 m/s velocity of air at process inlet

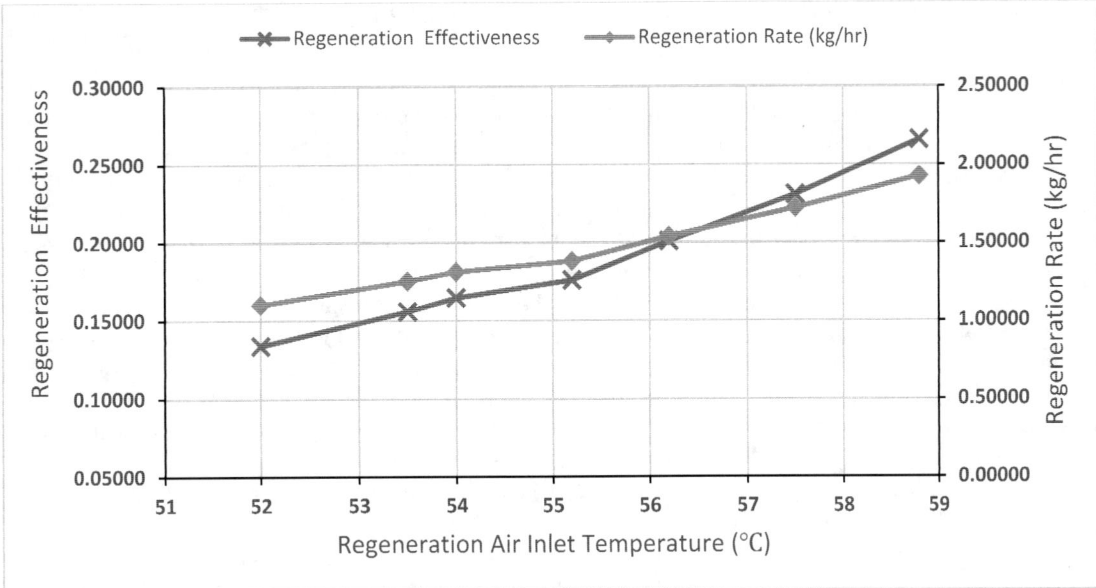

Figure 3.11 Regeneration Effectiveness and Regeneration Rate (kg/hr) variation with regeneration air inlet temperature at 2.5 m/s velocity of air at regeneration entry and fix 1.5 m/s velocity of air at process inlet

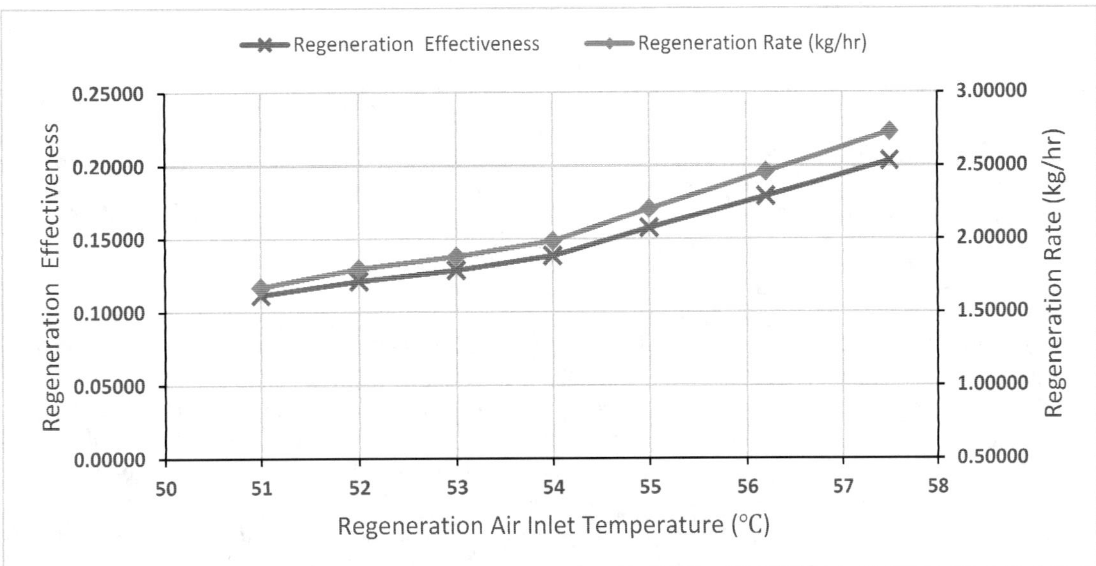

Figure 3.12 Regeneration Effectiveness and Regeneration Rate (kg/hr) variation with regeneration air inlet temperature at 4.5 m/s velocity of air at regeneration entry and fix 1.5 m/s velocity of air at process inlet

Figures 3.1, 3.2, 3.4 and 3.5 show that the value of refrigerating effect, heat rejection rate, VCOP and ECOP decreases with rising air temperatures at process inlet at different regeneration air inlet velocities (2.5 and 4.5 m/s) and fix 1.5 m/s process air inlet velocity. But, figure 3.3 shows that compressor work enhances with an increment in different process air inlet temperatures. It is because as ambient temperature increases, the rate of heat transfer from the condenser decreases. With an increase in ambient temperature, the air-cooled condenser is also affected and gets less cooled, increasing the load of the vapour compression refrigeration system. Thus due to this effect, less heat is absorbed by refrigerant from the air at the evaporator. It is clear that at a fixed process, when regeneration air inlet velocity decreases from maximum (4.5 m/s) to minimum (2.5 m/s) value of refrigerating effect, heat rejection rate, VCOP and ECOP increases. It is because lower regeneration air inlet velocity maximizes contact time between desiccant surface and regenerated air. Thus at lower air velocity at regeneration inlet, higher desiccant material is restored in the regeneration sector. Hence, ultimately this desiccant material in the adsorption sector adsorbs more of water vapour. Thus more latent load is removed at lower regeneration air inlet velocity, which eventually increases refrigerating effect [9,10]. Figures 3.6, 3.7 and 3.8 show that the value of dehumidification effectiveness, moisture removal capacity and DCOP decreases with an increment in air temperatures at the process inlet. It is due to, as the adsorption process is exothermic, so preferred by lesser temperatures. Lower temperature provides a higher presence of moisture content in process air. There is a significant variance of vapour partial pressure between desiccant material surface and process air, which creates a more diffusion of water vapour drops from perspective to surface of desiccant material [9,11]. Figures 3.9 and 3.10 shows that the value of regeneration effectiveness and regeneration rate increases with an increment in air temperatures at process entry. As ambient temperature increases, it also affects regeneration temperature, which leads to the improved exclusion of water vapour from the rotary desiccant wheel during the regeneration period. Figures 3.11 and 3.12 show the variation of regeneration effectiveness and rate (kg/hr) with regeneration air inlet temperature at 2.5 m/s and 4.5 m/s air velocity at regeneration entry and fix 1.5 m/s velocity of air at process inlet, respectively. When regeneration air inlet velocity decreases from maximum (4.5 m/s) to minimum (2.5 m/s) velocity, the value of regeneration rate decreases. As with increment in air velocity at regeneration inlet, there is an increment in regeneration mass flow rate of air which also offers desiccant wheel in regeneration segment with large hot regenerated air, as it removes water vapour from the desiccant surface in regeneration sector and thus ultimately increases regeneration rate. Also, figures 3.11 and 3.12 show that the value of regeneration effectiveness and regeneration rate increases with an increment in air temperatures at the regeneration inlet.

4. Error Analysis

The error investigation completed in the current research paper depends on the strategy of root sum square detailed by Kline and McClintock [12], the performance parameters described in this work refrigerating effect, heat rejection rate, compressor work, VCOP, ECOP, dehumidification effectiveness, moisture removal capacity, dehumidification COP, regeneration effectiveness and regeneration rate are found by calculation from noted variables such as air velocity, temperature, relative humidity and all of these noted variables is described by a known value of uncertainty. Comparative uncertainty values obtained for the considered factors are, 11.7% for refrigerating effect, 1.7% for heat rejection rate, 10% for compressor work, 11.8% for VCOP, 11.7% for ECOP, 6% for dehumidification effectiveness, 6.8% for moisture removal capacity, 7% for dehumidification COP, 14.8% for regeneration effectiveness and 14.9% for regeneration rate.

5. Conclusions

The primary objective of the current research work was to highlight the possibility of a hybrid rotary desiccant air conditioning unit utilizing waste heat from condenser for recovering at different air inlet temperatures at the process section, i.e. (28, 29.5, 31, 32.5, 34, 35.5 and 37°C) and different air inlet temperatures at regeneration section, i.e. (50-58.8°C) and different air velocities at regeneration inlet (2.5 and 4.5 m/s) and fix (1.5 m/s) air velocity at process inlet. It was concluded that optimum conditions of performance parameters are identified for effective working of hybrid desiccant air conditioning system under above operating states which are as follows:

- Refrigerating effect, heat rejection rate, VCOP and ECOP decreases with an increment in different air temperatures at the process inlet. Compressor work enhances with an increment in different process air inlet temperatures.
- The regeneration temperature should increase the refrigerating effect, heat rejection rate, VCOP and ECOP, and regeneration temperature. To lower the compressor work, the regeneration temperature should be increased.
- When the air velocity at regeneration inlet slows down from maximum (4.5 m/s) to minimum (2.5 m/s), the value of refrigerating effect, heat rejection rate, VCOP and ECOP increases.
- Dehumidification effectiveness, moisture removal capacity and DCOP reduces with an increment in air temperatures at the process inlet.
- Whereas regeneration effectiveness and regeneration rate increase with an increase in air temperatures at process inlet.
- The regeneration temperature should be high to enhance dehumidification effectiveness, moisture removal capacity, DCOP, regeneration effectiveness and rate.
- When air velocity at the regeneration inlet decreases from maximum (4.5 m/s) to the minimum (2.5 m/s), regeneration effectiveness increases. Whereas, when regeneration air inlet velocity decreases from maximum (4.5 m/s) to minimum (2.5 m/s), the value of regeneration rate decreases.

Nomenclature

COP	Coefficient of performance
VCOP	COP of vapour compression refrigeration system
ECOP	electric coefficient of performance
h_{ref1}	at the entry of compressor enthalpy of refrigerant in kJ/kg
h_{ref2}	at the outlet of compressor enthalpy of refrigerant in kJ/kg
h_{ref3}	at the outlet of condenser enthalpy of refrigerant in kJ/kg
h_{ref4}	at the entry of evaporator enthalpy of refrigerant in kJ/kg
MRC	Adsorption rate (kg/hr)
T_1	at process entry temperature of the air
T_2	at process exit temperature of the air
T_4	at regeneration entry temperature of the air
T_3	at regeneration exit temperature of the air
ω_1	at the process inlet humidity ratio of air
ω_2	at the process outlet humidity ratio of air
ω_4	at the regeneration inlet humidity ratio of air
ω_3	at the regeneration outlet humidity ratio of air
$\dot{V}$	volumetric flow rate of air (m^3/hr)

c_p air specific heat (kJ/kg K)
DW desiccant wheel

Greek symbols:

Δh_{vs} latent heat of vaporization of the water (kJ/kg)
ρ density of air (kg/m^3)
ω humidity ratio or specific humidity (kg/kg)

Subscripts:

P process
reg regeneration
dis discharge
suc suction

References

[1] Parmar, H., & Hindoliya, D. A. (2011). Desiccant cooling system for thermal comfort: a review. *International Journal of Engineering Science and Technology*, 3(5).
[2] Jia, C. X., Dai, Y. J., Wu, J. Y., & Wang, R. Z. (2006). Analysis on a hybrid desiccant air-conditioning system. *Applied Thermal Engineering*, 26(17-18), 2393-2400.
[3] Fatouh, M., Ibrahim, T. A., & Mostafa, A. (2009). Experimental investigation on a solid desiccant system integrated with a R407C compression air conditioner. *Energy conversion and management*, 50(10), 2670-2679.
[4] Mittal, V., & Khan, B. K. (2010). Experimental investigation on desiccant air-conditioning system in India. *Frontiers of Energy and Power Engineering in China*, 4(2), 161-165.
[5] Mandegari, M. A., & Pahlavanzadeh, H. (2010). Performance assessment of hybrid desiccant cooling system at various climates. *Energy Efficiency*, 3(3), 177-187.
[6] Belguith, S., Slama, R. B., Chaouachi, B., & Meddeb, Z. (2020, October). Performance analysis of hybrid solid desiccant-vapor compression air conditioning system: Application and comparative study. In 2020 11th *International Renewable Energy Congress (IREC)* (pp. 1-6). IEEE.
[7] Ansari, A. A., Goyal, V., Yahya, S. M., & Hussain, T. (2018). Experimental investigation for performance enhancement of a vapor compression refrigeration system by employing several types of water-cooled condenser. *Science and Technology for the Built Environment*, 24(7), 793-802.
[8] Dai, Y. J., Wang, R. Z., Zhang, H. F., & Yu, J. D. (2001). Use of liquid desiccant cooling to improve the performance of vapor compression air conditioning. *Applied Thermal Engineering*, 21(12), 1185-1202.
[9] Angrisani, G., Minichiello, F., Roselli, C., & Sasso, M. (2012). Experimental analysis on the dehumidification and thermal performance of a desiccant wheel. *Applied Energy*, 92, 563-572.
[10] Angrisani, G., Capozzoli, A., Minichiello, F., Roselli, C., & Sasso, M. (2011). Desiccant wheel regenerated by thermal energy from a microcogenerator: experimental assessment of the performances. *Applied Energy*, 88(4), 1354-1365.
[11] Yadav, A. V. A. D. H. E. S. H. (2012). Experimental and numerical investigation of solar powered solid desiccant dehumidifier (Doctoral dissertation, NATIONAL INSTITUTE OF TECHNOLOGY).
[12] Kline, S. J. (1953). Describing uncertainty in single sample experiments. Mech. Engineering, 75, 3-8.

IOP Conf. Series: Materials Science and Engineering 1224 (2022) 012013 doi:10.1088/1757-899X/1224/1/012013

Experimental Study for Cleanliness Evaluation of Tractor Engine Components

Pramod Kumar[1*], **Satendra Singh**[2]
[1,2] *Department of Mechanical Engineering, Vivekananda Global University, Jaipur, Rajasthan*

[1] *pramod_kumar@vgu.ac.in*, [2]*singh.satendra@vgu.ac.in*

Abstract

In present scenario most of the industries are competing for quality improvement. In each automotive company, to reduce the wear and tear between engine components and to increase the engine life, it is necessary to check the contamination level of engine components. To wash these components washing solvents or multi metal cleaner cum degreaser are used to wash the components in ordinary or ultrasonic washing machines. In this paper experimental work is carried out to find the life of two washing solvents used to wash the tractor engine components in ordinary and ultrasonic washing machines. This experiment is performed at different setting parameters date wise of both of the machines. Observations were taken for 15 days using two type of cleaning medium i.e. AFSO CLEAN-AD20 and METACLEAN 211. After data analysis it is concluded that if these washing solvents are used for both shifts in the industry than the solvents remain useful to clean the tractor engine components for 6 days as per specification limits.

Keywords: *Contamination level, Washing solvent, washing life, Millipore value, ultrasonic washing*

1. Introduction

It is very much clear that presently tractor manufacturing are using different types of washing solvents to remove the contaminants from various tractor engine components .It is necessary to wash or clean the components before forwarding to assembly line [1]. If components having contamination level higher than a standard value than it will impact the quality and will increase wear and tear between metal surfaces of components. Different methods are used by industries to evaluate a cleanliness level [2, 3]. In the case company, Millipore method is used to remove contaminates from the sample parts by brushing and flushing with a solvent and filtering the solvent to remove the contaminants and then determining the quantity and nature of the contaminants removed.

The brushing and flushing procedure does not yield complete removal of contaminants present but with good practice a 50% yield will be obtained discrepancies can occurs thus sufficient brushing and flushing [4, 5]. Also foreign particles can become trapped in the brushes, cleaning brushes before the check is essential.

2. Methodology

The method used in this project supplements the concept of ideal cleanliness of engine parts and related system. Instructions should be provided to apply this method using only nylon filter and tri-chloro ethylene. A consistent and precise technique is essential to obtain valid, re-producible results and special instructions for the application for this method to certain engine parts, are followed during the project.

2.1 Equipment

Millipore Equipment (Shown in Figure 1) including filter paper and sample disks filter paper size 0.45 Microns (Nylon Filter)-for filtering (T.C.E.) 5.0 Microns (Nylon filter)-for normal sample is used in this experimental study to calculate the life of washing solvents.

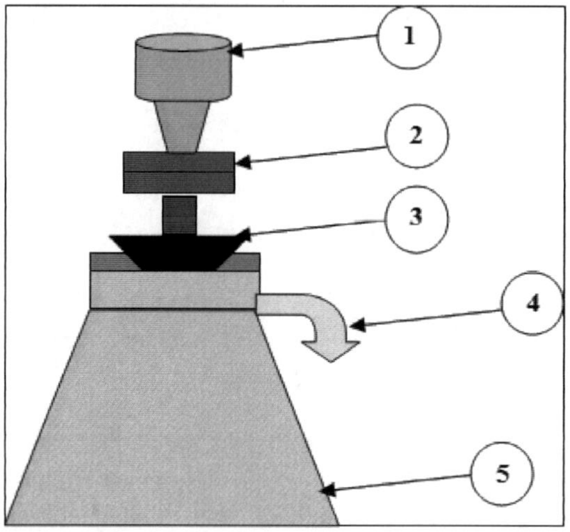

Fig.1: Filtration equipment for contamination.

1. Filter funnel
2. Holding Fixture for filter membrane
3. Vacuum pipe
4. Vacuum flask
5. Rubber stopper

Other items used with this equipment are as follows:

- Balance scale-reading upto 0.1 MG.
- 2x2" (51cm X 51cm) diameter plastic specimen plugs sanded smooth.
- Tri- Chloro Ethylene in a bottle for easy application. Normal usage is 200-400 ML per part. Incoming solvent is to be sample checked for cleanliness.
- Brushes-white Nylon bristle brushes to use unless otherwise specified round brushes for cleaning holes to be larger diameter than holes to be cleaned. Denature brushes are not acceptable.
- Receptacles to collect sample trays, jars, pans adapters as needed.

- All brushes trays, jars, other equipment used to make evaluation are to be precleared with Tri chloro-ethylene OR equivalent before use.
- All cleaning equipment's should be kept covered when not in use except when the equipment is kept in an enclosed separate room.
- Hands should be kept reasonable clean during all Millipore test evaluations so as not to influence results. Hands protection is recommended to minimize skin contact with Tri choloro ethylene or equivalent.
- Drying oven-automatic temperature controlled.

2.2 Parts or components selection for cleaning

Engine parts must be demagnetized prior to cleaning to aid the cleaning process. Limits of acceptance for residual magnetism shall be as outlined is inspection standard. Parts such as raw casting must be received from supplier at a cleanliness level as specified in inspection standard. Production parts shall be selected at random as per final wash but before final preservation, containerization or build up. Unless otherwise specified parts which are lubricated (from manganese phosphatized shall be inspected to this standard prior to treatment.

2.3 Method to use in washing and sample part

It is to rinse part with Tri choloro ethylene or equivalent catching fluid in the collection tray and brush water air, oil, fuel holes and bores using a brush larger than the holes size as specified in earlier.

Process for Brushing holes: When brushing holes, a spiral motion should be imparted to the brush taking care to avoid flipping particles from the bristles. Use a Nylon brush for holes and sash brush for bores. Note that brushing must be done in consistent manners with the following definition applying:

- Single stroke- movement in one direction only such as pulling or pashing a brush thru a hole bore.
- Double stroke-movement in both direction only such as pushing in and out of a blind hole or bore. All brushes are to be wetted in Tri choloro ethylene or equivalent.
- Flush the part and brushes with Tri choloro ethylene or equivalent catching all of the fluid in the collection tray.
- Brush and flush again as in above steps.
- Consistency in this washing and brushing operation is required to obtain reproducible results part. Care must be taken to prevent the solvent from contracting surface, which are not being evaluated.

Parts are submitted for refreshing after cleanliness checks are completed.

2.4 Filtering sampling solution

Sample is filtered using Millipore filtering apparatus, using 5-6 micron filter that has been previously weighted. After this steel pan or jars thoroughly washed and pour residue through the filter. If fluid sample is contained in one or more trays or jars of a given part sample, all fluid is to be filtered through the sample filter pad. When possible, note that vacuum of 15" (381 mm) Hg is to be maintained on the filter when introducing the sample. Normally only one part sample per filter pad.

2.4.1 Procedure of filtering solution

- Place the nylon filter pad on the analytical balance weigh and record the initial weight to the nearest tenth of milligram (0.0001) of the filter pad.
- Assemble the reweighted filter pad into the filter holder/apparatus.
- Pour the flushing solvent from the Millipore part into the filter apparatus.
- Rinse the container jar/steel pan two more times with clean solvent (TCE OR Equivalent) pouring the rinsing into the filter apparatus. Turn on the vaccum pump until the solvent has drained away from the filter apparatus.

- When the solvent is drained away flush the sides of the filter funneled with clean solvent and allow to drain to way.
- Turn off the vacuum pump, remove the filter pad and put into the drying oven (70^0 Celsius) in its unsealed Petridis.
- Remove the filter pad from the oven after the 30 minutes & allow to cool the room temperature.
- Weight the filter pad and record final weight to the nearest length of milligram.
- The contaminant level is difference between the initial weight of the pad and the final weight of the pad divided by the no. of part in the rest.
- Remove the ferrous residue by means of a magnet and the ferrous content of residue.
- Weight values may be excess of true dust values because of foreign impurities such as soap crystal residual oils brush bristles etc. if seem on the filter paper. These impurities should be noted down the evaluation.

3. Project Description

This experimental study is performed at a prestigious tractor engine manufacturing company situated in North Central Region of India. The case company has one of the oldest names in the field of tractor manufacturing sectors. The objective of the project is to find out the life of washing solvent/media in washing machine with respect to cleanliness level of Millipore value of Engine Components. During the study it was observed that the cleaning medium was not correctly mixed with water so it affects the contamination level of components. The study was carried out on two washing machines as ordinary washing and Ultra sonic washing machine .The washing solvent METACLEAN 211 and ASFOCLEAN – AD-20 in ordinary washing and Ultrasonic washing machine.

Table 1 Operating instructions for Ultrasonic machines (Parameters to be maintained)

Parameters	TANK				
	I	II	III	IV	Storage Tank
Qty. of Water	75	55	55	-	55
Qty. of AD-20	5%	5%	1%	-	5%
Cleaning Medium (Ltr.)	4	3	0.5	-	3
Temperature(^{0} C)	56	56	56	60	-
Cleaning cycle time (Min.)	5	5	5	5	-

Operating parameters in ultrasonic machine are given in Table 1. There are four washing tank in this machine. Fixed Quantity of water maintained in tanks.Percentage of AD-20 in three tank is used as 5%,5% and 1% respectively.Temerature maintaned in four tanks at 56°,56°,56° and 60°C respectively.

Table 2 Ordinary washing machine (setting parameters)

Parameters	Specifications
Temperature(^{0} C)	40-44
Concentration %	3-5-8
Pressure (Kg/cm²)	3-6
Washing Time(Min)	3

Operating parameters of ordinary washing machine are followed as per Table 2.Temperature is maintained in range of 40°-44°C, concentration percentage was limited to 3-5-8%, pressure was set at 3-6 Kg/cm², and washing time was 3 minutes.

Table 3 shows different setting parameters date wise in ordinary washing machine. These values were noted daily when cleaning was performed.

Table 3 Setting parameters date wise on ordinary washing machine (META CLEAN 211)

Date	Temperature °C	Liquid Concentration %	Pressure (Kg/cm ²)	Washing Time (Min)
05/06/2020	43	10	3.0	3
06/06/2020	44	8	4.2	4
07/06/2020	43	6	3.1	3
09/06/2020	44	10	4.0	5
10/06/2020	45	8	4.2	3
11/06/2020	44	6	3.3	4
12/06/2020	43	8	4.3	4
13/06/2020	45	5	3.0	4
15/06/2020	44	10	3.2	3
16/06/2020	43	10	4.0	3
17/06/2020	44	8	3.0	3
18/06/2020	43	8	3.2	5
19/06/2020	45	10	3.8	3
20/06/2020	44	8	4.0	4

Table 4 shows different setting parameters on ultrasonic washing machine such as AD 20 strength, temperature, contamination and cleaning time of the components. These data were noted on daily basis when the experiments were performed. AD 20 strength was measured in percentage with water and maximum percentage was 10% while minimum was 1%. Temperature was also noted in all four washing tanks. It was found 56-57 degree in first three tanks while in fourth tank it was variable. Contamination level was measured in mg/100ml. Contamination was continuously increasing from 15.2 mg/100ml to 38.2mg/ml. During the cleaning process, cleaning time was also noted and it was 5-6 min.

Seven components were cleaned on ultrasonic washing machine using AFSOCLEAN AD 20 as the cleaning medium. Observations were taken for 14 days from 05.06.2020 to 20.06.2020 as shown in Table 5. Contamination trend of these components were calculated using Millipore equipment. Weight of the contamination was noted in mg/pc on digital weigh machine. Date wise results shown in Table 5.These data were compared with standard values of contamination specified by the suppliers. For example- for connecting rod (C1) it is 10 mg/pc.It is clear from the Table 5 that the contamination level comes within specification limits upto 12 days. After 12 days it goes beyond the specifications.

Table 4 Setting parameters date wise for ultrasonic machine

Date	AD-20 Strength (% Tank)				Temperature (°C)				Contamination	Cleaning Time (Min)			
	I	II	III	IV	I	II	III	IV	Mg/100ml	I	II	III	IV
05/06/20	10	10	5	-	56	56	57	30-56	15.2	5	5	5	5
06/06/20	8	5	4	-	57	56	56	28-57	16.3	6	4	5	5
07/06/20	8	6	3	-	56	56	57	25-50	16.9	5	5	6	5
09/06/20	6	5	3	-	56	57	56	30-58	17.5	4	6	5	5
10/06/20	8	5	2	-	57	56	56	28-51	19.2	5	5	6	5
11/06/20	6	6	1	-	56	56	56	28-50	20.8	5	5	4	5
12/06/20	8	6	3	-	57	56	56	30-60	22.4	5	5	4	6
13/06/20	10	5	4	-	56	56	56	22-55	21.2	6	5	5	5
15/06/20	8	5	2	-	56	57	56	25-56	25.6	5	6	5	6
16/06/20	6	6	3	-	56	57	56	23-58	28.2	5	5	5	5
17/06/20	5	6	2	-	57	56	56	22-52	32.2	5	4	5	5
18/06/20	6	5	1	-	56	56	57	21-56	36.2	5	6	5	5
19/06/20	8	5	2	-	57	57	56	28-59	36.8	5	4	5	5
20/06/20	6	6	1	-	56	56	57	30-58	38.4	5	6	5	6

Table 5 Ultrasonic Washing Machine (Contamination Trend) Cleaning Medium –ASFOCLEAN –AD-20)

Component	Specification	Actual (mg/pc) Date wise													
	Mg/pc	5/6 /20	6/6 /20	7/6 /20	9/6 /20	10/6 /20	11/6 /20	12/6 /20	13/6 /20	15/6 /20	16/6 /20	17/6 /20	18/6 /20	19/6 /20	20/6 /20
C1	10	2.4	2.8	3.3	3.9	4.6	5.6	6.1	7.6	8.1	8.7	9.2	9.8	**10.5**	**12.8**
C2	2	0.8	1.1	1.2	1.4	1.4	1.4	1.5	1.6	1.7	1.7	1.8	1.9	**2.4**	**2.8**
C3	2	0.7	1.2	1.3	1.3	1.4	1.5	1.6	1.6	1.6	1.8	1.7	1.8	**2.5**	**3.1**
C4	2	0.9	1.2	1.3	1.4	1.5	1.4	1.6	1.7	1.8	1.7	1.8	1.7	**2.3**	**2.9**
C5	2	1.1	1.3	1.4	1.4	1.3	1.6	1.7	1.7	1.6	1.8	1.7	1.8	**2.3**	**2.9**
C6	2	0.7	1.9	1.1	1.2	1.5	1.4	1.5	1.6	1.7	1.7	1.7	1.9	**2.4**	**3.2**
C7	2	0.8	1.1	1.2	1.4	1.4	1.3	1.6	1.6	1.7	1.8	1.9	2.1	**2.8**	**3.4**

C1-Connecting rod, C2- RA. Exhaust, C3-R.A.Inlet, C4-Spring outer, C5-Spring inner, C6-V.S.Retainer Exhaust, C7-V.S. Retainer Inlet

For connecting rod it was 2.4 mg/pc on 05.06.2020 and after 12 days of observations it increased to 10.5 mg/pc which is out of specification limit. Similarly, for R.A. Exhaust the contamination was 0.8 mg/pc on the first day and it increased to 2.4 mg/pc after 12 days that shows out of the specification limit.

Table 6 shows the contamination trend on ordinary washing machine using METACLEAN 211 as cleaning medium. From the data it is clear that for protection sleeve, contamination level was 3.5 mg/pc on 05.06.2020 and it was found 5.8 mg/pc after 12 days. Similarly, it is clear from the observations that for all components, the cleaning media works satisfactory up-to 12 days only.

Table 6 Ordinary Washing Machine (Contamination Trend) Cleaning Medium METACLEAN 211

Component	Specification	Actual (mg/pc) Date wise													
	Mg/pc	5/6/20	6/6/20	7/6/20	9/6/20	10/6/20	11/6/20	12/6/20	13/6/20	15/6/20	16/6/20	17/6/20	18/6/20	19/6/20	20/6/20
C8	5	3.5	3.8	3.9	4.1	4.2	4.2	4.0	4.2	4.3	4.5	4.4	4.8	**5.8**	**7.1**
C9	20	12	12	13	14	14	15	17	17	19	19	20	20	**20**	**25**
C10	25	8.1	10	13	13	15	17	17	18	21	24	24	25	**26**	**31**

C8-Protection sleeve, C9-R.A.Housing, C10-Lid cover

4. Conclusion/recommendations:

It was observed from experimental study that if these washing solvents (AD-20 & METACLEAN 21) is used for both shift than these clean the components properly upto 06 days. After that the cleaning quality does not get upto standard level. If these washing media solvents will be used in single shift than it will clean the components upto 12 days max. If these are used beyond 12 days the component cleanliness level goes beyond the specification/standard.

The paper presented the experimental work for calculating life of washing media/solvent used in ultrasonic and ordinary washing machine. In the experiment components from tractor engine were collected for daily contamination checking and after washing in ultrasonic and ordinary washing machine, It was found that after 12 days the contamination of the component cross their specific value of contamination level.

Suggestions:

- The proportion of washing media with water should be correct.
- The tanks of the washing machine should be cleaned time to time.
- There should not be any dirty material around the washing machine.
- The time taken in washing machine the components should be correct.
- The tray in which the components are kept for washing should be washed.
- The contamination of the components washing in machine should be checked time to time.
- When the contamination of the components get their specification value. Then washing liquid should be changed.

References:

1. Kobayashi, Y., Ayame, T., Shibuya, k., Nakagawa, T., Kubota, Y., Gonda, K. and Ohuchi, N.(2016) "Stabilization of silica-coated silver iodide nanoparticles by ethanol-washing", *Pigment & Resin Technology*, Vol. 45 Issue: 2, pp.99-105.
2. Goldman, I.B., Aitken, D.F. and Charest, D. (1991) "Eliminating Solvents in the Cleaning of Circuit Assemblies: A Case History", *Circuit World*, Vol. 17 Issue: 3, pp.48-52.
3. Card, A., Moore, MA and Ankeny, M. (2006) "Garment washed jeans: impact of launderings on physical properties", *International Journal of Clothing Science and Technology,* Vol. 18 Issue: 1, pp.43-52.
4. Komal, (2017) "Fuzzy reliability analysis of the washing system in a paper plant using the TBFLT technique", *International Journal of Quality & Reliability Management*, Vol. 34 Issue: 8.
5. Brian N. Ellis, C.v Eng., M.I.E.R.E., F. Inst.C.T., (1980) "Water Soluble Fluxes for PCB Soldering and Contamination Control", *Circuit World*, Vol. 6 Issue: 3, pp.6-10.

IOP Conf. Series: Materials Science and Engineering 1224 (2022) 012014 doi:10.1088/1757-899X/1224/1/012014

Performance evaluation of hybrid desiccant air conditioning system regenerated by waste heat from air cooled condenser: Experimental Investigation

Taliv Hussain

Mechanical Engineering Department, Aligarh Muslim University, Aligarh, India.

* Author corresponding Email: hussaintaliv@gmail.com

Abstract. Heating ventilation and air conditioning units reject a large quantity of heat to the surroundings as waste. Thus proper reclamation of this unused heat could also enhance the COP of such arrangements and their lesser adverse effect on the environment. Therefore hybrid desiccant air conditioning unit using unused heat from condenser for regeneration can be more effective than the conventional air conditioning unit under hot and humid climatic conditions. In this paper, various performance parameters of hybrid desiccant air conditioning unit using unused heat from condenser is determined at different process air inlet temperatures (28, 29.5, 31, 32.5, 34, 35.5 and 37°C) and different process air inlet velocities, i.e. (1.5, 2.5, 3.5 and 4.5 m/s) and at fix (2.5 m/s) air velocity at regeneration inlet. Refrigerating effect, heat rejection rate, VCOP, ECOP, dehumidification effectiveness, moisture removal capacity and DCOP decreases with increment in different air temperatures at process inlet. While compressor work, regeneration effectiveness and regeneration rate increases with increment in different air temperatures at process inlet. On the other hand, refrigerating effect, heat rejection rate, VCOP, ECOP, dehumidification effectiveness, regeneration effectiveness, and regeneration rate decrease with increased air velocities at process inlet. Whereas compressor work, moisture removal capacity and DCOP increases with an increase in air velocities at the process inlet. Thus, optimization of these performance parameters is determined to achieve the maximum efficiency of the hybrid desiccant air conditioning system under the above operating states.

1. Introduction

The accessible energy must be employed appropriately, as there is a high requirement for electrical energy creation and circulation. Around 45% of electric power in every household is used to function refrigeration, heating, ventilation and air conditioning schemes in our nation. All these systems reject a tremendous quantity of heat to the environment as waste. Therefore, proper reclamation of this unused heat could also expand the COP of such structures and their minor adverse effect on the surroundings. Incorporating a desiccant structure with a vapour compression refrigeration system can distinctly eliminate latent and sensible load under hot and humid climates. It also increases COP and decreases the energy consumption of a vapour compression refrigeration system. Incorporating a desiccant scheme with a vapour compression refrigeration system will also lower desiccant regeneration temperature from (70–80)°C to (50–60)°C. Thus, the heat excluded by the condenser can contribute to renewal of desiccant in dehumidifier. Desiccant based air conditioning units have shown numerous benefits than that of conventional arrangements for dehumidification and cooling. Thus, their usage is also scattering for tertiary and domestic structures, mainly when renewal of desiccant can be acquired via utilizing accessible unused heat [1,2].

Few of the researchers worked in the area of hybrid desiccant air conditioning units. Ando et al. (2005) [3] proposed and experimentally investigated a four rotor desiccant cooling method equipped with two stage dehumidification. In this procedure, a restoration temperature of about 70°C could generate an adequate dehumidifying effectiveness at more air humidity. They attained heat from low grade energy like solar heat and unused heat as a replacement for electricity. Jalalzadeh-Azar et al. (2005) [4] constructed and examined cooling unit using a heat exchanger, reciprocating IC engine, desiccant dehumidifier and indirect/direct dehumidifier. Restoration of desiccant was executed via utilizing heat recovered from an IC engine. Lokapure and Joshi (2012) [5] aimed and constructed a heat exchanger in which COP of air conditioning unit was enhanced via 13% by recovering the waste heat. In this work, the principal stress was given on energy conservation by using the waste heat from the air conditioning system and thus improving COP. Angrisani et al. (2015) [6] investigated hybrid desiccant cooling unit with a desiccant wheel relating with small scale cogenerator. They scrutinized the performance by changing five operating conditions: rotational speed, restoration temperature, outside air temperature, volume airflow rates and specific humidity. They also examined different performance parameters based on primary, thermal and electric energy. They compared the hybrid desiccant cooling system with a micro cogenerator with additional electrical and thermal air conditioning systems. Jani et al. (2016) [7] discussed the principle of a solid desiccant cooling unit and its scientific uses and expansions. They highlighted numerous arrangements of desiccant cooling cycles, crossover/regular desiccant cooling cycles, various numerical models of a desiccant dehumidifier, execution assessment of the desiccant cooling framework and mechanical improvements. They also discussed the performance of solid desiccant cooling unit powered thru industrial waste/solar energy heat for regenerating the desiccant.

The main aim of research paper is to analyze various performance parameters of hybrid solid desiccant air conditioning system utilizing waste heat from condenser for restoration at different process air inlet temperatures (28, 29.5, 31, 32.5, 34, 35.5 and 37°C) and at different process air inlet velocities, i.e. (1.5, 2.5, 3.5 and 4.5 m/s) and fix (2.5 m/s) air velocity at regeneration entry. Thus identifying the optimum conditions of performance parameters for effective working of hybrid desiccant air conditioning system under above operating states.

2. Design of the experiment and Methodology

The experimental scheme contains a pretreatment segment, a solid desiccant wheel and a vapour compression refrigeration system (figure 2.1). The air pretreatment segment includes an electric heater and a humidifier. Experimental data has been collected for Hybrid VCRs in which the desiccant wheel is using waste heat from the condenser at different process air inlet temperatures (28, 29.5, 31, 32.5, 34, 35.5 and 37°C) and different process air inlet velocities (1.5, 2.5, 3.5 and 4.5 m/s) and fix (2.5 m/s) restoration air inlet velocity. The suction pressure (P_{suc}) and discharge pressure (P_{dis}) of refrigerant R-410a are measured with the help of Bourdon tube pressure gauges. The current (I) is measured with the help of an ammeter. The voltage (V) is noted with the use of a voltmeter. Data logger measures the temperature of refrigerant R-410a at the different points of VCRs. Data logger measures the experimental data at the entry and exit of process and regeneration sector of desiccant wheel, respectively. Humidity ratio of air has been calculated from relative humidity and dry bulb temperature with the help of a Psychrometric calculator.

The incoming air is heated through electric heater to the required temperature and then humidified thru a humidifier. After that, hot and humid air is blown over process segment of the rotary desiccant wheel. When processed air streams over the desiccant exterior in the adsorption segment, its vapour pressure is higher than that of the desiccant outward due to the more humid air state, which generates a significant difference. Thus adsorption sector takes away water vapour from moist air. This water vapour is adsorbed on the exterior of the desiccant and is condensed above it. Hence, adsorption occurs. Then the air passes over the vapour compression refrigeration system's cooling coil and is sensibly reduced to an ideal temperature. In the regeneration sector, when restoration air streams over the desiccant exterior in the restoration segment, it heats the desiccant surface. It increases its vapour pressure which upturns the vapour pressure difference within regeneration air and desiccant surface. Therefore, restoration air evaporates the moisture from a desiccant layer. Thus, regeneration occurs.

Figure 2.1 Photograph of an experimental setup for Desiccant Air Conditioning System utilizing waste heat of condenser for renewal of Desiccant wheel

2.1 Assessing device and Instruments

A Thermo hygrometer measures the relative humidity and dry bulb temperature of the air. An anemometer measures air velocity. Table 1 shows Assessing devices and instruments.

Table 1 Assessing device

Equipment	**Voltage**	**Sampling Rate**	**Range**	**Accuracy**
Tweex 8 Channel Data Logger	Single Phase 220V Standard supply	5 seconds to 255 minutes	-10 to +85 °C 0 to 100 %RH	±0.5°C ±2.5%
Digital Anemometer	3.0 V DC	-	0 to 30 m/s	±5%

IOP Conf. Series: Materials Science and Engineering 1224 (2022) 012014 doi:10.1088/1757-899X/1224/1/012014

2.2 Performance Factors

The different performance parameters of hybrid desiccant air conditioning unit utilizing the waste heat from condenser for renewal are shown in equations (1-10) at different process air inlet temperatures (28, 29.5, 31, 32.5, 34, 35.5 and 37°C) and different process air inlet velocities, i.e. (1.5, 2.5, 3.5 and 4.5 m/s) and at fix (2.5 m/s) air velocity at regeneration entry:

1. Refrigerating Effect = $(h_{ref1} - h_{ref4})$ (1)
2. Heat Rejection Rate = $(h_{ref2} - h_{ref3})$ (2)
3. Compressor Work = $(h_{ref2} - h_{ref1})$ (3)
4. VCOP of hybrid system [8]=Refrigerating Effect/ Compressor Work = $(h_{ref1} - h_{ref4})/(h_{ref2} - h_{ref1})$ (4)
5. ECOP of hybrid system [9] = Total Cooling Effect/ Total Electric Power Consumption (5)

6. Dehumidification Effectiveness (ηdeh): symbolizes the ratio between actual and ideal dehumidification ability of desiccant wheel [10].

$$\eta_{deh} = \frac{\omega_1 - \omega_2}{\omega_1} \quad (6)$$

7. Moisture Removal Capacity (MRC): signifies mass flow rate of dampness detached by desiccant wheel [10].

$$MRC = \rho_1 \times \dot{V}_P \times (\omega_1 - \omega_2) \quad (7)$$

8. Dehumidification Coefficient of Performance (DCOP): signifies ratio among thermal power linked to air dehumidification and thermal energy provided for restoration process [10].

$$DCOP = \frac{\rho_1 \times \dot{V}_P \times \Delta h_{vs} \times (\omega_1 - \omega_2)}{\rho_1 \times \dot{V}_{reg} \times C_p \times (T_4 - T_1)} = \frac{\dot{V}_P \times \Delta h_{vs} \times (\omega_1 - \omega_2)}{\dot{V}_{reg} \times (h_4 - h_1)} \quad (8)$$

The latent heat of vaporization of water Δhvs has been assessed using subsequent empirical cubic function [10].

$$\boldsymbol{\Delta h_{vs} = -0.614342 \times 10^{-4} \times T_1^3 + 0.0158927 \times 10^{-2} \times T_1^2 - 0.236418 \times 10 \times T_1 + 0.250079 \times 10^4}$$

9. Regeneration Effectiveness (ηreg): implies latent load held by a desiccant wheel about thermal regeneration power needed for adsorption procedure of desiccant wheel [11].

$$\eta_{reg} = \frac{\omega_3 - \omega_4}{\omega_4} \quad (9)$$

10. Regeneration Rate (RR) signifies the mass flow rate of dampness detached by a wheel from the process side to the regeneration side [11, 12].

$$RR = \rho_1 \times \dot{V}_R \times (\omega_3 - \omega_4) \quad (10)$$

3. Graphs and Discussions

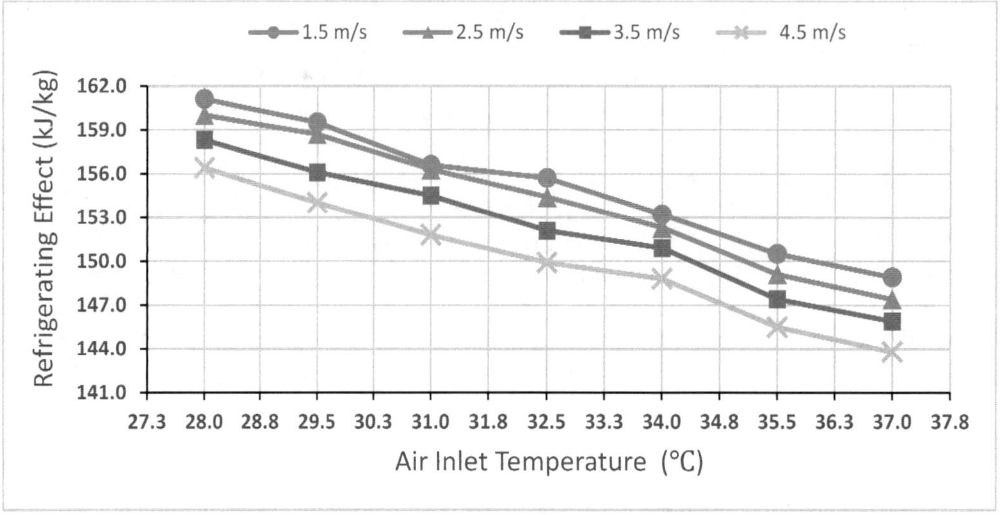

Figure 3.1 Variation of Refrigerating Effect with Air Inlet Temperatures at different process air inlet velocity and 2.5 m/s air velocity at regeneration inlet

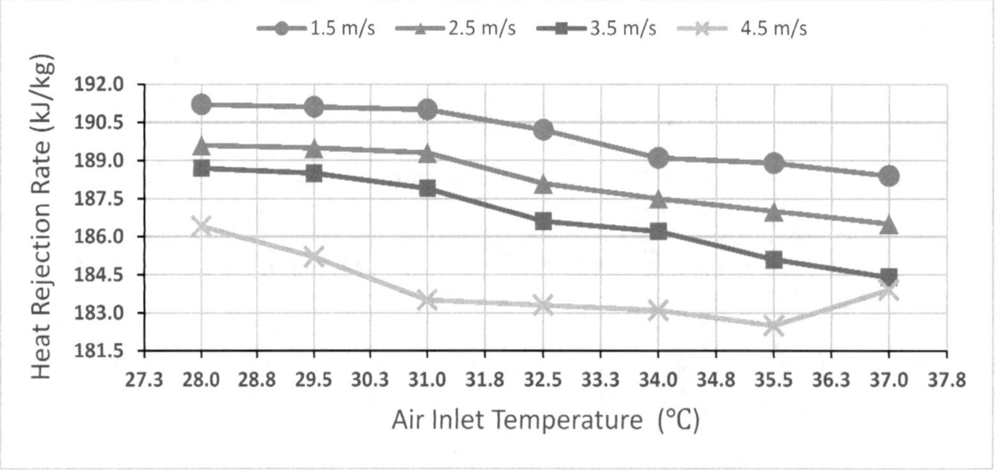

Figure 3.2 Variation of Heat Rejection Rate with Air Inlet Temperatures at different process air inlet velocity and 2.5 m/s air velocity at regeneration inlet

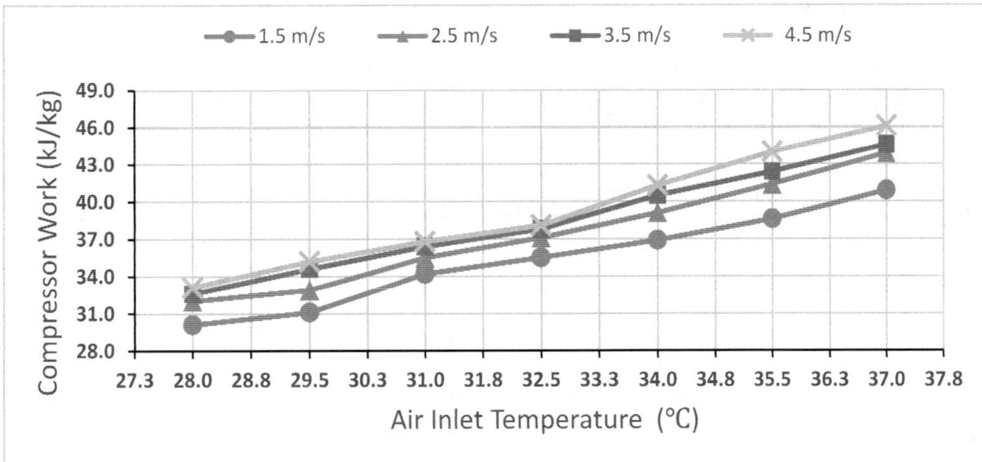

Figure 3.3 Variation of Compressor Work with Air Inlet Temperatures at different process air inlet velocity and 2.5 m/s air velocity at regeneration inlet

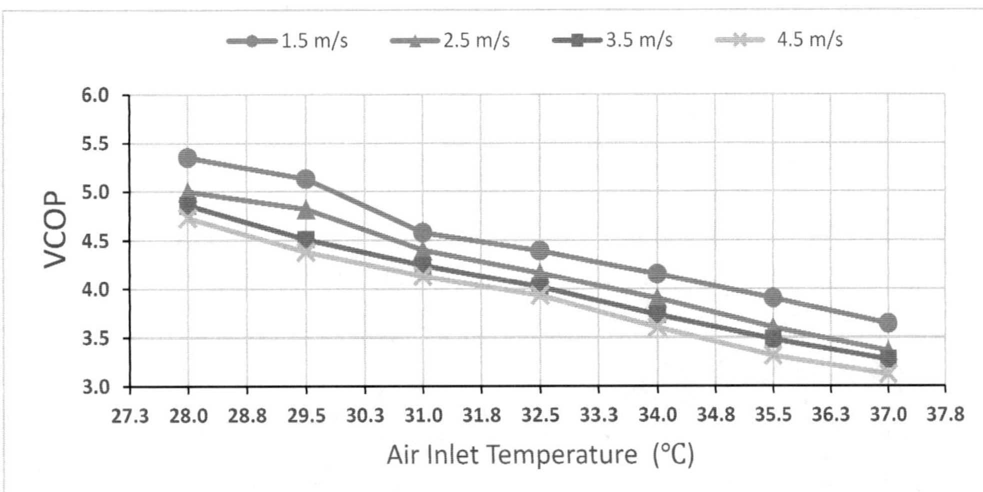

Figure 3.4 Variation of VCOP with Air Inlet Temperatures at different process air inlet velocity and 2.5 m/s air velocity at regeneration inlet

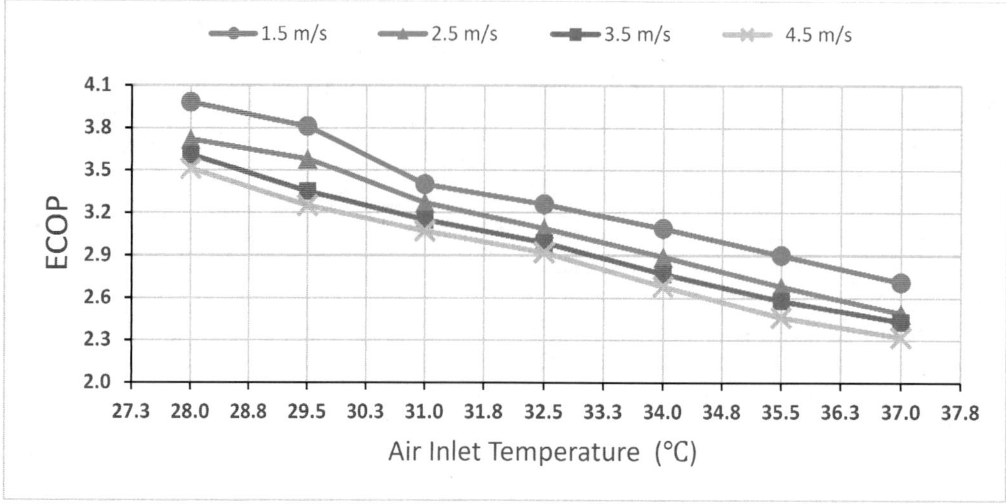

Figure 3.5 Variation of ECOP with Air Inlet Temperatures at different process air inlet velocity and 2.5 m/s air velocity at regeneration inlet

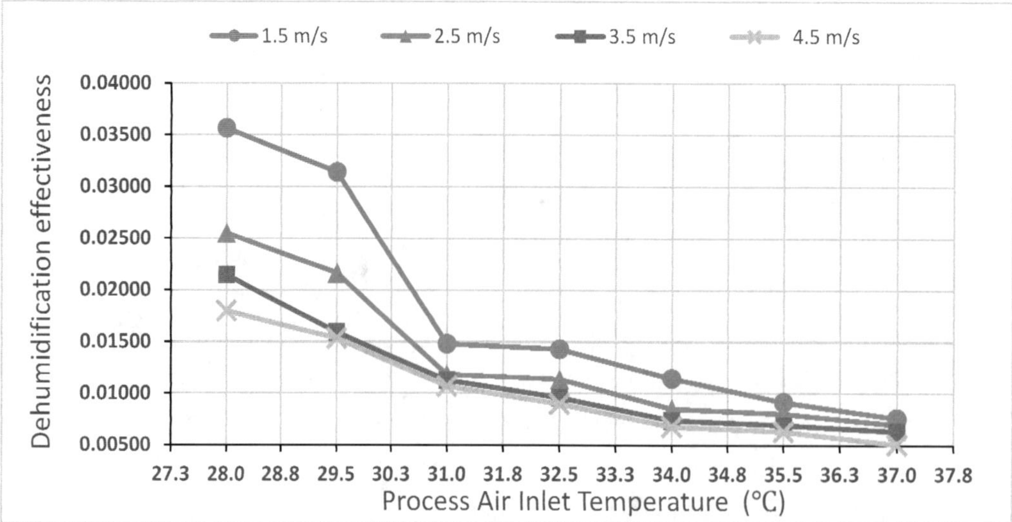

Figure 3.6 Variation of Dehumidification Effectiveness with Air Inlet Temperatures at different process air inlet velocity and 2.5 m/s air velocity at regeneration inlet

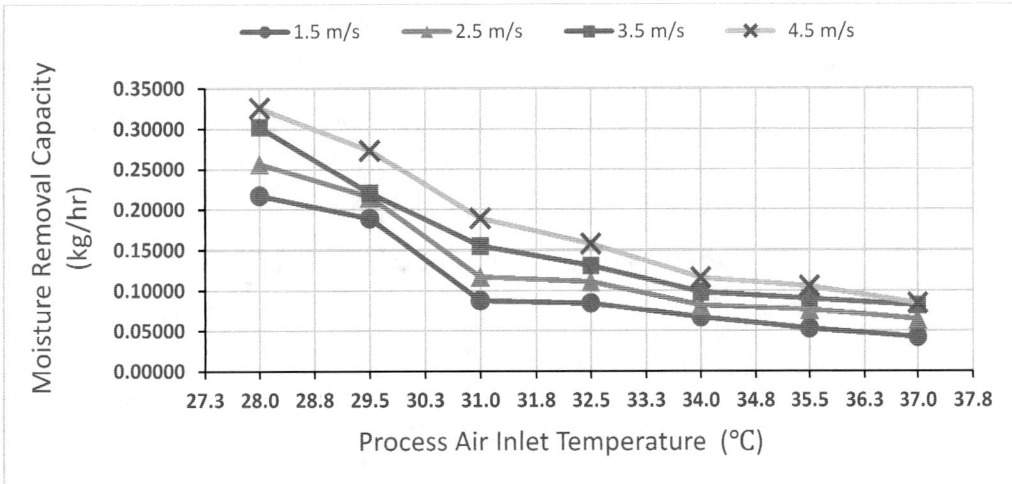

Figure 3.7 Variation of Moisture Removal Capacity (kg/hr) with Air Inlet Temperatures at different process air inlet velocity and 2.5 m/s air velocity at regeneration inlet

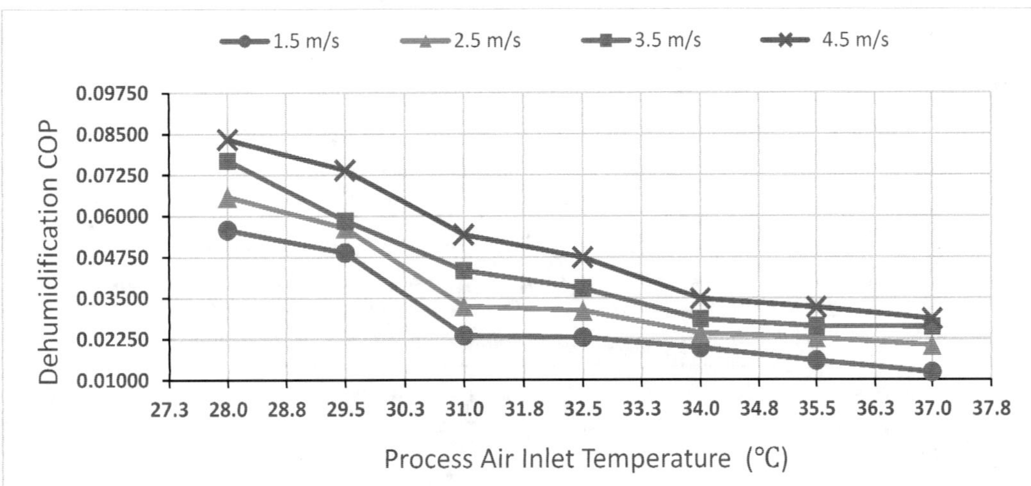

Figure 3.8 Variation of Dehumidification COP with Air Inlet Temperatures at different process air inlet velocity and 2.5 m/s air velocity at regeneration inlet

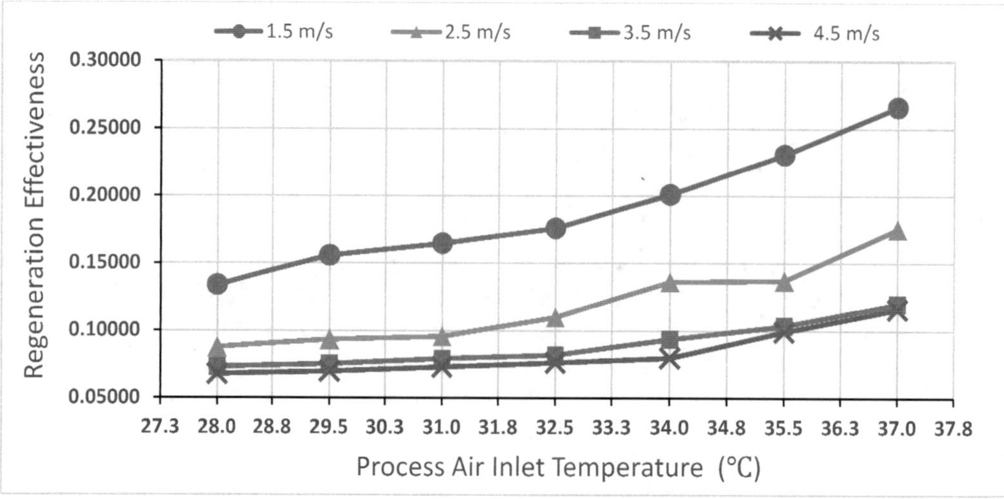

Figure 3.9 Variation of Regeneration Effectiveness with Air Inlet Temperatures at different process air inlet velocity and 2.5 m/s air velocity at regeneration inlet

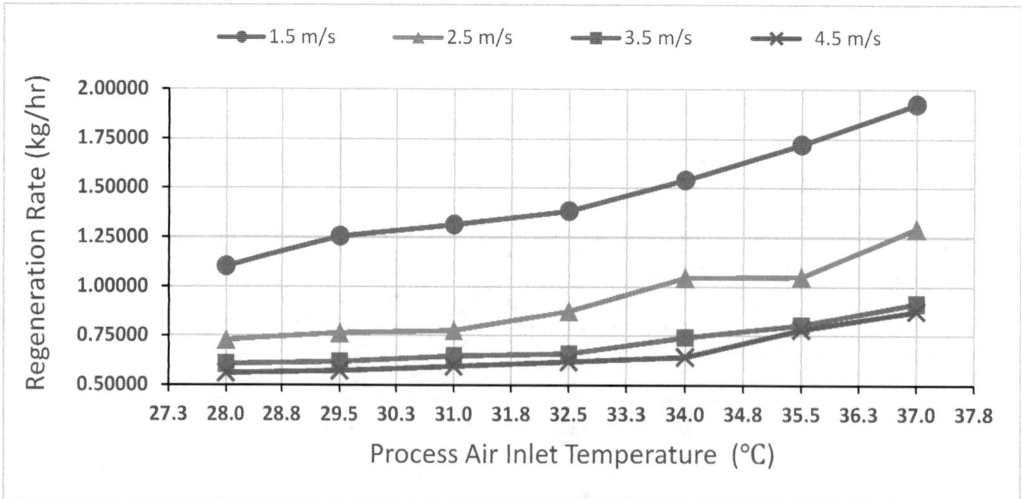

Figure 3.10 Variation of Regeneration Rate (kg/hr) with Air Inlet Temperatures at different process air inlet velocity and 2.5 m/s air velocity at regeneration inlet

Figures 3.1, 3.2, 3.4 and 3.5 shows for all the different process air inlet velocities, the value of refrigerating effect, heat rejection rate, VCOP and ECOP decreases with increment in the distant air temperatures at the process a fix 2.5 m/s air velocity at regeneration inlet. But, figure 3.3 shows that compressor work upsurges with rise in various process air entry temperatures. It is so, as ambient temperature increases, rate of heat transfer from the condenser decreases. With the rise in ambient temperature, the air-cooled condenser is also affected and gets less cooled, increasing the load of the vapour compression refrigeration system. Thus due to this effect, less heat is absorbed by refrigerant from the air at the evaporator. Also, figures 3.1, 3.2, 3.4 and 3.5 shows that for all different process air entry temperatures, value of refrigerating effect, heat rejection rate, VCOP and ECOP declines with rise in various air velocities at process inlet and a fix 2.5 m/s air velocity at regeneration inlet. More incredible the air velocity at the process inlet minimizes the contact time between the surface of the cooling coil and incoming air. Thus, the refrigerant absorbs less heat from the incoming air due to lower contact time [9,12]. Figures 3.6, 3.7 and 3.8 show that the value of dehumidification effectiveness, moisture removal capacity and DCOP decreases with increased process air inlet temperatures. There is lower ambient relative humidity at higher ambient temperatures, which leads to drop in dehumidification effectiveness. At entry process, the air specific humidity also decreases due to lesser relative humidity, which causes air vapour pressure to fall adsorption potential [6,10]. Figures 3.9 and 3.10 show that the value of regeneration effectiveness and regeneration rate increases with air temperatures at process inlet. As ambient temperature rises, it also affects regeneration temperature, which also increases, which clues to enlarged removal of water vapour from the desiccant wheel during the renewal period. Thus the regeneration effectiveness and regeneration rate also increase with increment in the ambient temperature.

4. Error Analysis

The error investigation completed in the present work depends on root sum square strategy detailed by Kline and McClintock [13], the performance parameters described in this work i.e. refrigerating effect, heat rejection rate, compressor work, VCOP, ECOP, dehumidification effectiveness, moisture removal capacity, dehumidification COP, regeneration effectiveness and regeneration rate are found via calculation from noted variables such as velocity of air, relative humidity and temperature is described via a recognized value of uncertainty. The comparative uncertainty values obtained for the considered factors are, 9.5% for refrigerating effect, 1.8% for heat rejection rate, 8.3% for compressor work, 9.8% for VCOP, 9.5% for ECOP, 6.1% for dehumidification effectiveness, 6.9% for moisture removal capacity, 7.2% for dehumidification COP, 14.1% for regeneration effectiveness and 14.3% for regeneration rate.

5. Conclusions

The key objective of the current investigation was to explore the possibility of a hybrid solid desiccant air conditioning unit using the waste heat from condenser for regeneration at different process air inlet temperatures (28, 29.5, 31, 32.5, 34, 35.5 and 37°C) and at different process air inlet velocities, i.e. (1.5, 2.5, 3.5 and 4.5 m/s) and at fix (2.5 m/s) air velocity at regeneration entry. It was concluded that best conditions of performance parameters are identified for effective working of hybrid desiccant air conditioning system under above functioning situations, which are as follows:

- Refrigerating effect, heat rejection rate, VCOP, and ECOP decrease with increment in different air temperatures at process inlet and different air velocities at process inlet at a fixed 2.5 m/s renewal air entry velocity.
- Compressor work upsurges with rise in various air temperatures at process inlet, different air velocities at process inlet at a fixed 2.5 m/s restoration air entry velocity.

- The regeneration temperature must be high to increase the refrigerating effect, heat rejection rate, VCOP and ECOP, and regeneration temperature. To lower the compressor work, the regeneration temperature should be increased.
- The maximum value of refrigerating effect, heat rejection rate, VCOP and ECOP is 161.1 kJ/kg, 191.2 kJ/kg, 5.35 and 3.98 respectively at 28°C process air entry temperature, 1.5 m/s processes and 2.5 m/s air velocity at regeneration inlet.
- The minimum value of compressor work is 30.1 kJ/kg at 28°C air temperature at process inlet, 1.5 m/s process and 2.5 m/s air velocity at regeneration inlet.
- Refrigerating effect, heat rejection rate, VCOP, ECOP, dehumidification effectiveness, regeneration effectiveness and regeneration rate decreases with increment in various air velocities at process inlet and a fix 2.5 m/s air velocity at regeneration entry.
- Compressor work, moisture removal capacity and DCOP increases with an increase in air velocities at the process inlet.
- Dehumidification effectiveness, moisture removal capacity and DCOP decreases with increment in air temperatures at process inlet. In contrast, regeneration effectiveness and regeneration rate increase with rise in process air inlet temperatures.
- Regeneration temperature should be high to enhance dehumidification effectiveness, moisture removal capacity, DCOP, regeneration effectiveness, and regeneration rate.

Nomenclature

c_p	air specific heat (kJ/kg K)
COP	Coefficient of performance
VCOP	COP of vapour compression refrigeration system
ECOP	electric coefficient of performance
h_{ref1}	at the entry of compressor enthalpy of refrigerant in kJ/kg
h_{ref2}	at the outlet of compressor enthalpy of refrigerant in kJ/kg
h_{ref3}	at the outlet of condenser enthalpy of refrigerant in kJ/kg
h_{ref4}	at the entry of evaporator enthalpy of refrigerant in kJ/kg
MRC	Adsorption rate (kg/hr)
T_1	at process entry temperature of the air
T_2	at process exit temperature of the air
T_4	at regeneration entry temperature of the air
T_3	at regeneration exit temperature of the air
ω_1	at the process inlet humidity ratio of air
ω_2	at the process outlet humidity ratio of air
ω_4	at the regeneration inlet humidity ratio of air
ω_3	at the regeneration outlet humidity ratio of air
$\dot{V}$	volumetric flow rate of air (m^3/hr)
DW	desiccant wheel

Greek symbols:

Δh_{vs}	latent heat of vaporization of the water (kJ/kg)
ρ	density of air (kg/m^3)
ω	humidity ratio or specific humidity (kg/kg)

Subscripts:

P	process
reg	regeneration
dis	discharge
suc	suction

References

[1] Harriman, L.G. (2002).The Dehumidification Handbook, Second Edition. Munters Corporation, AMESBURY, MA.

[2] Azhar, M., & Siddiqui, M. A. (2017). Optimization of operating temperatures in the gas operated single to triple effect vapour absorption refrigeration cycles. *International journal of refrigeration*, 82, 401-425.

[3] Ando, K., Kodama, A., Hirose, T., Goto, M., & Okano, H. (2005). Experimental study on a process design for adsorption desiccant cooling driven with a low-temperature heat. Adsorption, 11(1), 631-636.

[4] Jalalzadeh-Azar, A. A., Slayzak, S., Judkoff, R., Schaffhauser, T., & DeBlasio, R. (2005). Performance Assessment of a Desiccant Cooling System in a CHP Application with an IC Engine. *International Journal of Distributed Energy Resources*, 1(NREL/JA-550-36974).

[5] Lokapure, R. B., & Joshi, J. D. (2012). Waste heat recovery through air conditioning system. *International Journal of Engineering Research and Developmen*t, 5(3), 87-92.

[6] Angrisani, G., Roselli, C., & Sasso, M. (2015). Experimental assessment of the energy performance of a hybrid desiccant cooling system and comparison with other air-conditioning technologies. *Applied energy*, 138, 533-545.

[7] Jani, D. B., Mishra, M., & Sahoo, P. K. (2016). Solid desiccant air conditioning–A state of the art review. *Renewable and Sustainable Energy Reviews*, 60, 1451-1469.

[8] Ansari, A. A., Goyal, V., Yahya, S. M., & Hussain, T. (2018). Experimental investigation for performance enhancement of a vapor compression refrigeration system by employing several types of water-cooled condenser. *Science and Technology for the Built Environment*, 24(7), 793-802.

[9] Dai, Y. J., Wang, R. Z., Zhang, H. F., & Yu, J. D. (2001). Use of liquid desiccant cooling to improve the performance of vapor compression air conditioning. *Applied Thermal Engineering*, 21(12), 1185-1202.

[10] Angrisani, G., Minichiello, F., Roselli, C., & Sasso, M. (2012). Experimental analysis on the dehumidification and thermal performance of a desiccant wheel. Applied Energy, 92, 563-572.

[11] Goodarzia, G., Thirukonda, N., Heidari, S., Akbarzadeh, A., & Date, A. (2017). Performance evaluation of solid desiccant wheel regenerated by waste heat or renewable energy. Energy Procedia, 110, 434-439.

[12] Yadav, A. V. A. D. H. E. S. H. (2012). Experimental and numerical investigation of solar powered solid desiccant dehumidifier (Doctoral dissertation, NATIONAL INSTITUTE OF TECHNOLOGY).

[13] Kline, S. J. (1953). Describing uncertainty in single sample experiments. Mech. Engineering, 75, 3-8.

Dehumidification Analysis of Rotary Solid Desiccant Wheel System for different Surface Materials

Taliv Hussain, Ashad Ahmad, Nabeel Hidayat, Danish Ahamad, Md Salik Anwar and Sadique Istekhar Ahmed

Mechanical Engineering Department, Aligarh Muslim University, Aligarh, India.

* Author corresponding Email: danishahamad1425@gmail.com

Abstract. In recent decades, desiccant dehumidification has been given much interest in the air conditioning industry area. Compared to the conventional vapour compression air conditioning system, desiccant dehumidification has various advantages, including separate humidity and temperature control. In this paper, the analysis of the Rotary Solid Desiccant Wheel System has been performed. Here, we have studied three diffusion coefficients: Ordinary, Knudsen and Surface diffusion coefficients for Rotary Solid Desiccant Wheel System. Also, the analysis has been performed for three different solid desiccant materials: Silica Gel, Molecular Sieve and Activated Alumina to incorporate and to study the effect of pore size on moisture transport and to examine the impact of Tortuosity factor on moisture transport phenomenon. This analysis can correctly select the best desiccant material suitable for the desiccant dehumidification industry.

1. Introduction

Desiccants are unique materials that exhibit an excellent capability to adsorb water vapour upon contact with moist air at average temperatures. However, desiccants utilized for space conditioning must be able to hold significant quantity of water. Desiccant dehumidification cooling does not involve cooling to the dew point. It can adsorb moisture at almost any level of humidity. The air has to be cooled to a dew point in a conventional air conditioning system for dehumidification. In some cases, humid air is overcooled to achieve low humidity, which decreases energy efficiency. Industrial solid desiccant ingredients can hold up water to more than half of their weight. Silica gel, molecular sieve and activated alumina are the most general marketable solid desiccants in use. The desiccants having porous materials with tiny microscopic pores and large surface area are called solid desiccants. The porous characteristic of desiccants is the indication that desiccants have a remarkable ability to hold water vapour. Desiccants undergo large numbers of adsorption/desorption cycles over their lifetime. Adsorption and desorption are heat and mass transfer processes between humid air and desiccant [1].

At present, much experimental and simulation research has been done in the desiccant constructed dehumidification industry. Chant (1991) [2] studied solid side resistance differently. They solved the diffusion coefficient equation for moisture passage in the solid side, assuming a parabolic concentration profile (PCP) inside the constituent part. Based on the PCP model, developed a heat and mass transfer model used for desiccant wheel with moist laminar airflow. The experimental data matched with simulation results. Dai et al. (2001) [3] obtained wave analysis to evaluate rotary

desiccants' performance using a psychometric chart. According to wave, the shape was proposed to improve dehumidification performance and specific essential parameters, i.e., heat capacity, rotation speed, adsorption heat, desiccant matrix thickness, regeneration temperature and desiccant isothermal shape were deliberated in information using a Psychrometric chart. Niu and Zhang (2002) [4] established a two dimensional transient heat and mass transfer model designed for the desiccant wheel, allowing for heat conduction and diffusion in axial and thickness side. They analyzed the optimum rotary speed of the desiccant wheel affected by the passage wall thickness for two main utilizations, i.e. dehumidification of air and recovery enthalpy. Zhang et al. (2003) [5] established a one dimensional attached heat and mass transfer model that is predictable for usage in designing and manufacturing a honeycombed desiccant wheel rotary type. They found a humped curve about air humidity ratio alongside air channel in the process of regeneration side. The bulge of air specific humidity moves from duct entry to duct leaving and gradually increases till bulge reaches duct leaving the side in the regeneration process. Ge et al. (2008) [6] studied that investigators have made a mathematically model to the attached heat and mass transfer process inside desiccant wheel. They described in their research paper the fundamental principle of heat and mass transfer mechanism and technique of model formation of desiccant wheels.
Here, we have studied three diffusion coefficients, i.e. Ordinary, Knudsen and Surface diffusion coefficient for Rotary Solid Desiccant Wheel System having three altered solid desiccant materials, such as Silica Gel, Molecular Sieve and Activated Alumina. Thus, we studied effect of pore size on moisture transport and examined the impact of the Tortuosity factor on the moisture transport phenomenon. The proper selection of the best desiccant material as per the suitable application in the desiccant dehumidification industry can be attained through this analysis.

2. Experimental Arrangement

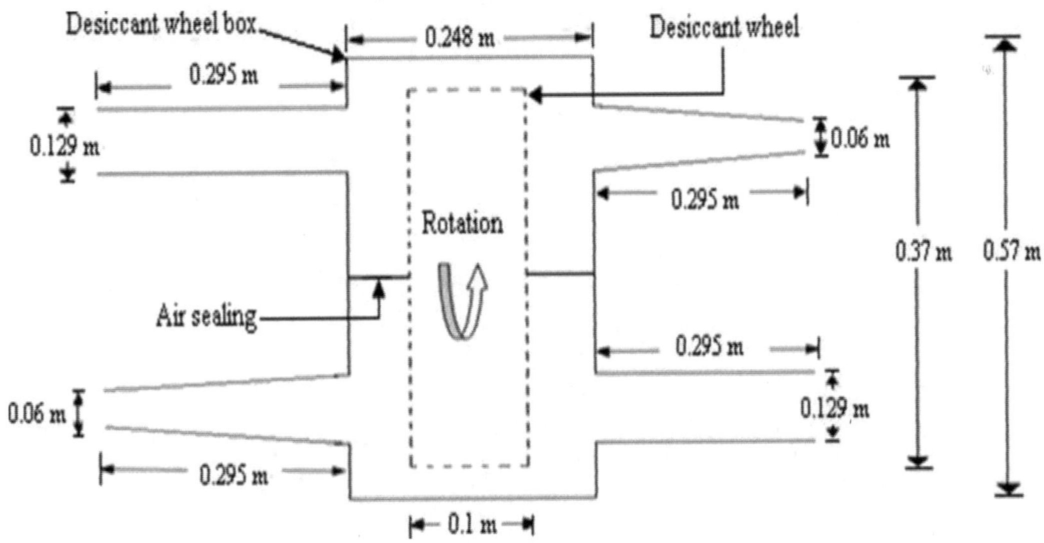

Figure 1. Plan diagram of a rotary desiccant wheel

Experiments have been performed on the rotary desiccant wheel by measuring dry bulb temperature and relative humidity with the help of a digital Thermo hygrometer. A detailed plan diagram of rotary desiccant wheel is presented in figure 1. Line diagram of experimental arrangement is displayed in figure 2.

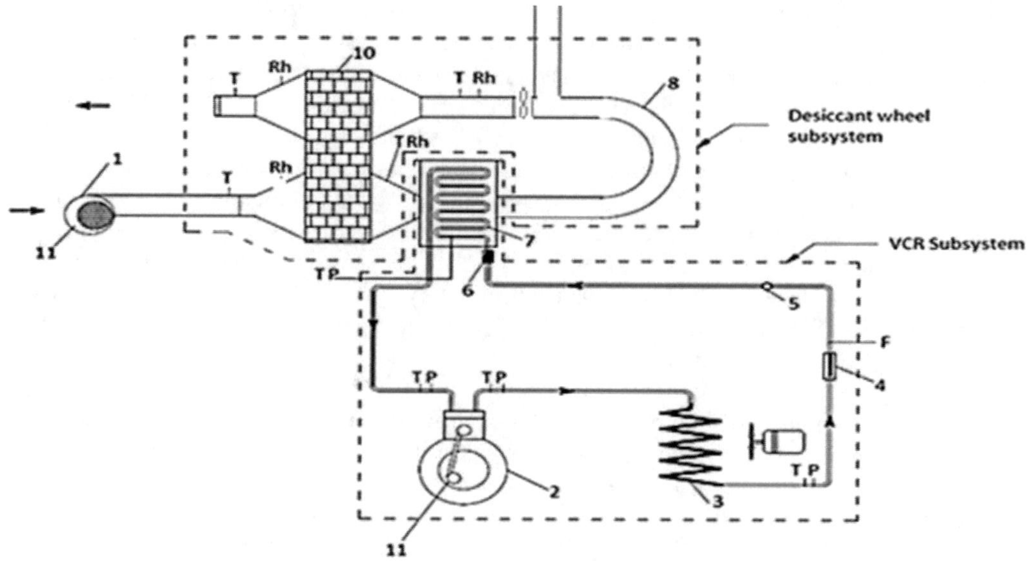

Figure 2. Line diagram of the actual setup

T- Temperature	1) Blower	5) Sight Glass	9) Anemometer
P- Pressure	2) Compressor	6) Expansion Valve	10) Desiccant wheel
Rh- Relative humidity	3) Condenser	7) Cooling Coil	11) Variable speed derive
F- Flow Rate	4)Refrigerant Flow Meter	8) Duct Connection	

2.1. Assessing device and Instruments

Here Relative humidity and dry bulb temperature (DBT) of air are measured with help of a Thermo hygrometer. An anemometer measures air velocity. Table 1 shows the assessing device and instruments used.

Table 1. Assessing device

Equipment	Range	Accuracy
Thermo Hygrometer	-10 to +60 °C 0 to 100 %RH	±0.5°C ±2.5%

3. Methodology

Mass transfer around material particles and moist air contains external convection and internal diffusion. Mass transfer is complicated inside the solid particle due to the desiccants of porous nature. In porous solids, three diffusion mechanisms can occur, i.e. i) Ordinary diffusion ii) Knudsen diffusion iii) surface diffusion [7]. The relations for diffusions coefficients are shown in equations (1-3).

3.1 Ordinary diffusion coefficient

The ordinary diffusion coefficient can be deliberated by Niu et al. [4] :

$$D_0 = 1.758 \times 10^{-4} \frac{T_a^{1.685}}{P_{atm}} \quad (1)$$

3.2 Knudsen Diffusion coefficient

In desiccant pores Knudsen diffusion occurs. It depends on a parameter that cannot be neglected, i.e., pore diameter. Knudsen diffusion can be deliberated by Niu et al. [4] :

$$D_K = 97r(\frac{T_a}{M})^{0.5} \quad (2)$$

3.3 *Surface Diffusion coefficient*

The surface diffusivity can be deliberated by following equation given by Niu et al. [4] :

$$D_s = \frac{1}{\zeta} \times 1.6 \times 10^{-6} e^{(-0.974 \times 10^{-3} \frac{q_{st}}{Ta})} \quad (3)$$

Where ζ is tortuosity factor that accounts for increase in diffusion length due to tortuous path of real pores.

4. Results and Discussion

4.1. Diffusion Coefficients Analysis for Silica Gel as a Desiccant Material

Table 2 shows the observation table for mass transfer analysis at P_{atm}=1 bar

Table 2 Observation table for mass transfer analysis

T_1 (°C)	RH_1 (%)	T_2 (°C)	RH_2 (%)	T_3 (°C)	RH_3 (%)	T_4 (°C)	RH_4 (%)
27.1	74	30	45	26.3	63	44.5	12
28.1	67	30.4	45	26.2	62	45.1	12
30.5	57	32.5	45	26.3	63	46.2	11
32.2	50	34.6	45	26.6	63	47.5	11.4
33.2	49	35.5	45	26.7	61	48.2	11.2
33.7	44	35.8	40	26.9	61	49.4	10.5
34.1	45	34.7	41	27.1	59	50.5	9.8
34.8	46	35.1	42	27.1	58	51.8	9.6

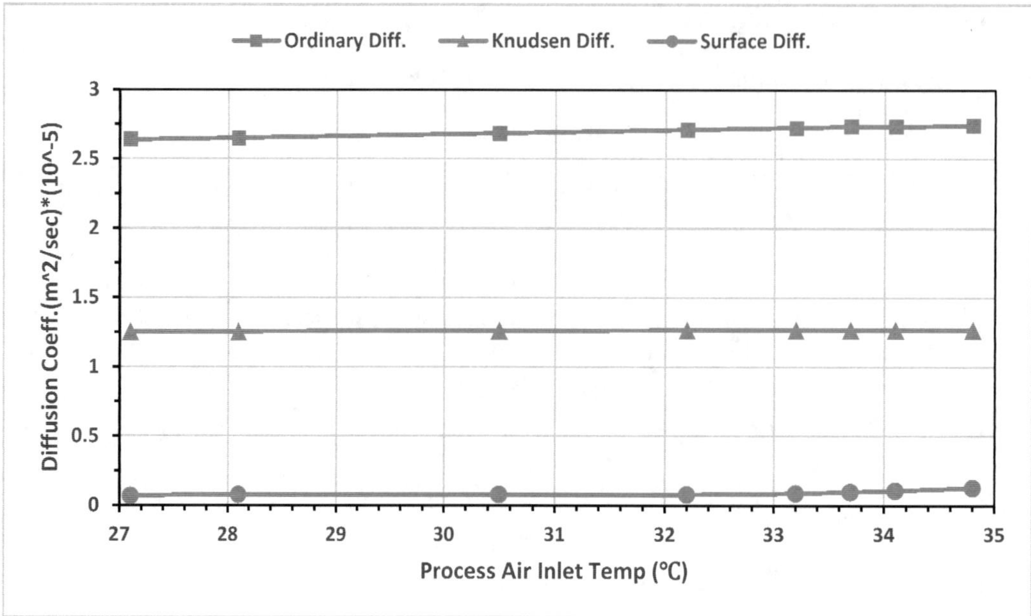

Figure 3. Variation of Diffusion coefficients with process air inlet temperatures

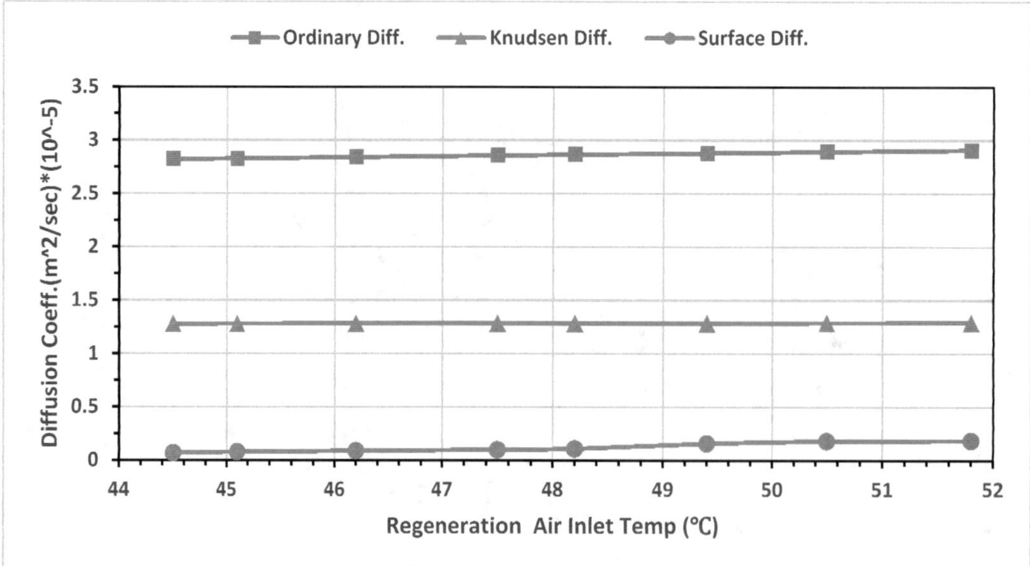

Figure 4. Variation of Diffusion coefficients with regeneration air inlet temperatures

4.2. Knudsen diffusion coefficient Analysis for Silica gel, Molecular sieve and Activated alumina

Desiccants	Avg. pore diameter (nm)
Silica gel	2-5
Molecular sieve	0.4
Activated Alumina	4.5

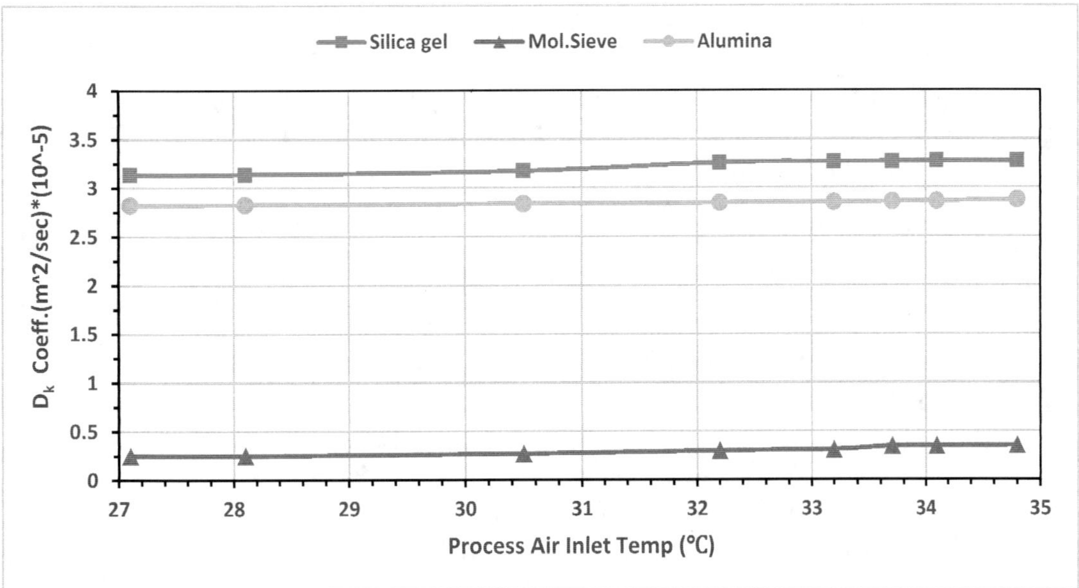

Figure 5. Variation of Knudsen diffusion coefficient for Silica gel, Molecular sieve and Activated alumina with process air inlet temperatures

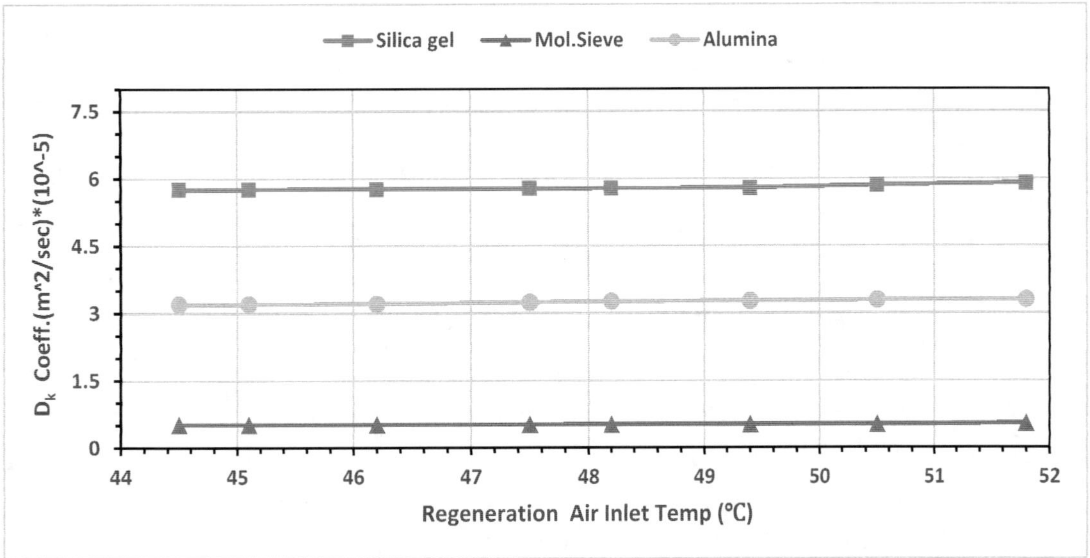

Figure 6. Variation of Knudsen diffusion coefficient for Silica gel, Molecular sieve and Activated alumina with regeneration air inlet temperatures

4.3. Surface diffusion coefficient Analysis for Silica gel, Molecular sieve and Activated alumina

Desiccants	Tortuosity factor
Silica gel	2.5
Molecular sieve	6
Activated Alumina	0.7

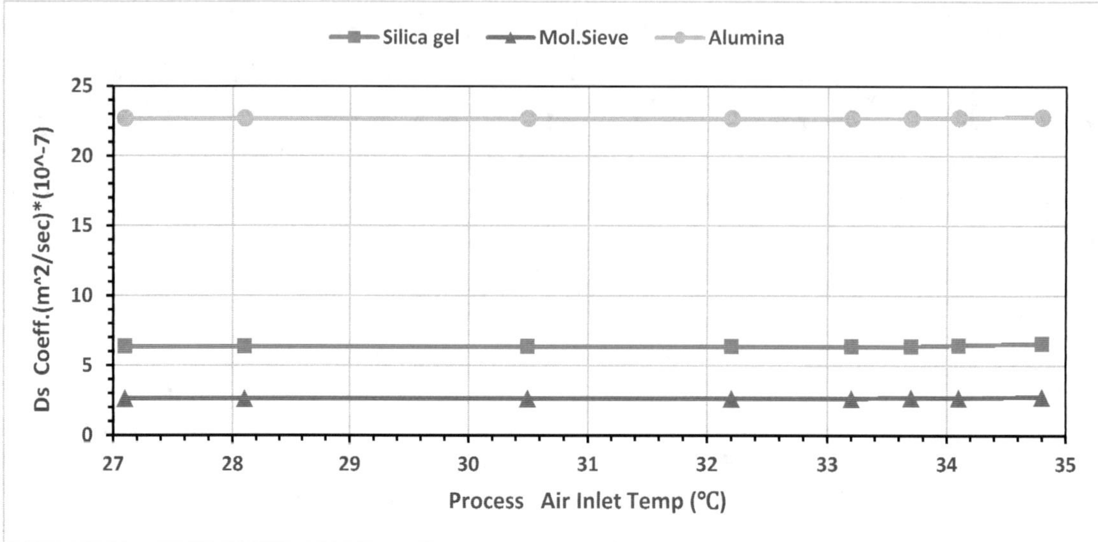

Figure 7. Variation of Surface diffusion coefficient for Silica gel, Molecular sieve and Activated alumina with process air inlet temperatures

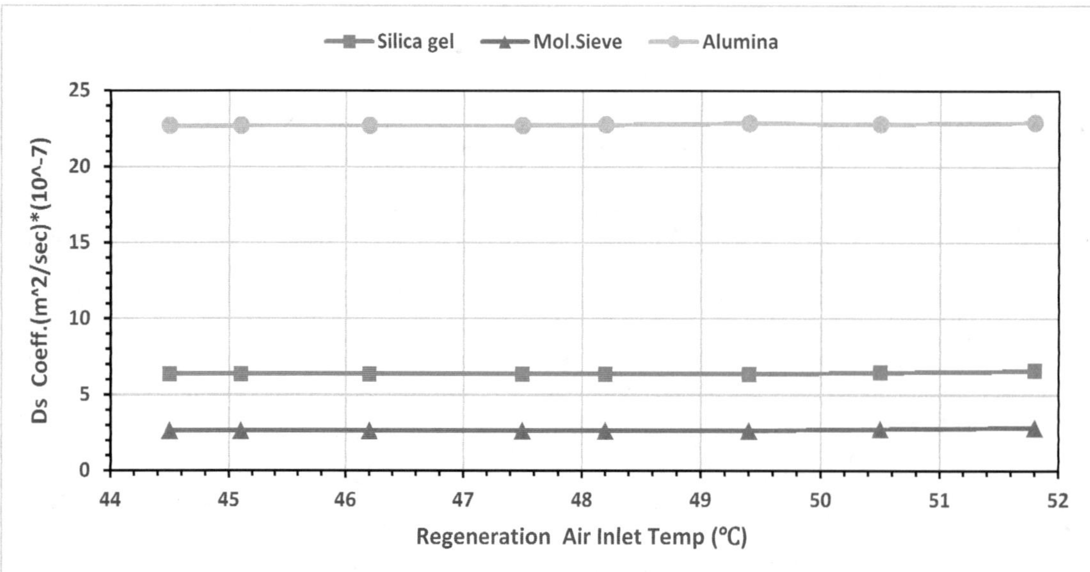

Figure 8. Variation of Surface diffusion coefficient for Silica gel, Molecular sieve and Activated alumina with regeneration air inlet temperatures

Figures 3 and 4 shows that all three diffusion coefficients, i.e. Ordinary diffusion, Knudsen diffusion & Surface diffusion offer a steep increment with temperature. However, they are increasing with air inlet temperatures at the process and regeneration sectors. Also, figures 3 and 4 show that for the exact temperature of the air at process and regeneration sector $D_{Surface}<D_{Knudson}<D_{Ordinary}$. Thus, the mass transfer analysis of silica gel shows that the surface diffusion coefficient is the most dominant moisture transport mechanism for the rotary desiccant wheel system. The smaller the diffusion coefficient, greater the effectiveness of rotary desiccant wheel system [8]. Figures 5 and 6 shows the Knudsen diffusion coefficient variation for Silica gel, Molecular sieve and Activated alumina with air inlet temperatures at process and regeneration sectors. It shows that as the air temperature rises at the process and regeneration sectors, the Knudsen diffusion coefficient value increases for all three desiccant materials, confirming the increase in moist air's diffusivity with desiccant. As pore radius increases, the Knudsen diffusion coefficient increases. Since the pore radius of Silica gel is the largest among the three desiccant materials, it has the highest value of the Knudsen diffusion coefficient. Figures 7 and 8 shows variation of Surface diffusion coefficient for Silica gel, Molecular sieves and Activated alumina with air inlet temperatures at process and regeneration sectors. It shows that as the air temperature rises at process and regeneration sectors, the Surface diffusion coefficient value increases, which confirm the increase in diffusivity of moist air with desiccant (but very steep variation). As the Tortuosity factor increases, the Surface diffusion coefficient decreases. Since the Tortuosity factor of Activated alumina has the most negligible value of 0.7, it has the highest value of Surface diffusion coefficient [9].

5. Conclusions

The main aim of this paper was to study three diffusion coefficients, i.e. Ordinary, Knudsen and Surface diffusion coefficient for Rotary Solid Desiccant Wheel System having three different solid desiccant materials, i.e. Silica Gel, Molecular Sieve and Activated Alumina. The effect of pore size on moisture transport and the impact of the Tortuosity factor on the moisture transport phenomenon were examined. Thus it was concluded:

- In a nutshell, this complete moisture transfer process between desiccant material and moist air can be understood as smaller the diffusion coefficient more will be effectiveness of rotary desiccant wheel system.
- Surface Diffusion is dominant mechanism of moisture transport for rotary desiccant wheel system.
- Considering the effect of pore size, Silica gel is the best desiccant material amongst all three materials.
- Considering the effect of the Tortuosity factor, Activated Alumina is the best choice for desiccant dehumidification compared to Silica gel and molecular sieve.

Nomenclature

D_0	Ordinary diffusion coefficient (m^2/s)
D_K	Knudsen diffusion coefficient (m^2/s)
D_s	Surface Diffusion coefficient ((m^2/s)
T_a	Air Temperature (°C)
T_1	process air inlet temperature
T_2	process air outlet temperature
T_4	regeneration air inlet temperature
T_3	regeneration air outlet temperature
RH_1	process air inlet relative humidity
RH_2	process air outlet relative humidity
RH_4	regeneration air inlet relative humidity
RH_3	regeneration air outlet relative humidity
q_{st}	Adsorption heat of regular density of the material

Greek symbols:

ζ	tortuosity factor

References

[1] Xing, H. Y. H. (2000). Desiccant dehumidification analysis (Doctoral dissertation, Massachusetts Institute of Technology).
[2] Chant, E. E. (1991). Transient and steady state simulations of an advanced desiccant enhanced cooling cycle (Doctoral dissertation, Georgia Institute of Technology).
[3] Dai, Y. J., Wang, R. Z., & Zhang, H. F. (2001). Parameter analysis to improve rotary desiccant dehumidification using a mathematical model. International journal of thermal sciences, 40(4), 400-408.
[4] Niu, J. L., & Zhang, L. Z. (2002). Effects of wall thickness on the heat and moisture transfers in desiccant wheels for air dehumidification and enthalpy recovery. International Communications in Heat and Mass Transfer, 29(2), 255-268.
[5] Zhang, X. J., Dai, Y. J., & Wang, R. Z. (2003). A simulation study of heat and mass transfer in a honeycombed rotary desiccant dehumidifier. Applied Thermal Engineering, 23(8), 989-1003.
[6] Ge, T. S., Li, Y., Wang, R. Z., & Dai, Y. J. (2008). A review of the mathematical models for predicting rotary desiccant wheel. Renewable and Sustainable Energy Reviews, 12(6), 1485-1528.
[7] Mills, A. F. (1995). Heat and mass transfer, Richard D. Irwin, Boston.
[8] Ahlberg, J. E. (1939). Rates of water vapor adsorption from air by silica gel. Industrial & Engineering Chemistry, 31(8), 988-992.
[9] Sladek, K. J., Gilliland, E. R., & Baddour, R. F. (1974). Diffusion on surfaces. II. Correlation of diffusivities of physically and chemically adsorbed species. Industrial & Engineering Chemistry Fundamentals, 13(2), 100-105.

Aqueous Two-Phase Separation (ATPS) Methods for Oleic acid extraction from Neem leaves

Shubham Gawade, Sandeep P. Shewale and Amravati Gode

School of Chemical Engineering, MIT Academy of Engineering, Alandi (D), Pune, India-412105

Corresponding author's e-mail address: agode@mitaoe.ac.in

Abstract. The aqueous two-phase separation system (ATPS) signifies an environmentally responsible approach for the extraction of bioactive compounds from a plants basis, as it is a liquid-liquid fractionation technique centred on the inconsistency of two aqueous solutions. In this investigation, various experimental parameters are optimized as the speed of agitation (200, 300 400 and 500 rpm) and solvent ratio (1:1, 2:3 and 3:2) with 20 % (w/w) of Ammonium Sulphate (AMS) salt composition and 30 % (w/w) of Polyethelyene Glycol (PEG). The obtained extract contains alkaloids, flavonoids, tannins, glycosides, acids and total phenolic compounds (TPC). The extracted Oleic acid by the ATPS method was measured with gallic acid equivalent (GAE) of TPC extracted from neem leaves powder. The determined concentration of oleic acid in the practice of TPC is 8.033 mg of GAE/g from the optimized experimental parameter. The optimized results can be cast off for a commercial process on an industrialized scale. Also, the mathematical modelling investigation was done to intent the critical impeller speed (Njs) with the Zwittering model. The identified model calculates the essential speed of agitation (rpm) for maximum extraction yield.

Keywords: Oleic acid, Total phenolic compounds (TPC), Aqueous Two-Phase Separation System (ATPS), Gallic Acid Equivalent (GAE), Critical Impeller Speed (Njs).

1. Introduction

Herbs are used for flavouring, food, medicine, or perfume from ancient times. Culinary use naturally differentiates herbs are implying the leafy green portions of spice also a product from a different part of the plant containing seeds, roots, bark, and fruits. Furthermore, medicinal contents available in the plant's parts are used for the production of some pharmaceutical products such as aspirin, colchicine, ephedrine, morphine, physostigmine, pilocarpine, quinidine, reserpine and vincristine, etc [1-2]. The approach of isolating the bioactive components from the medicinal plants and used for the manufacturing of some pharmaceutical products are becoming prominent. Generally, organic solvents such as methanol, ethanol and diethyl ether are usually helped for extracting the bioactive compounds as of plant basis by the outmoded extraction arrangements. These solvents are relatively expensive, needs distinctive processing conditions and most importantly disposal of the solvents is a major concern as they are not environmental responsive [3].

Conventional extraction processes are time-consuming and need more solvent for carrying out an operation, also after extraction, the added cost of purification and solvents recovery makes the process uneconomical.

Whereas, ATPS two-stage extraction is developing as a successful and flexible green system for the downstream handling of biomolecules. Fluid two-stage frameworks are low unpredictability frameworks with high adaptability [4]. That is, an expansive assortment might be acquired utilizing substances that pursue the Green Chemistry guideline on ecotoxicity, biodegradability, bioaccumulation and constancy, limiting waste and amplifying yields. Furthermore, they conform to the guideline of changeover of naturally safe structures to permit work under air weight. Since the 1970s numerous classified and out examinations have announced the filtration of proteins and other biologic materials utilizing ATPS, and numerous specialists have considered different operations of ATPS for the extraction and cleaning of organic products [5]. However, the utilization of such frameworks for the recuperation of phenolic mixes from plant materials is extremely constrained. In addition, there is broad writing about the thermodynamic properties of ATPS be that as it may, to the best of our insight, their application to crude unpurified examples has been very constrained [6].

Subsequently, the late nineteenth-century fluid two-phase extraction has been identified to the entire world. Aqueous two-phase can be framed through an extensive diversity of characteristics or else engineered water-solvent polymers& salt blends [7]. Watery two-phase extraction is developed for protected, sparing partition and cleaning of biomolecules, for example, proteins and catalyst extraction. Fluid extraction has numerous favourable circumstances; it is biocompatible, has low interfacial surface pressure among stages and it has high water content, the procedure can incorporate and the ability for strengthening [8-9].

Likewise, the level of corruption for biomolecules is low. In any case, two-polymer and polymer-based salt frameworks have developed quickly and a considerable measure of effort has been placed keen on concentrate this strategy utilizing these sorts of aqueous two-phase separation systems (ATPS). Aqueous two-phase extraction is known as an operative, adaptable and significant developing green method for the subsequential treating of biomolecules. This strategy has points of interest completed traditional extraction systems similar to, simplicity of scaling-up, condition benevolent, minimal effort, fit for nonstop activity and is effective for some sorts of trials exceptionally for the fixation and refinement of biomolecules. The utilization of partiality in ATPS can affect the developed recuperation earnings and developed refinement bends of bio consistent items such as it is an essential phase recuperation strategy [10]. Water as the foremost constituent of together stages in ATPS practices a moderate setting for bioactive molecules to distinct and polymers steady to the assembly and biotic doings through further liquid-liquid extraction approaches could impairment natural goods since of the development circumstances and biological solvents such method reduces the purity of active ingredient present in the extract.

There are two fundamental sorts of ATPS: polymer-polymer and polymer–salt frameworks. The mind-boggling expense of some shaping stage polymers (e.g dextran) limits the use of these frameworks, just legitimized when the expense of the result of intrigue is extensive. Consequently, the choice of the more temperate polymer–salt frameworks is profoundly suggested [10-11].

The novelty of the proposed work is that during the extraction itself two different layers of aqueous solution and salt is obtained, which can help further to reduce the cost of separating components. Also, the systems can be designed by partying a diversity of components in water and two-polymer and polymer-salt systems have developed quickly. The said work majorly focuses on the extraction of oleic acid from neem leaves to powder using ATPS (water + polymer + salt) based on PEG and ammonium sulphate. The aim is to optimize various experimental parameters (time, ATPS composition, particle size) for the removal of Oleic acid from neem leaves powder and its additional practices as natural antioxidants.

IOP Conf. Series: Materials Science and Engineering 1224 (2022) 012016 doi:10.1088/1757-899X/1224/1/012016

2. Material and Methods

The Neem powder was obtained from Hari Parshuram Aushdhalya, Pune., Polyethelyene Glycol (PEG) was procured from SRL Chemicals Pvt Ltd, Mumbai. Folin Ciocalteu's reagent was procured from Qualigens Fine Chemicals, Mumbai. Ammonium Sulphate (AMS) was procured from S.D. Fine Chemicals Pvt Ltd, Mumbai.

2.1. Batch extraction

Batch extractions are a modest method for the extraction of bioactive compounds. The stages in this method are equipped with a 50 ml glass reactor with a four-bladed glass turbine impeller and the combination to be divided is supplementary. Subsequently collaborating, phase parting is proficient each by resolving below gravity. The stages are disconnected and investigated to improve the alienated constituents of the preliminary mixture. The object product would be focused at any of the stages and the pollutants in the additional form. In various cases, reclamation and attentiveness of the product that produces beyond 90% can be attained with a particular extraction stage.

One particular phase removal does not give adequate retrieval, recurrent extractions can be supported obtainable in a sequence of communicating and parting components [12]. The fluid dividers into two stages, each covering added 80% liquid. When basic biomolecules are supplementary to these combinations, biomolecules and cell wreckages are dividers among the stages; by choosing suitable circumstances, cell remains can be limited to one stage as the object bioactive molecule barriers. The segregating of biomolecules among segments mostly be contingent on the equilibrium connection of the arrangement. The partition coefficient is demarcated as [12-13].

$$K = \frac{C_{AT}}{C_{AB}}$$

Where C_{AT} is the equilibrium attentiveness of constituent A in the upper phase and C_{AB} is the equilibrium adsorption of A in the lesser phase. If constituent A helps the greater stage the worth of K will be better. In numerous aqueous arrangements, K is continual finished with an extensive collection of deliberations, as long as the molecular possessions of the stages are not transformed. The theoretic yield in the topmost stage, Y_T, can be premeditated relative to the capacity ratio of the stages, R (up to volume / below volume), and the partition coefficient K of the object molecule as follows [12-13]:

$$Y_T = \frac{V_T C_{AT}}{V_o C_o} = \frac{V_t C_{AT}}{V_t C_{AT} + V_B C_{AB}} = \frac{1}{1 + [\frac{1}{KR}]}$$

Similarly, the theoretic yield in the bottommost stage, Y_B is known by,

$$Y_B = \frac{1}{1 + [\frac{1}{KR}]}$$

Consequently, by changing anyone like K or R we can effortlessly upsurge or reduction the profit of the object particle [14]. Additional constraint recycled to describe two-phase partitioning is the concentration factor, δc, distinct as the ratio of produce attentiveness in the favoured stage to the original product attentiveness.

$$\delta_{C,T} = \frac{C_{AT}}{C_{Ao}} \quad \text{(Product partitions to the higher phase)}$$

$$\delta_{C,B} = \frac{C_{AB}}{C_{Ao}} \quad \text{(Product partitions to the lower phase)}$$

2.2. Determination of the binodal

By settlement, the constituent mainly in the lower layer is represented as abscissa and the predominant element in the upper stage is represented as ordered. The three systems are explained realistically.

2.2.1 Turbidometric titration

In the tubing, with suitable backup solutions, formulate systems through different configurations of recognized weight. Note the added size due to titration, for example, if 5 g methods are organized, use 10 ml tubes. As an example, shows the systems for different systems that can use PEG-phosphate and PEG-dextran, and the essential designs. This can be replicated in a worksheet to permit easy intention. Note down the mass of the tube and titration drop by drop, with suitable dilution till the scheme is zeroed, i.e, a stage is formed.

This can be done through the scheme is continuously mixing or accumulation a droplet, collaborating, adding a second drop and continues the same process. To confirm that it has formed a single-stage system, the schemes must be centrifuged (for example 1000-2000 g, 5 min). Record the concluding mass of the tube and estimate the mass of the additional dilution just before the formation of a phase. Since the number of graduate systems is relative to the total of points in the binodal, superior precision is obtained with a superior number of schemes [15].

2.2.2 Cloud point method

Balance 5 g of a standard solution of constituent X into a 25 ml narrowed flask. Then add drop by drop, a reserve solution of the Y component up to the principal indication of turbidity, which is the cloud point. Note the weight of the Y component necessary for the mixture to become cloudy. This provides the first point in the binodal. Also, add a known weight of diluent lower the cloud point and duplication as indicated [15-16].

2.2.3 Determination of the Tie line

Measuring the connection line for polymeric schemes comprising an optically active composite, for example, PEG-dextran, PEG-Ficoll and ethylene oxide-propylene as well as oxide-Reppal PES 100.

2.2.3.1 Polymeric Methods Comprehending One Optically Active Polymer

Formulate a standard curve for the lively constituent, in the variety of 0 to 10% (p / v), i.e. inside the linear series, through the identical sections arrange a second standard curve for the refractive measure of the index. If the scheme is arranged in a shield, the average curves for the clean constituents must be completed with a similar safeguard, since the salts similarly subsidize the refractive index. Get ready the phased scheme for investigation, assembly sure that the phase components mix well; let the phases separate.

To ensure complete separation, centrifuge at low speed (for example 1000-2000 g, 5 min). The system proportions should be appropriate to permit deduction of at least 5 g upper and lower stage for phase concentration investigation and an additional quantity for density extents. Distinct the higher and subordinate stages building certain not to origin stage relations. Make the suitable dilutions, for example, watery 5 g of phase with the suitable solvent to 25 ml in a volumetric container. Extent the visual revolution for every stage and estimate the individual concentrations. The concentration of another constituent is resolute by determining the refractive index of the developed and subordinate stage and deducting the influence of the refractive index acquired from the optically lively constituent [17].

2.2.3.2 Polymer-Salt Systems

Formulate a standard curve aimed at salt conductivity within the linear series (in% w / v). Arrange the stage schemes as indicated above and eliminate 5 g of samples of the higher too subordinate phases and diluted with liquid and lyophilized, then note the dehydrated weight. Eliminate an additional section from the higher and subordinate stage, diluted with water and extent the conductivity of every stage. Estimate the salt absorption and deduct the mass involvement of the dehydrated mass of the section [16-18].

3. Results and Discussion

In this investigation, various experimental parameters were optimized such as speed of agitation (200 rpm, 300 rpm, 400 rpm & 500 rpm) and solvent ratio (1:1, 2:3 and 3:2). Extract samples were pipette out at specific time intervals like every 15 min and further standard Folin–Ciocalteu's method was used for analysis purposes. The obtained outcomes indicated that the TPC concentration of TPC in the extraction stage at a specific time. By changing the parameters speed of agitation and solvent ratio in the batch reactor at different times the results were optimized and used for further study.

3.1 Speed of agitation

About 5 g of powder of neem leaves was weighted and fed to 50 ml of batch reactor along with AMS salt (20 %w/w) and 50 ml of PEG (30 %w/w) at a temperature of 30°C till the extraction rate was a steady-state. The maximum speed of agitation produces high turbulence in the batch reactor and increases the rate of mass transfer [11]. The results of various experiments were performed for multiple agitation speeds is shown in figure 3.1 The experimental results shows that the concentration of TPC at 200 rpm significantly low as compared to 500 rpm, but there could be a marginal difference of TPC concentration of 400 and 500 rpm, speed of agitation., therefore for the further study, speed of agitation (400 rpm) was used. The circulation of TPC compounds from the neem leaves powder in the solvents could expand with the accumulative agitation speed. An added rise in the agitation speed has no substantial effect on final extraction yield; it clues that external mass transfer fighting is inconsequential at 400 rpm

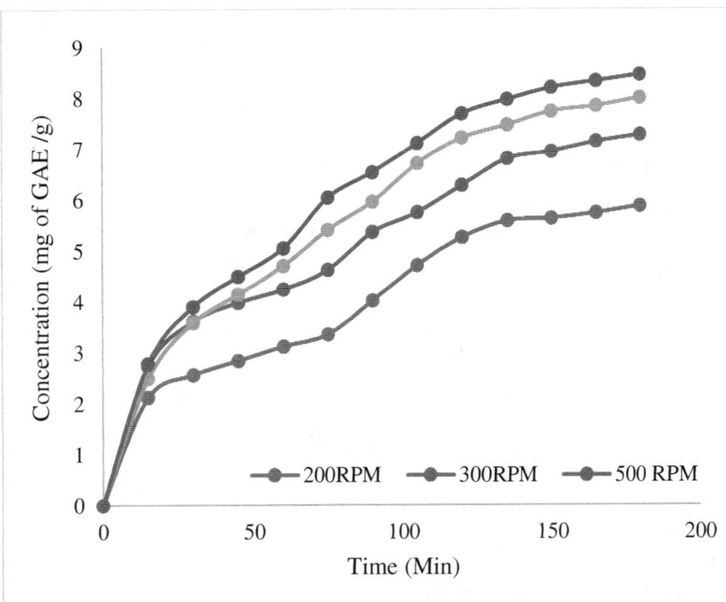

Fig. 3.1 Concentration of TPC obtained from Batch at temp 30^{0}C speed of agitation 200, 300, 400 & 500 rpm)

3.2 Effect of Solvent Ratio (AMS: PEG)

The TPC concentration values in the extract were considered for changed extraction times at different solvent ratios (1:1, 2:3 & 3:2) and the same is shown in Figure 3.2. About 5 g of powder of neem leaves was weighted and fed to 50 ml of batch reactor along with AMS salt (20 %w/w) and 50 ml of PEG (30 %w/w) at a temperature of 30°C till the extraction rate was a steady-state. There was an increase in TPC concentration and experiential for solvent ratio 01:01. A substantial quantity of solvent favours an additional concentration gradient and cuts diffusional resistance that rises the rate of extraction rate. There was a rise in TPC concentration, for the solvent ratio of 1:1 to 3:2. A substantial quantity of solvent tends to the added concentration gradient and drops diffusional resistance that rises the rate of extraction [11].

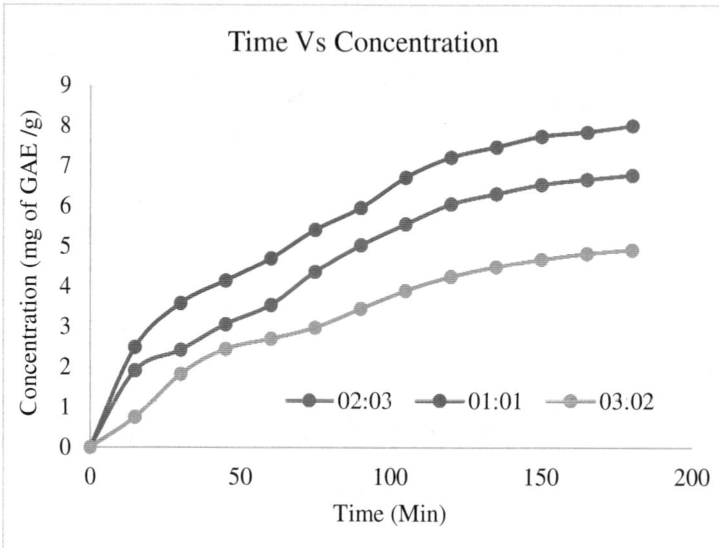

Fig.3.2 Effect of Solvent Ratio (AMS: PEG)

4. Conclusion

The investigation of the ATPS method was beneficial for the sub sequential treating of biomolecules. Also, the experimental parameters were optimized such as speed of agitation and solvent ration and the optimized parameters as 400 rpm and 01:02 solvent ration respectively with AMS salt (20 % w/w) & PEG (30 % w/w). The obtained extract also contains alkaloids, flavonoids, tannins, glycosides, acids and total phenolic compounds (TPC). The extracted Oleic acid by ATPS technique was measured with gallic acid equivalent (GAE) of TPC extracted from neem leaves powder. The determined concentration of oleic acid in the practice of TPC is 8.033 mg of GAE/g from the optimized experimental parameter. The optimized results can be used for a commercial process on an industrial scale.

References

[1] Alasalvar C, Karamac M, Amarowicz R, Shahidi F, 2006. Antioxidant and antiradical activities in extracts of hazelnut kernel (*Corylus avellana* L.) and hazelnut green leafy cover. *JAgric Food Chem* 54: 4826-4832.

[2] Babitskaia V, Shcherba V, Lkonnikova N, 2000. Melanin complex of the fungus *Inonotus obliquus*. *Prikl Biokhim Mikrobiol* 36(4): 439-444.

[3] Dallora N, Klemz J, Filho P, 2007. Partitioning of model proteins in aqueous two-phase systems containing polyethylene glycol and ammonium carbamate. *Biochem Eng J* 34(1): 92-97.

[4] Karakatsanis A, Liakopoulou M, 2007. Comparison of PEG/ fractionated dextran and PEG/industrial grade dextran aqueous two-phase systems for the enzymic hydrolysis of starch. *Eng* 80(4): 1213-1217.

[5] Alencar L.V.T.D., Passos L.M.S., Martins M.A.R., Barreto I.M.A., Soares C.M.F., Lima A.S., Souza R.L., 2020.The complete process for the selective recovery of textile dyes using an

aqueous two-phase system.*Sep. Purif. Technol.* 253: https://doi.org/10.1016/j.seppur.2020.117502.

[6] Ying H, Diyun C, Shuqi C, Minhua S, Yongheng C, Yixiong P, Gutha Y, 2021. A green method for recovery of thallium and uranium from wastewater using polyethylene glycol and ammonium sulfate based on the aqueous two-phase system. *J. Clean. Prod.* 297: https://doi.org/10.1016/j.jclepro.2021.126452.

[7] Leitner M, Vandelle E, Gaupels F, Bellin D, Delledonne M, 2009.NO signals in the haze: Nitric oxide signalling in plant defence.*Curr Opin Plant Biol* 12(4): 451-458.

[8] Liu SP, Zong ZM, Wei Q, Wei XY, 2010. Study on organic compounds in aqueous two phase system phase forming and distribution. *Chem Ind Times* 24(12): 21-24.

[9] Ahareh AS, Gholamreza P, Javad RS, Shahla S, Naghmeh H, 2020.Separation of erythromycin using aqueous two-phase system based on acetonitrile and carbohydrates. *Fluid Phase Equilib.*505: https://doi.org/10.1016/j.fluid.2019.112360

[10] Perla JV, Salvador VG, Iran AT, Yolanda SM, Leticia GC, Artemio PL, Diana GR, 2020. Separation of bioactive compounds from epicarp of 'Hass' avocado fruit through aqueous two-phase systems. *Food Bioprod. Process.*123:238-250.

[11] Sandeep Shewale & Virendra K. Rathod (2018) Extraction of total phenolic content from Azadirachta indica or (neem) leaves: Kinetics study, Preparative Biochemistry & Biotechnology, 48:4, 312-320, DOI: 10.1080/10826068.2018.1431784

[12] Singh SB, Jayasuriya H, Dewey R, Polishook JD, Dombrowski AW, Zink DL, Guan Z, Collado J, Platas G, Pelaez F, Felock PJ, Hazuda DJ, 2003. Isolation, structure, and HIV-1 integrase inhibitory activity of structurally diverse fungal metabolites. *J IndMicrobiol Biotechnol* 30(12): 721-731.

[13] Srinivas ND, Barhate RS, Raghavarao KSMS, 2002. Aqueous two-phase extraction in combination with ultrafiltration for down-stream processing of Ipomoea peroxidase. *J Food Eng* 54(1): 1-6.

[14] Uma DB, Ho CW, Wan Aida WM, 2010. Optimization of extraction parameters of total phenolic compounds from Henna (*Lawsoniainermis*) leaves. *Sains Malaysiana* 39(1): 119-128.

[15] Wang SY, Wu JH, Cheng SS, Lo CP, Chang HN, Shyur LF, Chang ST, 2004. Antioxidant activity of extracts from *Calocedrusformosana* leaf, bark, and heartwood. *J Wood Sci* 50: 422-426.

[16] Zhao YX, Miao KJ, Zhang MM, Wei ZW, Zheng WF, 2009. Effects of nitric oxide on production of antioxidant phenolic compounds in *Phaeoporus obliquus*. *Mycosystema* 28(5): 750-754.

[17] Zheng WF, Zhao YX, Zhang MM, Wei ZW, Miao KJ, Sun WG, 2009. Oxidative stress response of *Inonotus obliquus* induced by hydrogen peroxide. *Med Mycol* 47: 814-823.

[18]Zheng WF, Miao KJ, Liu YB, Zhao YX, Zhang MM, Pan SY, Dai YC, 2010. Chemical diversity of biologically active metabolites in the sclerotia of *Inonotus obliquus* and submerged culture strategies for up-regulating their production. *Appl MicrobiolBiotechnol* 87(4): 1237-1254.

Study of Mechanical Properties of Concrete by Replacement of Natural Coarse Aggregate with Recycled Coarse Aggregate with Metakaolin

Om Prakash Singh, Mahendra Kr. Singar, Ravi Kant Pareek

Department of Civil Engineering, Vivekananda Global University, Jaipur, Rajasthan,

Corresponding author email id: om.prakash.singh@vgu.ac.in

Abstract. This research work investigates the compressive strength of concrete specimens cured at the age of 7 & 28 days, the split tensile strength, and the flexural strength of concrete specimens cured at 28 days. Natural coarse aggregate (NCA) has been replaced with 0 percent, 25 percent, 40 percent, 60 percent, and 100 percent by RCA of construction & destruction (C&D) waste of 30-year-old building and Ordinary Portland cement (OPC) has been replaced with 10 percent metakaolin (MK).To understand the fresh properties of concrete of slump test was performed and to find satisfactory results. To analyze the hardening properties of MK with recycled coarse aggregate (RCA) base mix concrete, compressive strength was measured at the age of 7 and 28 days, and split tensile strength, the flexural strength was measured at the age of 28 days. Based on the experiment's study, it can be concise that the 40 percent substitute of NCA with RCA carried good results with 10 percent MK. This research presents the effect of MK as a cement substitute on the properties of fresh and hardened with RCA.

Keywords: NCA, RCA, Metakaolin (MK), Construction & Demolition (C&D), Compressive Strength, Flexural Strength, Split Tensile Strength

1. Introduction

A large quantity of C&D waste has been produced due to the economic growing of infrasture. The general way to organize of such waste is by landfill. Improper handling of C&D waste as time passes, landfills will cause many environmental issues. The continuous utilization of NCA will reason reserve scarcity. The potential solution for these problems is the recycling of aggregates. Still, due to strength reduction associated with RCA, its usage is restricted for low-strength concrete. Another hand, the incorporation of MK as a cement replacement in RAC could pave a path towards a sustainable future. This research presents the effect of MK as a cement substitute on the property of fresh and hardened RCA as well as NCA.

The RCA is extracted from first-hand cast laboratory cubes with known average compressive strength. The matter of waste disposal has become harsh both in industrialized and developing countries, including India. This one is related to a huge increase in the number of waste products, its continuing preservation of dumping grounds, and increasing the cost of transport and removal. The broad extraction of natural aggregate and the huge volumes of C&D waste going to landfill sites are having an effect on the environment & creating serious challenges to the public and environmentalist ambitions of a recyclable future. As a result, this idea of recycling waste and reusing it in some manner has garnered attention. C&D waste is the aforesaid issue in many countries. As a result, recycling technology has made advances in recycling the demolished concrete structure.

The pattern in reduction of aggregate assets can be in India. The scenario in India isn't fearfully severe, there are a few parts of a country wherein stone dust aggregate are inaccessible within a few kilometers. The harshness of an issue in the future requires significant thinking mostly on part of the Indian community, especially as the volume of the concrete building is expected to striking in the coming decades. The use of cement in concrete, along with C&D waste, is indeed a major source of concern for environmentalists. By high carbon emissions the cement manufacturing part is one of the major contributors to climate change. Studies are being conducted to reduce the use of cement by replacing mineral admixtures counting such MK, as a portion of cement. With the huge growth in construction waste more in the previous decade, social and environmental concerns about recycling have grown. The recycling of waste concrete has much increased according to current technological advancements. The characteristics of natural and RA are examined and the behavior of RAC results of MK.

1.1 World Scenario

The use of C&D waste in buildings is not really a new concept. Recycled concrete waste was found to be among the best ways to reduce construction waste. The research of RAC dates all the way back more than three decades. In the developed world C&D waste disposal is a serious problem in so many nations. Due to the lack of disposal sites, a problem had developed. There are major trash dumps in the area, and the rates of waste creation are extremely high. The rate of generation in industrialized countries is quite high. Because traditional recycling methods are ineffective, Sub base filling, land reclamation, and other waste, etc. are insufficient to address the issue of disposal of garbage. According to the extensive review of the literature, one primary obstacle to using RA is the ever-increasing demand for aggregates as infrastructure construction increases. Reduced specific gravity, low packing density, and lower impact crushing, and abrasion resistance are issues with RA. It's been shown that typical structural concrete can be easily achieved by using RA.

1.2 Indian Scenario

In India, there is a severe shortage of infrastructural amenities such as houses, roads, hospitals, and enormous quantities of building materials are supposed to create every one of these facilities. In the 10th and 11th five-year plans, the Indian Planning Commission dedicated approximately half of the financial plan to infrastructure development. Increasing infrastructural development of motorways, airports, other infrastructure, and rising demand for housing, has resulted in a scarcity of building materials and an increase in their cost. Most of the waste produced by destroyed constructions is dumped as landfills. Land-based disposal is producing a scarcity of dumping sites. The Indian standard standards of concrete requirement create no necessities for utilization of RA in concrete. The lack of a codified rule does not involve a limit on the utilization of RA indirectly and directly. As a result, it is start recycling and reusing C&D waste in arrangements to conserve the environment, energy, and money.

2. Literature Review

The resources of natural aggregates of engineering quality can be used. Meta-kaolin has been safely and successfully used as a construction aggregate in applications that as cement concrete, road embankment, road and walkways. The consequence of properties of concrete both fresh and hardened with metakaolin that as partial replacement of the NCA by C&D waste that previously research. Metakaolin is a de-hydroxy method of the clay mineral kaolinite. Metakaolin is an appreciated admixture purpose of utilization of cement. Good stone rocks kaolinite are recognized for example china clay used in the product acknowledged as porcelain. The size of metakaolin particles is lesser than compared to cement. Metakaolin is a relatively material in the concrete industry that is growing strength decreasing sulphate attack and air-void connection. Pozzolanic reactions disparity the structure of concrete produce through powerful the existing calcium hydroxide of added calcium silicate hydrates subsequent in an enlarged strength and reduced porosity.

The examined the significance of including metakaolin in concrete and mortar was established. Concrete and mortar grieve destruction of workability outstanding to the mixture of metakaolin advanced the critical to improving a superplasticizer. The strengths of compressive and flexural were improved at the age of 7 days. The various parameters of durability were improved after the mixture of 10 to 15 percent metakaolin (1).

In this studies the mechanical properties of RAC with and without metakaolin. The different properties of concrete such comp. strength, splitting tensile and flexural strength is calculated. The variable of test is replacement relations for NA (0%, 20 %, 40 %, 60 %, 80 %, and 100 %) with RA and use of 20 percent metakaolin to replacement of cement. The result of combined RA and metakaolin on the concrete properties of is discovered in this study (2).

In this studied MK partial replacement of cement by 10 percent with RCA had 12 percent and 10 percent increase in mechanical properties for 25 and 30 grades. An increase of content cement improved the mechanical properties of RCA concrete. The totalling of metakaolin as an admixture in concrete completed from RCA reduced the workability of the concrete as the percentage totalling of MK increases. The concrete produced from RCA exceeded the conventional mix; it determined which the usage of MK as an admixture in concrete made from RCA aggregates better-quality both the strength tensile and compressive of concrete (3).

In this studied different result of the RCA ratio of water absorption, RAC mixtures display lesser value of slump of related to NAC mixes. The procedure MK at various substances decreases RAC mixes workability mainly due to the minor of particles that essential content of water. The RCA weaker, the compressive strength of the RAC mixes fallen upto 30 percent associated to the conventional mix with NCA. The processes MK can provision to progress the 28 day's comp. strength mix of the RAC. The enhancement of strength outstanding for addition of MK that depends on MK.Up to 20 percent strength improvement that attained related to NAC when 20 percent is used of MK. This is subsequently of the reaction of pozzolanic of MK. Addition metakaolin that affect in dropping the absorption of water RAC mixes (4).

The level of replacement of cement by metakaolin is founding 13 percent. The increase in strength due to replacement of cement by metakaolin is 41 percent at the age of 7 days. The workability of concrete is established to reduction as the level of additional of natural aggregate by RA is enlarged. The Compressive strength increases upto 50 percent replacement of natural aggregate by RA and past that it reductions. The results found from non-destructive test are lesser than the obtained, excluding at 100 percent replacement level (5).

The study examines the properties of sandcretes formed the adding of 10 to 20 percent w/c MK in the estimated of binder. The adding of MK has created an enhancement of the properties of the material. The specimens of meta-kaolin in the binder performance improved strength associated to the reference. The progression in strength was extra noticeable in examples with advanced concentration of binder and low ratio of W/B. The strength rise up was experimental in the case of sandcrete samples concluded little binder concentrations (6).

3. Research Significance

It was mentioned in the previous section that brief evaluation writing on the subject was available and that it was not in the proper sequence on the RCA with MK finished. Considering the use of RCA with MK, this research has been considered to evaluate the strength and durability, and mathematical prediction with MK. It is proposed the most suitable mix of components for enhancing concrete durability in terms of RCA and cement additions being suggested. MK has been used the partial substitute for OPC in various amounts. In addition, all RCA mixes were submitted to comp. strength, flexural strength, and split tensile strength tests.

3.1 Experimental Program

Materials

In the preparation of concrete mixes are divided into two groups. OPC of grade 43 as according to IS 269:2015, natural sand, NCA, and RSA with a maxi. size of 20 mm was utilized. The C&D waste was collected near Gandhinagar Railway Station, Jaipur and manually crushing in at the concrete laboratory of department of Vivekananda Global University, Jaipur. The aggregates were crushed manually and size reduced as per requirement, which was a maximum of 20 mm. As shown in table no.1 show the physical characteristics of the RCA used in this investigation. MK was used as a constituent in cement-based materials in various ratios.

1. Cement

The cement used is Ultratech OPC of 43 grades, which has a specific gravity of 3.15 as per IS: 269: 2015.

Table 1 Characteristics of OPC

Sr. No.	Properties	Units	Result Obtained	Permissible range specified (IS:269:2015)
1	Sp. Gravity	-	3.11	3.10–3.15
2	Fineness	cm^2 / gm	2385	2250 (mini.)
3	Soundness	mm	4	10 (max.)
4	Consistency	%	33.45	30–35
5	Setting Time of cement	Minutes	63 410	30 (min.) 600 (max.)
6	Comp. Strength	MPa	24 (3-Days) 36.5 (7-Days) 46.1 (28-Days)	23.00 (minimum) 33.00 (minimum) minimum)

2. Metakaolin (MK)

The development of building products offering scientific and ecological advantages is the primary dispute of the novel millennium single such material is MK that is categorized as current era of additional material of cementious. Complementary cementitious material is a fine material used to substitute parts of clinker in cement in mix of concrete. MK is distinctive that as is neither a by-product of industrial it is obtained from a naturally occurring mineral and is produced specifically for cement purposes. MK is generated by thermal treatment, i.e. by calculations of kaolin clays in certain limit of temperature. The heating method from the mineral kaolinite (Al2O3·2SiO2·2H2O), the primary component of kaolin clay, and dissolves the structure of material resulting an alumino-silicate (Al2O3·2SiO2). The equation for this di-hydroxylation process is shown below: Al2O3·2SiO2·2H2O → Al2O3·2SiO2 + 2H2O

Table 2 Chemical Properties of Metakaolin

Sr. No.	Name of Ingredient	Percent
1.	SiO_2	52–54%
2.	Al_2O_3	41–43%
3.	Fe_2O_3	<2.24%
4.	TiO_2	<2.89%
5.	SO_4	<0.45%

6.	P_2O_5	<0.2%
7.	CaO	<0.2%
8.	MgO	<0.1%
9.	Na_2O	<0.05%
10.	K_2O	<0.40%
11.	L.O.I.	<0.5%

3. NCA and RCA

The designing a concrete mix, different physical characteristics of NA and RA are determined. During this project, NCA with A size of 20 mm and the grading criteria as per IS: 383-2016 was purchased from locally available quarries. The demolished structure provides the RCA. The waste concrete from the demolished structure is carried to a nearby crusher, where it is converted into aggregates into 20 mm.

Table 3 Properties of NCA and RCA

Sr. No.	Description of test	Test Result of NCA	Test Result of RCA
1	Fineness Modulus	6.94	6.82
2	Specific Gravity	2.66	2.48
3	Aggregate Impact Value	15.41%	28.16%
4	Aggregate Crushing Value	15.67%	26.54%
5	Water Absorption	0.58%	5.42%

4. Natural Fine Aggregate (Sand)
   In this work sand was used FA (sand) confirms to Zone-II of IS 383:2016 purchased from locally available in Jaipur Banas River.

Table 4 Test Result of Fine Aggregate

Sr. No.	Description of test	Test Result
1	Silt Content	1.65%
2	Moisture Content	1.61%
3	Absorption Capacity	2.89%
4	Finesse Modulus	2.84%
5	Unit Weight	1562kg/m^3
6	Specific Gravity	2.61

5. Water

The water used for this research is municipal water of Jaipur, found in the Vivekananda Global University, Campus.

Specimen Preparation

All ten mixes of concrete are produced during the present experimental research each mix has a proper reference mix ID.

Group I: Total five concrete mixes 0%, 25%, 40%, 60%, and 100% of RCA replacements were cast as given the targeted compressive strength of 25-grade concrete.

Group II: Total five concrete mixes 0%, 25%, 40%, 60%, and 100% of RCA replacement with 10 percent MK were cast as given of 25-grade concrete.

The mixes are prepared in the related guidelines as per IS10262: 2019. In table no. 3 and 4 the proportions for various mixtures that have been taken.

Table 5 Mix Group I mix proportion of concrete (kg per meter cube)

S. No.	Mix ID	Cement	NCA	RCA	FA	Water	W/C Ratio
1	M011	352.00	1055.00	0.00	880.00	201.00	0.40
2	M012	352.00	791.25	263.75	800.00	207.00	
3	M013	352.00	633.00	422.00	880.00	212.00	
4	M014	352.00	422.00	633.00	880.00	219.00	
5	M015	352.00	0.00	1055.00	880.00	226.00	

Table 6 Mix Group II mix proportion of concrete (kg per meter cube)

S. No.	Mix ID	Cement	MK	NCA	RCA	FA	Water	W/C Ratio
1	M021	352.00	35.20	1055.00	0.00	880.00	201.00	0.40
2	M022	352.00	35.20	791.25	263.75	880.00	207.00	
3	M023	352.00	35.20	633.00	422.00	880.00	212.00	
4	M024	352.00	35.20	422.00	633.00	880.00	219.00	
5	M025	352.00	35.20	0.00	1055.00	880.00	226.00	

4. Results

The following are the different test results discussed below.

4.1 Fresh Concrete Tests

Slump Test: The horizontal free flow of NCA and RCA concrete mixes was investigated using the slump test. A rigid base plate and a slump cone were used. A slump test is carried out to check the consistency and workability of freshly made concretes. Slump tests conducted for all experiments are presented below.

4.2 Hard Concrete Test

Compressive Strength: In the laboratory, for each mix, three cubes with dimensions of 15cm x 15cm x 15cm were utilized to determine comp. strength at age of 28 days of curing. The comp. strength test is accepted absent in agreement as per IS 516:1959.

Flexural Strength: The property of materials, used to conclude the bend strength. The three beams with dimensions of 10cm x 10cm x 50cm were used to determine flexural strength after 28 days of curing period. The flexural test is performed as per IS 516:1959.

Split Tensile Strength: This test used to determine the tensile strength. Split tensile strength was calculated by cylinder of size 100mm X 300mm. The split tensile testing is showed as per IS 516: 1959. The split tensile strength of each mix is calculated using an average of three cylinders.

Table 7 Slump value in mm of Group-I

Sr. No.	**Mix ID**	**Slump value (mm)**
1	M011	77
2	M012	72
3	M013	71
4	M014	68
5	M015	64

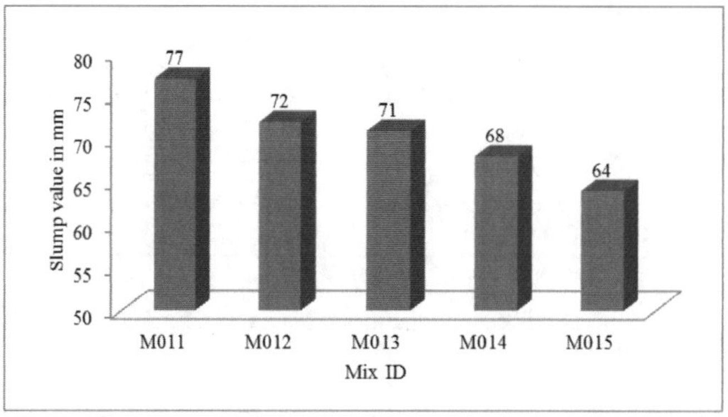

Fig.1. Bar Chart Slump value in mm of Group-I

Table 8 Slump value in mm of group-II

Sr. No.	Mix ID	Slump value (mm)
1.	M021	77
2.	M022	75
3.	M023	74
4.	M024	71
5.	M025	76

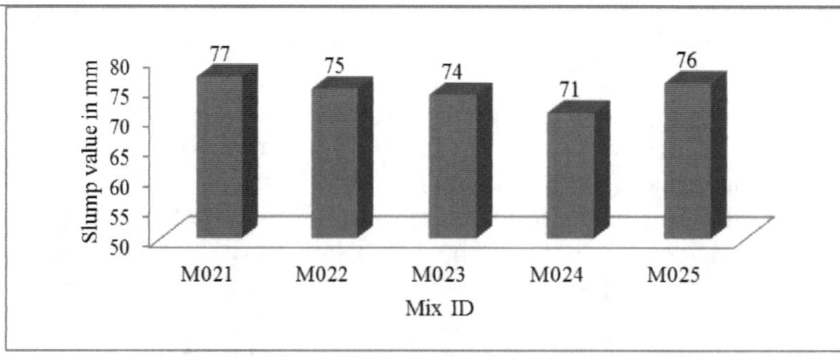

Fig.2. Bar Chart Slump value in mm of Group-II

Table 9 Compressive Strength Test of Group-I

Compressive Strength in MPa			
Sr. No.	Mix ID	7-Days	28-Days
1.	M 011	24.62	36.82
2.	M 012	20.94	32.24
3.	M 013	20.46	31.81
4.	M 014	20.74	31.02
5.	M 015	18.92	28.29

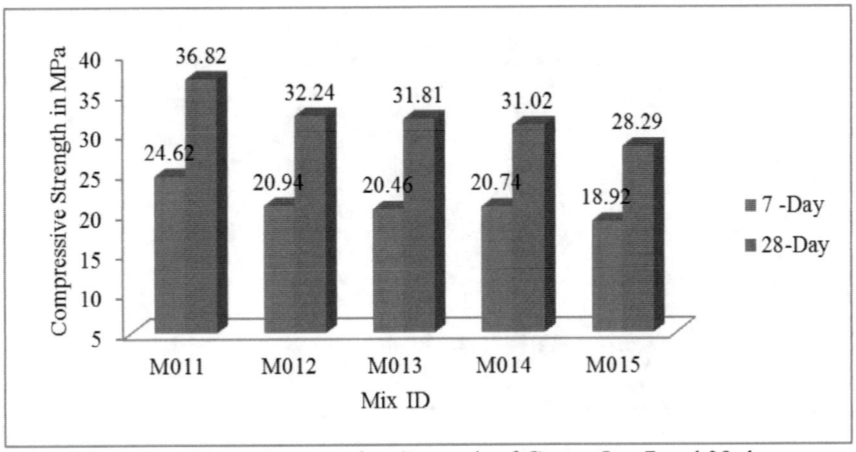

Fig.3. Bar Chart Compressive Strength of Group-I at 7 and 28 days

Table 10 Compressive Strength Test of Group-II

Compressive Strength in MPa			
Sr. No.	Mix ID	7-Days	28-Days
1	M 021	24.71	36.91
2	M 022	22.07	33.12
3	M 023	21.12	32.14
4	M 024	20.83	31.51
5	M 025	19.76	29.42

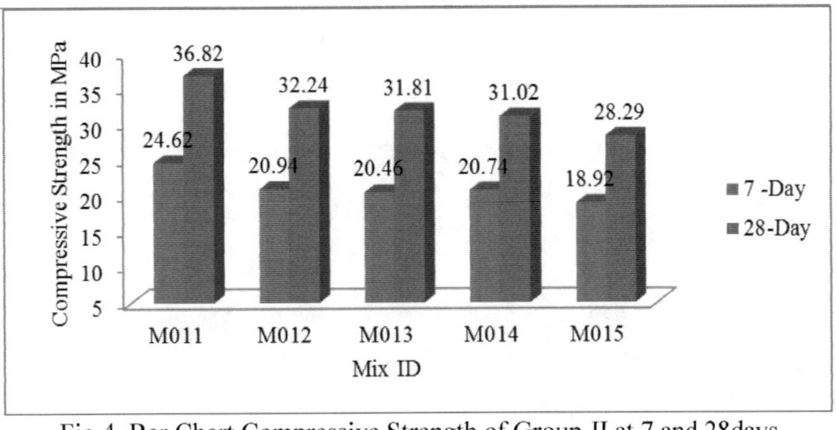

Fig.4. Bar Chart Compressive Strength of Group-II at 7 and 28days

Table 11 Split Tensile Strength of Group-I

Sr. No.	Split Tensile Strength in MPa at 28-Days	
1.	Mix ID	28-Days
2.	M 011	4.61
3.	M 012	3.78
4.	M 013	3.76
5.	M 014	3.47
6.	M 015	3.42

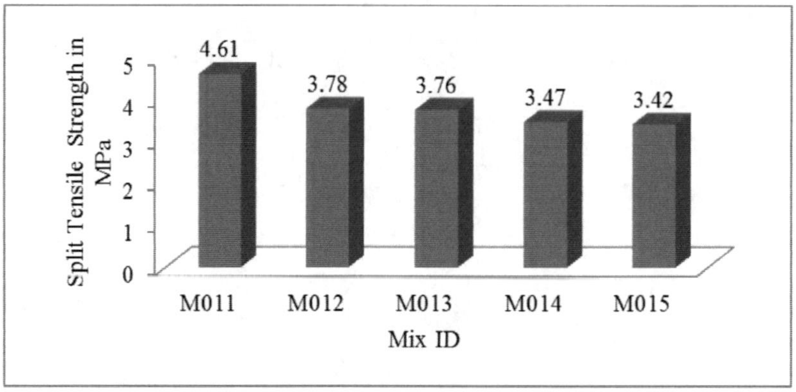

Fig.5. Bar Chart Split Tensile Strength of Group-I at 28-Days

Table12:- Split Tensile Strength of Group-II

Split Tensile Strength in MPa at 28-Days		
Sr. No.	Mix ID	28-Days
1.	M 021	4.75
2.	M 022	4.11
3.	M 023	3.89
4.	M 024	3.72
5.	M 025	3.56

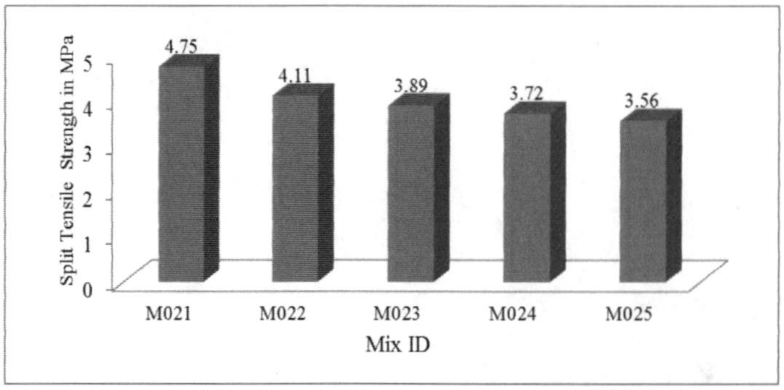

Fig.6. Bar Chart Split Tensile Strength of Group-II at 28-Days

Table 13 Flexural Strength Test of Group-I

Sr. No.	Mix ID	28-Days
1	M 011	4.23
2	M 012	3.72
3	M 013	3.68
4	M 014	3.59
5	M 015	3.28

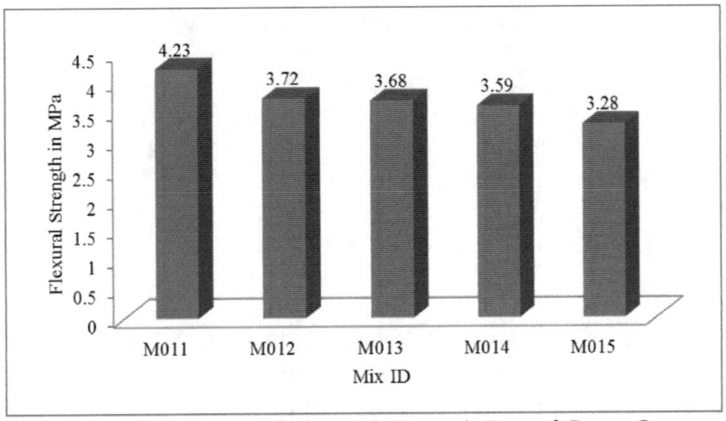

Fig.7. Bar Chart of Flexural Strength Test of Group-I

Table 14 Flexural Strength Test of Group-II

Sr. No.	Mix ID	28-Days
1	M 021	4.59
2	M 022	3.81
3	M 023	3.87
4	M 024	3.71
5	M 025	3.54

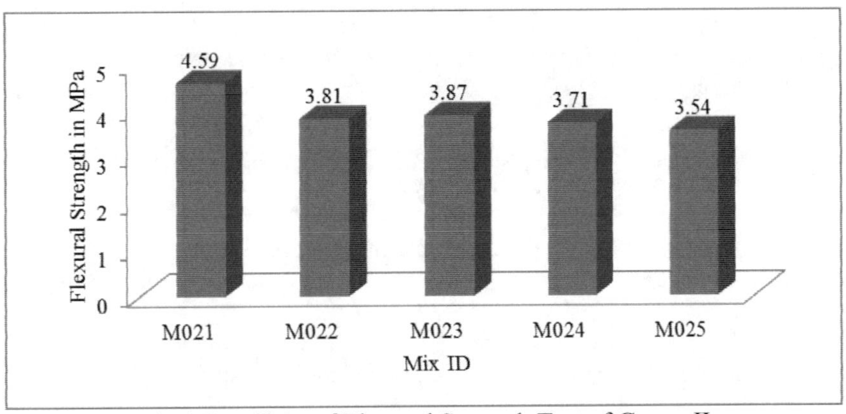

Fig.8.Bar Chart of Flexural Strength Test of Group-II

5. Conclusion

By investigation, some significant conclusions can be made aiming to reuse the RCA instead of NCA as partial substitute of cement. The mechanical properties of RAC with and without MK are explored in this work. In this study, the mixed effect of RCA and MK on concrete is investigated. Based on slump they brief that using RCA in mixed concrete with and without MK a little affects the workability. The Workability of concrete decreases as the substitute of NCA with RCA increases. An acceptable result was experiential with RCA and not effect was seen on workability. The comp. strength at 28 days is 36.91 MPa for mix ID M021. Comp. strength decreases slowly with increasing the RCA replacing of NCA. The effect of MK enhanced the strength of the RCA concrete mix. The compressive strength of mix ID of M 015 was 23.17 percent lesser vale than of control concrete mix. mix ID from M 021 to M 025 had a satisfactory split tensile strength but mix ID M011 to M 015 had a lesser value of strength. The split tensile strength of mix ID of M 015 was 25.81 percent lesser vale than of the control concrete mix. The Split tensile strength decreased with growing the content of RCA. Generally RCA base concrete failed to create a adequate pledge between aggregate and cement. To short out this problem use MK in appropriate percentages weight of cement. Mix ID from M 021 to M 025 had acceptable flexural strength apart from mix ID of M 012 to M 015 had a lesser value of flexural strength. The flexural strength of mix ID M015 was 22.46 percent lesser vale than of control concrete mix. The split tensile strength reduced with growing the content of RCA. To conquer this complexity use MK in percentile by weight of cement.

REFERENCE:

1. Bheel, N., Awoyera, P., 021). Synergic effect of MK and groundnut shell ash on the behavior of fly ash-based self-compacting geopolymer concrete. construction and building materials, 311, 125327.
2. Jadhav, R. M., Jadhao, P. D., & Pande, S. G. (2015). A Study on behavior of MK base RAC. Int. J. Struct. Civ. Eng. Res, 4, 50-62.
3. Dahunsi, B. I. O., Adebanjo, A. U., & Labiran (2021). Effects of Concrete Grades on Strength Characteristics of MK Modified RAC. Nigerian Journal of Technological Development, 18(3), 184-193.
4. Younis, K. H., Alzeebaree, Khoshnaw, G. J., & Ibrahim, T. K. (2021). Performance of RAC incorporating MK. In IOP Conference Series: Earth and Environmental Science (Vol. 856, No. 1, p. 012029).
5. Kumar, R., Mehta, P. K., & Srivastava, V. (2014). Effect of MK and recycled fine aggregate on workability and comp. strength of concrete. J. Environ. Nanotechnol, 3(3), 82-87.
6. Asteris, P. G., Kolovos, K. G., A., Plevris, V., & Konstantakatos, G. (2019). Investigation of the mechanical behaviour of MK-based sandcrete mixtures. European Journal of Environmental and Civil Engineering, 23(3), 300-324.
7. Younis, K. H., Amin, A. A., Ahmed, H. G., & Maruf, S. M. (2020). RAC including Various Contents of Metakaolin: mechanical behavior. Advances in materials science and engineering, 1-17.
8. Jadhav, R. M., Jadhao, P. D., & Pande, S. G. (2015). A Study on behavior of metakaolin base recycled aggregate concrete. Int. J. Struct. Civ. Eng. Res, 4: 50-62.
9. Singh, N., & Singh, S. P. (2016). Carbonation and electrical resistance of self-compacting concrete made with recycled concrete aggregates and metakaolin. Construction and Building Materials, 121, 400-409.
10. Kapoor, K., Singh, S. P., & Singh, B. (2017). Permeability of self-compacting concrete made with recycled concrete aggregates and metakaolin. Journal of Sustainable Cement-Based Materials, 6(5): 293-313.
11. Kumar, R., Mehta, P. K., & Srivastava, V. (2014). Effect of Metakaolin and recycled fine aggregate on workability and compressive strength of concrete. J. Environ. Nanotechnol, 3(3): 82-87.

Effect of Steel-Nylon Fibre as Reinforcing Material on Strength Characteristics of Concrete

Ramoo Ram[a], Dr Ravi Kant Pareek [b] , Kulbir Nain [c] , and Akhilesh Nautiyal [d,*]
[a b] Department of Civil Engineering, Vivekananda Global University, Jaipur, Rajasthan, India
[c] Department of Environment Science and Engineering, Guru Jambheshwar University of Science and Technology, Hisar, India
d Department of Civil Engineering, National Institute of Technology (NIT), Hamirpur, Himachal Pradesh, India
* Corresponding e-mail address: akhil.nauti09@gmail.com

Abstract. Concrete is reinforced using different kind of admixtures in order to increase its properties and making it cost effective. Fibre is one type of reinforcement, which is a composite material made up of cement or mortar and fragmented, separate, distributed, arbitrarily located, and disconnected reasonable fibres. Concrete is broadly utilized as a part of structural designing with its high compressive strength, minimal effort and bottomless crude material. However, regular cement has a few weaknesses, for instance, low malleability, restricted flexibility, little protection from splitting, weak hardness, high fragility, and other factors limits its use." The cement may get split because of basic and natural properties; however, a large portion of the breaks are shaped because of inherent inward smaller scale breaks as well as the material's inherent flaws to oppose ductile powers. Such condition of the structure at the tensile end subjects to the cracks under a certain amount of load. Initially there become small splits or cracks and with due respect of time these cracks gets widened. After drying of cement, Cracks may be developed due to shrinkage. The small-scale splits frame because of drying shrinkage or different reasons for volume change if there should be an occurrence of plain concrete. Along these lines many endeavours have been made to expand the rigidity of cement by method for utilizing traditional strengthened steel bars and furthermore by applying limiting systems.

1. Introduction

Reinforcement of Concrete using fibres is not a new concept. A lot of research regarding this has been conducted in the past. The idea of using fibre in such a way is as old as Indus Valley Civilization. Ancient developments utilized lignocelluloses resource like straw fibres (Wheat fibre etc.) in sun-dried mud blocks to make a composite in order to prevent cracks in the structure. A French Planner, named Joseph Lambot, inventor of Ferro-cement, in 1847 turned out with including constant strands into the solid, as wires or wire networks.

Fibres prevalently used to control the cracks rather than structural strengthening. The characters of fibre fortified solid change with fluctuating solid, fibre materials, geometries, dissemination, introduction, and densities. The consideration of both steel and synthetic fibres is regularly utilized as a part of development extends so as to consolidate the advantages of the two strands; auxiliary changes gave by steel fibres and the protection from plastic shrinkage enhancements gave by manufactured strands." "Manufactured strands are not a swap for steel in cement concrete, but rather are of advantage in diminishing the occurrence of little splits inside the section realized by plastic shrinkage and settlement amid the curing procedure. On-ground flooring, air terminal and parkway asphalts, seismic tremor safe and hazardous safe buildings, mine and tunnel linings, connect deck overlays, water powered structures, shake slant modification, and so on are all examples of its current area of application. Fibers are commonly employed as a piece of concrete to control shrinkage in plastic parts and drying shrinkage breaking. At the point when strands are utilized for auxiliary strengthening this is known as essential fortifying. In exhibit think about steel fibres and nylon 6 strands are utilized as secondary support. The aggregate volume of fibre is taken 0.75 % of aggregate volume of concrete. In this examination rates of steel fibre-nylon fibre are utilized as 100-0%, 75-25%, 50-half and 25-75%.

2. Materials

The strength of Concrete (grade M25) was tested using steel and nylon fibres. End hook steel fibres (NOVOCON HE0630) with a length of 30 mm and a diameter of 0.60 mm were employed, with a 50 aspect ratio. The second fibre was nylon 6, which had a length of 45 mm and a diameter of 0.3 mm, as well as a 150 aspect ratio. With steel-nylon proportions of 100-00 percent, 75-25 percent, 50-50 percent, and 25-75 percent, the total volume of fibre was 0.75 percent of the total volume of concrete.
Properties of Steel- nylon fibre:

Parameters	Steel	Nylon
Lenght (mm)	30	17
Density (g/cm^3)	7.75	1.18
Elastic Modulus (GPa)	210	5.52
Tensile Strength (MPa)	1200	1000

3. Results and Discussion

3.1. Slump cone test

Slump cone tests were used to measure the workability and consistency of fresh concrete. In general, the workability of concrete diminishes as the fiber content rises. The slump test data are reported and shown in fig 1."

In comparison to the reference mix, Table No. 1 shows that adding fiber to concrete reduces slump. Mix MN100S0 experienced the most slump. MS50N50 was found to have the highest slump loss. Because superplasticizer was used to retain workability, there was no noticeable slump loss.

Table 1. Slump value for different mixes

Mix	Steel-nylon	Slump value (in mm)
MS_0N_0	0-0 %	79
$MS_{100}N_0$	100-00 %	67
$MS_{75}N_{25}$	75-25 %	59
$MS_{50}N_{50}$	50-50 %	47
$MS_{25}N_{75}$	25-75 %	55

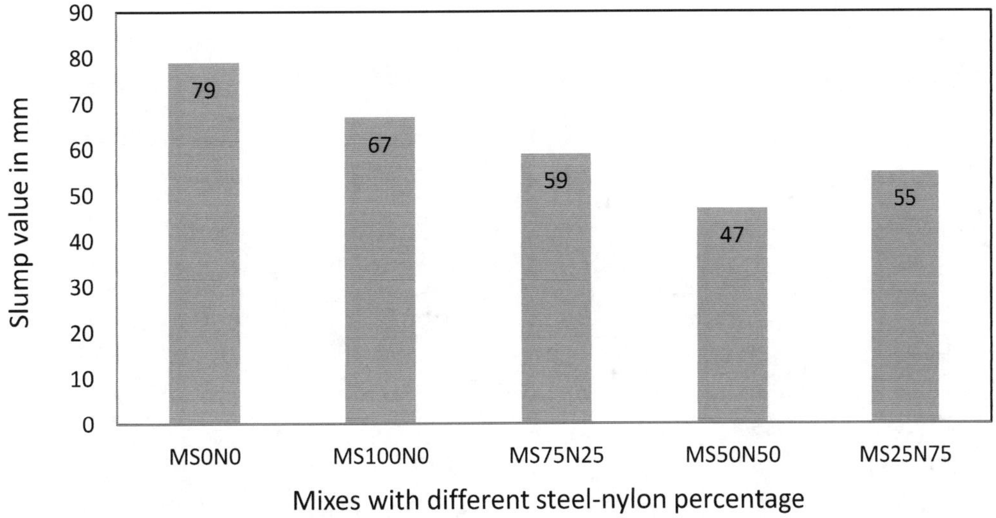

Figure 1. Slump value for all mixes

3.2. Compressive strength

For the compressive strength test, cube specimens of 150 x 150 x 150 mm were cast for the M25 concrete grade. They were allowed to cure for 7 days and 28 days after 24 hours. These cubes were evaluated on a compression testing equipment after 7 and 28 days of curing, according to I.S. 516-1959. The findings of the compressive strength test at the 7th and 28th days are presented in tables 2 and 3 and graphed.

"

Table 2. Compressive strength of hybrid reinforced concrete at 7 days

Mix	% addition of fibres (steel-nylon	Time (in days)	Collapse load (KN)	Compressive strength (measured in MPa)	Avg. compressive strength (measured inMPa)
MS_0N_0	0-0	7	490	21.72	
			460	20.30	20.97
			470	20.90	
$MS_{100}N_0$	100-0	7	735	32.82	
			720	32.18	32.40
			700	32.21	
$MS_{75}N_{25}$	75-25	7	650	29.06	
			600	26.82	28.15
			640	28.58	
$MS_{50}N_{50}$	50-50	7	585	26.11	
			620	27.05	27.21
			615	28.49	
$MS_{25}N_{75}$	25-75	7	535	24.08	

			560	24.89	24.34
			555	24.06	

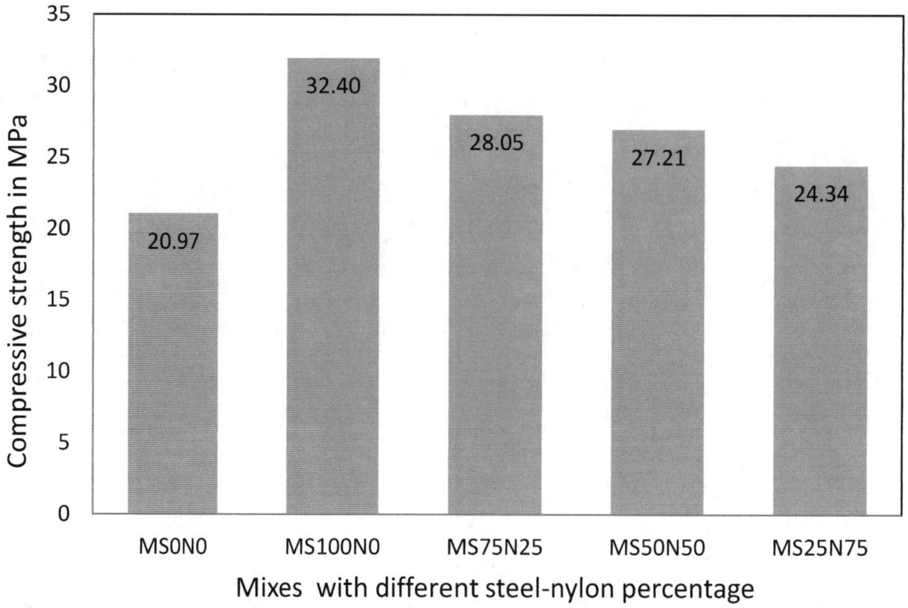

Figure 2. Compressive strength at 7 days

Table 3. Compressive strength of hybrid reinforced concrete at 28 days

Mix	% addition of fibres (steel-nylon)	Time (in days)	Collapse load (KN)	Compressive strength (measured in MPa)	Avg. compressive strength (measured inMPa)
MS_0N_0	0-0	28	680	30.29	
			700	32.09	32.11
			760	33.94	
$MS_{100}N0$	100-0	28	985	44.30	
			950	42.93	43.88
			985	44.42	
$MS_{75}N_{25}$	75-25	28	900	40.09	
			920	41.20	41.37
			930	42.82	
$MS_{50}N_{50}$	50-50	28	835	36.85	
			860	38.45	38.11

			870	39.02	
$MS_{25}N_{75}$	25-75	28	730	32.85	
			740	32.46	33.29
			760	34.56	

Figure 3. Compressive strength at 28 days

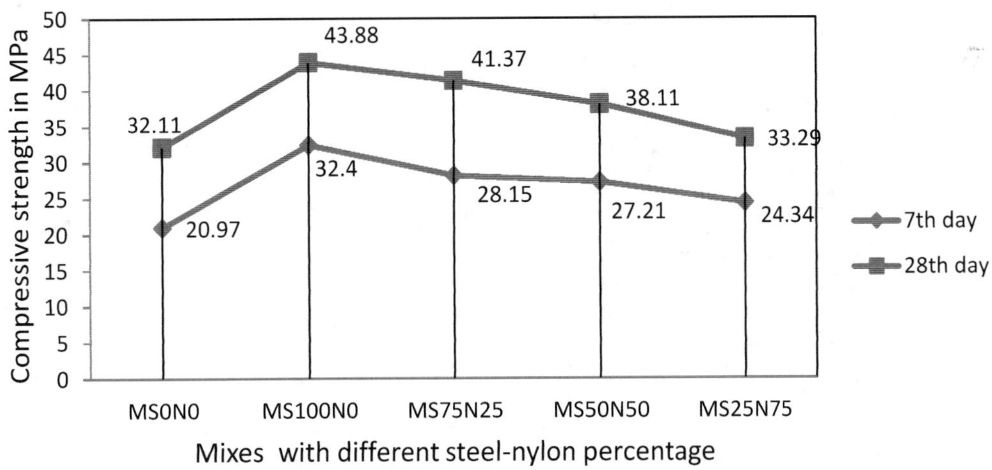

Figure 4. Compressive strength at 7th & 28th days

3.3. Split tensile strength
The test was performed according to IS 5816:1999. For the tensile strength test, cylindrical specimens having a diameter of 100 mm and a length of 200 mm were cast. The de-moulded specimens were placed in a curing tank and left to cure for 7 and 28 days, respectively. The findings of the split tensile strength test at 7th and 28th days are presented in tables 4 and 5, respectively, and graphs 5, 6 and 7 are drawn.

Table 4. 7-day split tensile strength of hybrid reinforced concrete.

Mix	% addition of fibres (steel-nylon)	Age in days	Failure load (KN)	Split tensile strength (MPa)	Avg. Split tensile strength (MPa)
MS_0N_0	0-0	7	140	2.08	2.12
			140	2.09	
			150	2.19	
$MS_{100}N_0$	100-0	7	190	2.77	2.72
			190	2.78	
			180	2.62	
$MS_{75}N_{25}$	75-25	7	175	2.4	2.43
			170	2.37	
			180	2.53	
$MS_{50}N_{50}$	50-50	7	160	2.22	2.37
			180	2.55	
			165	2.33	
$MS_{25}N_{75}$	25-75	7	155	2.23	2.19
			150	2.2	
			145	2.13	

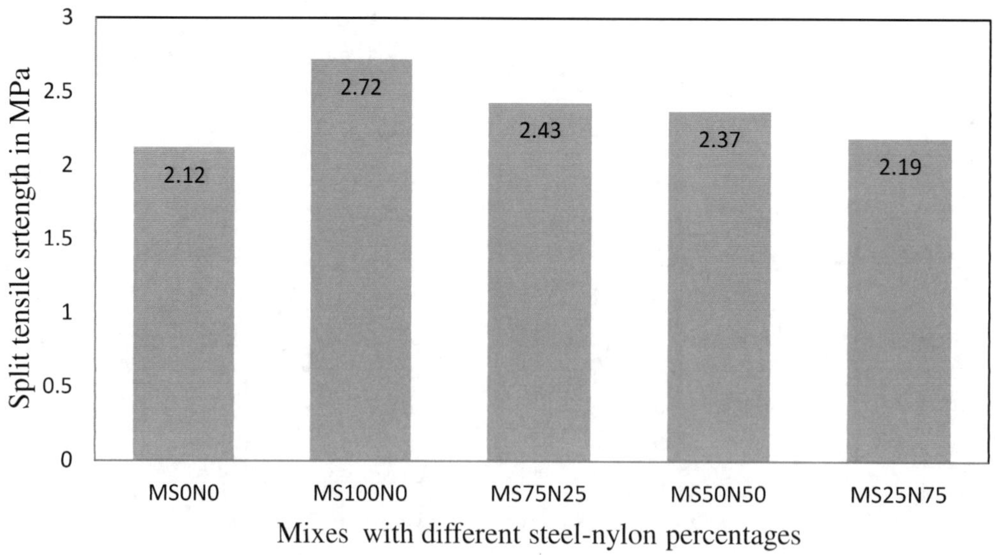

Figure 5. Split tensile strength at 7 days

Table 5. Split tensile strength of hybrid reinforced concrete at 28th day

Mix	% addition of fibres (steel-nylon	Time (in days)	Collapse load (KN)	Split tensile strength (Measured in MPa)	Average Split tensile strength (measured in MPa)
MS_0N_0	0-0	28	190	2.79	

			190	2.79	2.93
			220	3.21	
$MS_{100}N0$	100-0	28	235	3.28	
			255	3.57	3.38
			235	3.28	
$MS_{75}N_{25}$	75-25	28	220	3.09	
			240	3.37	3.18
			220	3.09	
$MS_{50}N_{50}$	50-50	28	210	3.01	
			230	3.29	3.15
			220	3.15	
$MS_{25}N_{75}$	25-75	28	215	3.12	
			205	2.98	3.10
			220	3.19	

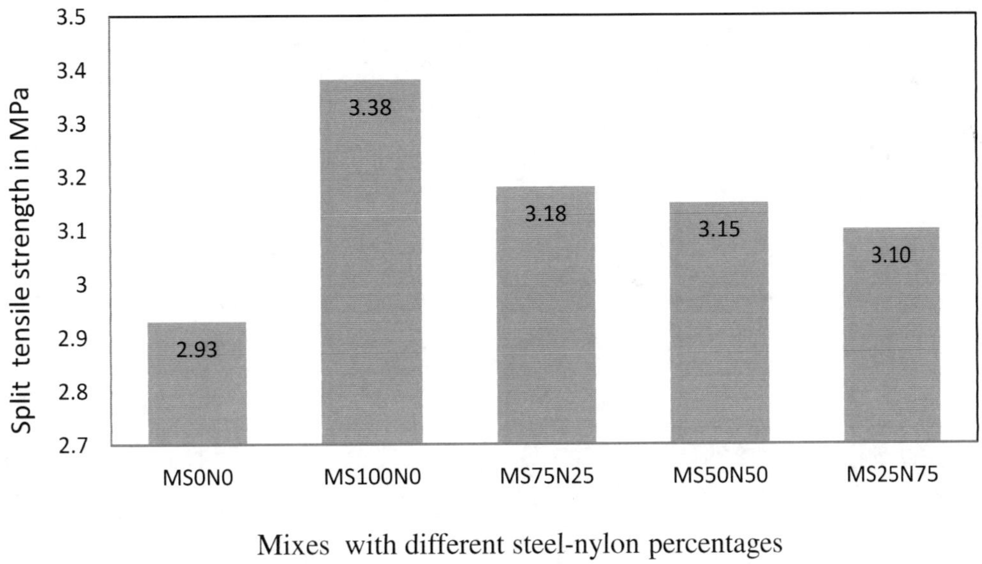

Figure 6. Split tensile strength at 28 days

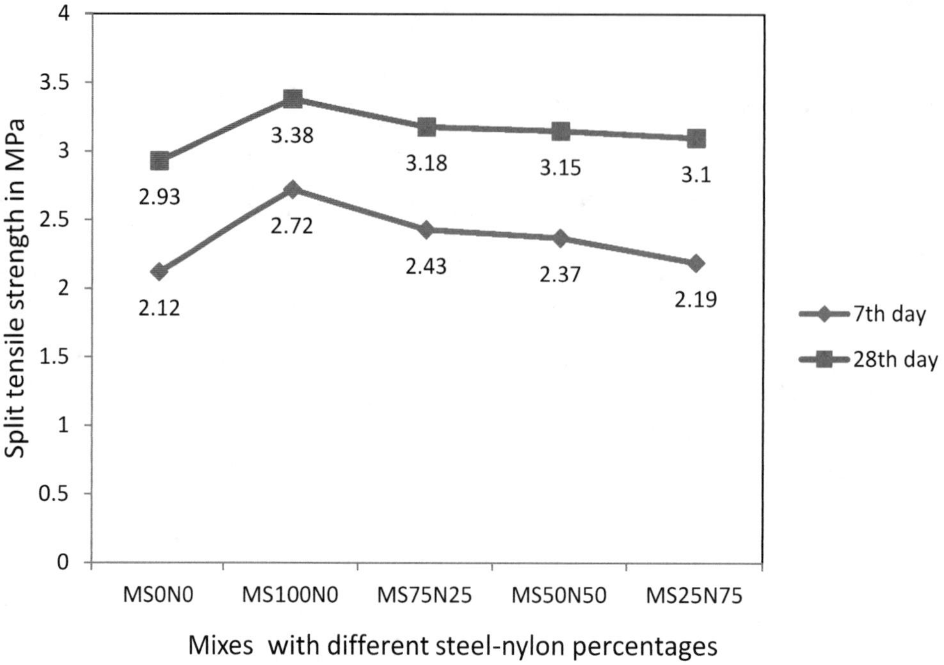

Figure 7. Split tensile strength at 7th & 28th day

3.4. Flexural strength

The flexural strength of concrete is measured by loading a 50cmx10cmx10cm beam for 7 and 28 days in a flexural testing equipment in line with IS: 9399:1979. " Tables 5 and 6 give the results of the flexural strength test at the 7th and 28th days, respectively, and graphs 7, 8, and 9 are drawn.

Table 6. Hybrid reinforced concrete flexural strength after 7 days

Mix	% addition of fibres (steel-nylon)	Time (in days)	Collapse load (KN)	flexural strength (measured in MPa)	Average flexural strength (measured in MPa)
			9.4	5.1	
MS_0N_0	0-0	7	9.0	5.5	5.3
			8.4	5.3	
$MS_{100}N_0$			11.0	6.8	
	100-0	7	10.0	6.2	6.4
			10.0	6.2	
$MS_{75}N_{25}$			11.0	6.7	
	75-25	7	11.0	6.7	6.8
			11.5	7.0	
$MS_{50}N_{50}$			9.5	5.9	

	50-50	7	9.5	5.9	6.1
			10.5	6.5	
$MS_{25}N_{75}$			10.0	6.2	
	25-75		9.0	5.6	5.9
		7	9.5	5.9	

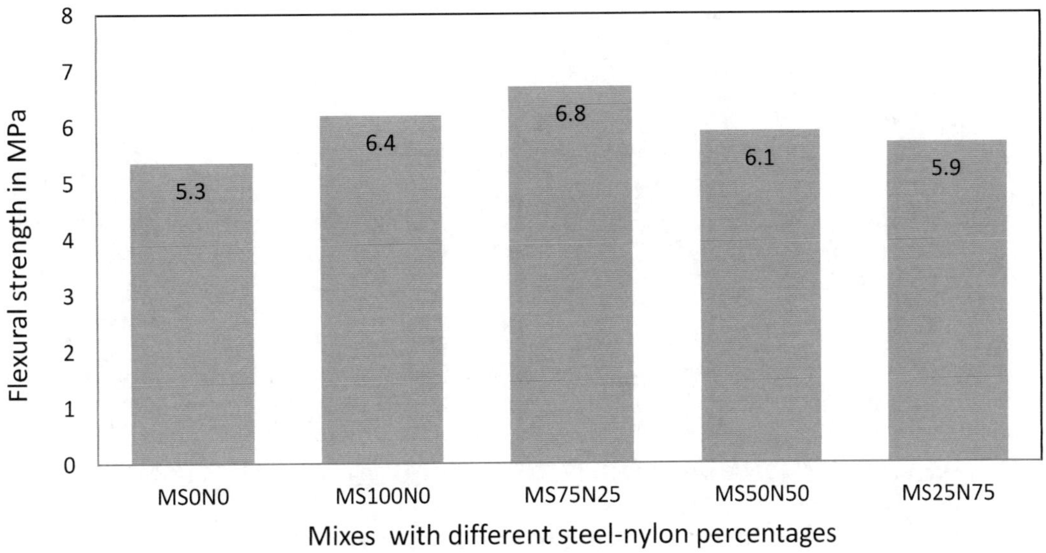

Figure 8. Flexure strength at 7th day

Table 7. Hybrid reinforced concrete flexural strength on the 28th day

Mix	% addition of fibres (steel-nylon	Time (in days)	Collapse load (KN)	flexural strength (Measured in MPa)	Average flexural strength (Measured in MPa)
MS_0N_0	0-0	28	10.5	6.4	
			11.0	6.7	6.5
			10.5	6.4	
$MS_{100}N_0$	100-0	28	12.0	7.4	
			13.0	8.0	7.6
			12.0	7.4	
$MS_{75}N_{25}$	75-25	28	13.0	7.8	
			12.0	7.2	7.7
			13.5	8.1	

$MS_{50}N_{50}$	50-50	28	12.0	7.3	
			11.5	7.0	7.1
			11.5	7.0	
$MS_{25}N_{75}$	25-75	28	11.0	6.6	
			12.0	7.2	6.9
			11.5	6.9	

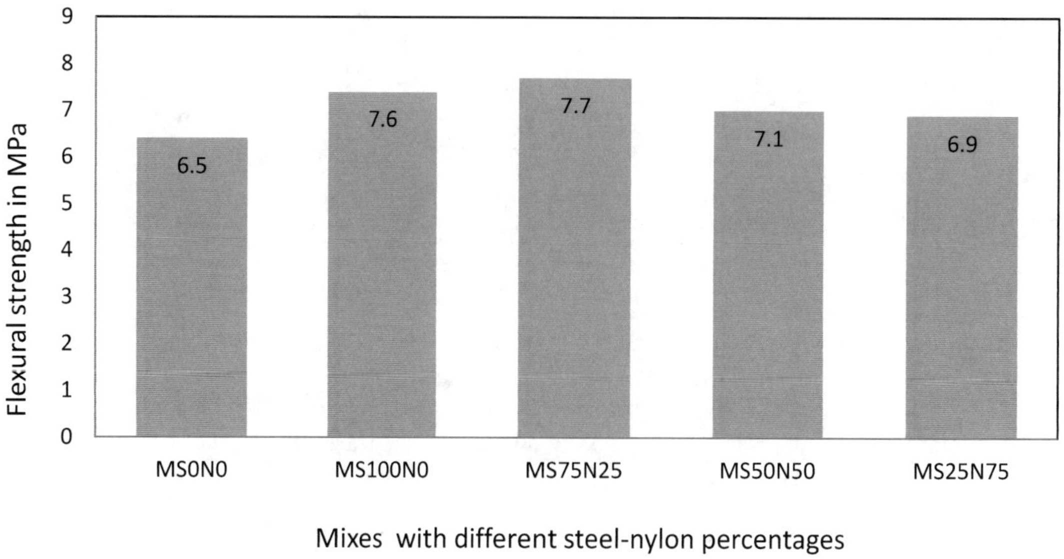

Figure 9. Flexural strength at 28 days

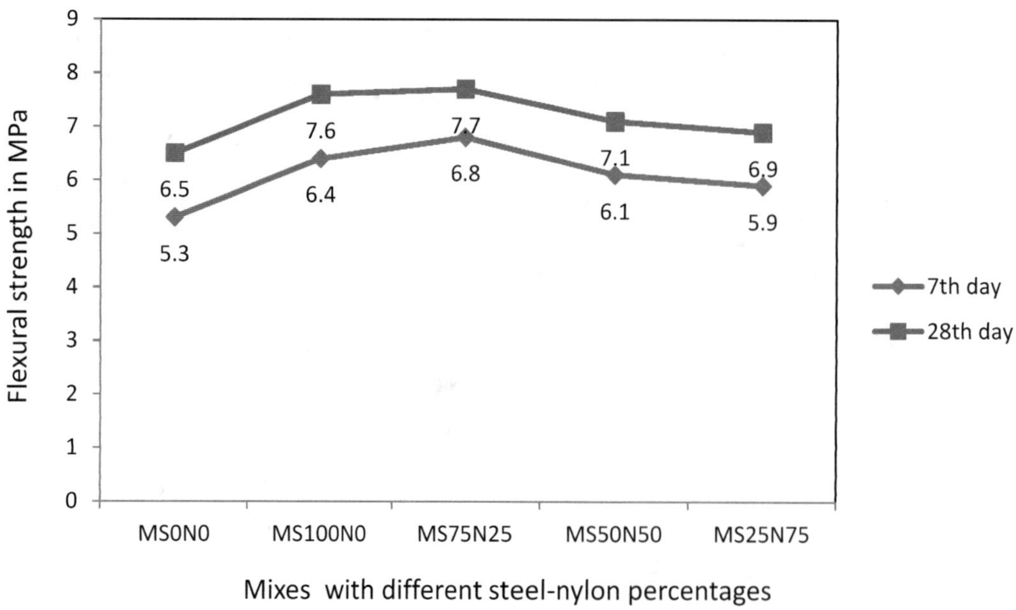

Figure 10. Flexural strength at 7th & 28th day

4. Conclusions

- In comparison to the reference mix, slump decreases when fibre is added to concrete.
- The steel nylon fibre-reinforced concrete outperformed its polypropylene-steel fibre, companion in the upgrading of compressive strength
- Steel fibre integration increases tensile strength more than nylon fibre incorporation due to its .
- Increasing the number of nylon fibres in the concrete improves the splitting tensile strength just slightly.
- It's been determined that adding nylon fibres improves flexural strength.
- The use of steel and nylon fibres in combination improves flexural strength."

5. `References

(1) Ningaraj C. Birajdarand M. M. Magdum"Durability Of Hybrid Fibre Reinforced Concrete". International Journal of Engineering Sciences & Research Technology.vol 6 issue 7,July 2017
(2) Navilesh andVivekanand A Gutteder "A Study on Hybrid Fiber Reinforced Concrete". International Research Journal of Engineering and Technology (IRJET) Volume: 04 Issue: 06 , June -2017
(3) R T Annavahini and S Christian Johnson"Experimental Investigation on Hybrid Fibre Reinforced Concrete". International Journal of Innovative Research in Science,Engineering and Technology. Vol. 6, Issue 6, June 2017
(4) K. Mounika and A. SuchithReddy"Performance of Supplementary Cementitious Material based Hybrid Fibre Reinforced Concrete". International Journal of Innovative Research in Science, Engineering and Technology. Vol. 6, 15Issue 6, June 2017
(5) Abid S. Mulla and Akshata A. Mulgund "An Investigation On Mechanical Properties Of Hybrid Fiber Reinforced Concrete". International Research Journal of Engineering and Technology Volume: 03 Issue: 05 , May-2016
(6) LakshmiPriya and T.Ch.Madhavi "Flexural Behaviour of Hybrid Fiber Reinforced Concrete Beams Strengthened by Glass FRP Laminate". International Journal of Innovative Research in Science, Engineering and Technology. Vol. 5, Issue 5, May 2016
(7) Vinayadeep and Brijbhushan "Strength aspects of hybrid fiber reinforced concrete". International Research Journal of Engineering and Technology (IRJET) Volume: 02 Issue: 08 ,Nov-2015.
(8) SudheerJirobe and Brijbhushan "Experimental investigation on strength and durability properties of hybrid fiber reinforced concrete". International Research Journal of Engineering and Technology. Volume: 02 Issue: 05 ,Aug-2015.
(9) K.Ravi and A.V.Karvekar " Performance evaluation of hybrid fibre reinforced concrete subjected to freezing and thawing effect". International Journal of Research in Engineering and Technology. Vol 3 issue 11, nov 2014.
(10) C Chandra et al, "Flexural behavior of solo and hybrid fibre concrete- a comparative study", International Journal of Engineering Sciences & Research Technology, Vol. 2 Issue7, July 2013.
(11) Pey-Shiuan Song, et al, "Investigating mechanical properties of steel- nylon hybrid fiber-reinforced concrete by using experimental design", Journal of C.C.I.T., Vol.42, No.1, May, 2013.
(12) Parveen and Ankit Sharma, "Structural behaviour of fibrous concrete using polypropylene fibres", International Journal of Modern Engineering Research (IJMER), Vol.3, Issue.3, May-June, 2013 pp-1279-1282.
(13) Karthik M.P. and Maruthachalam D., "Mechanical properties of hybrid fibre reinforced concrete with available of rural fibres", International Journal of Engineering Sciences & Research Technology, April, 2013.

(14) Kavita S Kene et al, “Experimental study on behavior of steel and glass fiber reinforced concrete composites”, Bonfring International Journal of Industrial Engineering and Management Science, Vol. 3, Issue 4, Jul-Aug 2013, pp.1408-1411.
(15) Ankur, Singh Varinder, Pareek Ravi Kant "Effect of rice husk ash and plastic fibers on concrete strength", International Journal of Civil and Structural Engineering, Vol 6, issue 1, July 2015

Mathematical Expressions Model to forecast for Chloride Ion Penetration and Comp. Strength of Recycled Coarse Aggregate Concrete Incorporating Meta-kaolin

Om Prakash Singh[1], Mahendra Kr. Singar[2], Sandeep Kumar[3]
Department of Civil Engineering, Vivekananda Global University, Jaipur, Rajasthan

Corresponding author email id: om.prakash.singh@vgu.ac.in

Abstract. This research investigates the mathematical modeling of compressive strength (CS) of concrete cured cube at 3, 7 & 28 days and chloride penetration resistance of cylindrical concrete specimens cured at 28 days incorporating of Meta-kaolin (MK). The experimental data results were shown various concrete specimens close to recognize compressive strength (CS) and penetration of chloride ion. Ordinary Portland cement (OPC) has been partial replacement of 0%, 5%, 8%, 12%, 16%, 18% and 20% by Meta-kaolin (MK). Compressive strength (CS) results discover through the cubes and rapid chloride penetration test (RCPT) presented. Two predictive regression models established, first for compressive strength (CS) of concrete specimen at the days of 3, 7, and 28 and second for charge passed Q at 28 days. To predictive the mathematical models have been established and have good precision, correlation by experimental results of data.

Keywords:RCA, Metakaolin (MK), Compressive Strength (CS), RCPT, Cement

1. Introduction

The continued growth of infrastructure in developing countries requires large quantities of construction materials. Developing countries like India produce vast quantity of C&D waste year wise. Proper utilize of the waste is a big difficulty because the one side, it requires a major amount of break for disposal while, on either side, it pollute the environment. Natural resources such as stone and sand must be protected. An important issue is indeed the continuing use of natural resources including such river sand. It increases the depth of the riverbed, producing draughts and changing climatic conditions. As a consequence, the sustainable construction thought was introduce into the construction field in response to rising distress for our planet's future as it is a large purchaser of natural assets (1). In construction waste, the proportion of concrete remains is highest (1). The impartial of this investigation was to review the assets of concrete construct by RCA from C&D structure. It'd be extra appropriate for construction if the loss of strength in concrete induces from RCA strength be address with the addition of pozzolana. To settle on the majority constructive amount of metakaolin (MK), Ordinary cement was partially replaced with metakaolin up to 20 percent in the experiment (2). NCA and RCA with MK are being used to evaluate the experimental compressive strength and predicted compressive strength. NCA and RCA with MK are being used to evaluate the experimental RCPT value and predicted RCPT value.

2. Experimental Investigation

To appreciate persuade of the MK filling in different properties of concrete of comp. strength (CS) and durability to chloride penetration. The major collection of the properties of dissimilar adding material to will be an appearance of resulting concrete and mix design M-25 concrete grade.

2.1 Materials

The properties materials to development of concrete specimens

a) Ultra-tech Ordinary Portland cement of 53 grade of cement used as per IS 269:2015.
b) Sand of river (fine aggregate) used as confirmed to zone-II as per IS 383:2016.
c) C&D waste of 30 years old available in Jaipur 20 mm and 10 mm coarse aggregates.
d) Meta-kaolin (MK) used as partial substitution of cement.
e) BS 17 3P water reducing super plasticizer admixture used as water reducing admixture.

Table 1 property physical & chemical of MK

S. No.	Properties	Value
	Physical	
1	Sp. Gr.	2.51
2	Sp. area (cm2/gm)	160000 - 170000
3	Colour	White
	Chemical	
4	Al_2O_3	31-34
5	SiO_2	62-65
6	Fe_2O_3	1.12
7	CaO	0.15-0.41
8	MgO	0.22-0.81
9	Na_2O	0.45-1.25
10	K_2O	0.51-1.3
11	Loss on Ignition	<1.42

Table 2 Analysis of sieve for natural F.A.

S. No.	Size of sieve	Fine Aggregate (% passing)
1	4.75 mm	94.3
2	2.36 mm	82.4
3	1.18 mm	72.3
4	600 micron	59.1
5	300 micron	39.7
6	150 micron	8.6
7	< 150 micron	-

2.2 Mix Design

Mix design of concrete procedure of M-25 grade is execute as per IS 10262:2019 and IS 456:2000. As shown in table no.03 show of mix design per cubic meter of M-25 of concrete. The total different concrete specimens have developed for M-25 grade.

2.3 Value of RCPT test (ASTM C1202)

The RCPT value is being conduct on different cylindrical samples of concrete. The charge is passed during specimen via RCPT machine. The charge value in coulomb is further that suggest concrete is extra preface and a reduced amount in case of vice versa.

Table 3 The value of mix design of M-25 grade per cubic meter

Mix ID	Cement	MK	%-MK	FA (Sand)	RCA (10mm)	RCA (20mm)	Water	w/c ratio
M-01	382.16	0.00	0	692.86	458.55	687.83	192	0.43
M-02	363.05	19.11	5	692.86	458.55	687.83	192	0.42
M-03	351.59	30.57	8	692.86	458.55	687.83	192	0.41
M-04	336.30	45.86	12	692.86	458.55	687.83	192	0.41
M-05	321.01	61.15	16	692.86	458.55	687.83	192	0.40
M-06	313.37	68.79	18	692.86	458.55	687.83	192	0.40
M-07	305.73	76.43	20	692.86	458.55	687.83	192	0.39

3. RESULTS

The value of comp. strength (CS) of different mixes of concrete with MK at age of 3, 7 and 28 days are respectively discuss and value of RCPT investigation as per ASTMC1202 manner for a range concrete of cylindrical samples with/without MK substance at 28 days period. The predictive regression model developed for comp. strength (CS) at 3, 7, 28 days and charge passed at 28 days in different specimens of concrete.

Table 4 Comp. Strength (CS) Result

Sr. No.	Mix ID	Comp. Strength (CS) in MPa		
		3-Days	7-Days	28-Days
1	M-01	15.30	20.53	32.24
2	M-02	14.11	22.43	32.98
3	M-03	14.25	22.26	33.23
4	M-04	14.46	22.90	33.89
5	M-05	14.52	23.04	34.12
6	M-06	15.02	23.79	35.26
7	M-07	15.23	23.83	35.79

3.1 Compressive Strength (CS)

The value of comp. strength (CS) of many concrete cubes at age of 3, 7 and 28 days are respectively measured as per IS 516:1959 and table no 4 below show the comp. strength (CS) of diverse concrete samples at 3, 7 and 28 days with/without MK. The bar graph in fig. 1 shows the comp. strength (CS) of the entire concrete samples at 3, 7 and 28 days are respectively period of curing.

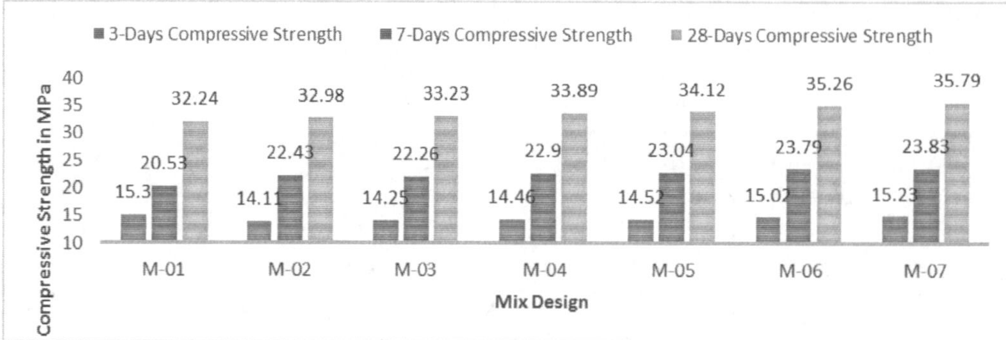

Fig. 1 Bar Graph of Comp. Strength (CS)

3.2 RCPT Test Results

RCPT investigation significances on different concrete samples incorporate with/without MK be as below shown in table no 5 & fig. 2 show the variant of RCPT outcome charge accepted at 28 days in every concrete.

Table 5 Total charge in coulomb at 28-days via RCPT

Sr. No.	Mix ID	Charge coulomb at 28-days
1	M- 01	3244.65
2	M- 02	3192.81
3	M- 03	3052.71
4	M- 04	2956.43
5	M- 05	2851.72
6	M- 06	2738.83
7	M- 07	2716.79

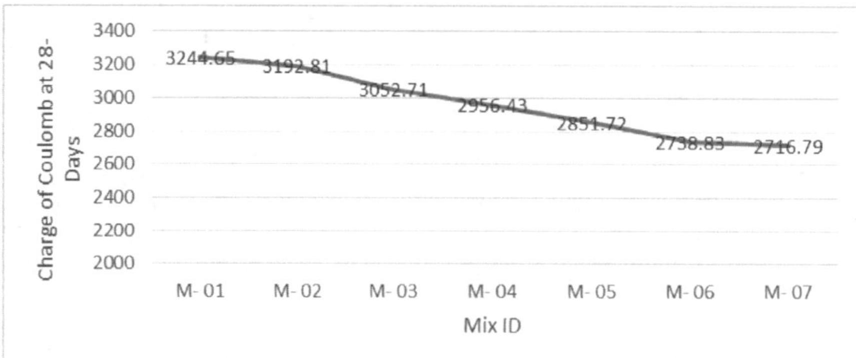

Fig. 2 Differences of RCPT test results at 28 days

3.3 extrapolative Regression Model

Double extrapolative regression replicas developed, first for comp. strength (CS) of concrete cubes at 3, 7 and 28 days and second for charge passed Q at 28 days in specimens of cylindrical concrete via value of RCPT.

3.3.1 Regression Models for Comp. Strength (CS)

Regression functions are being developed for MK, water-cement ratio and cement comp. strength (CS) as independent variables.

Comp. Strength (CS) = f (Cement, MK, Water Cement Ratio)
i. e. CS = f (C, MK, W/C)

Compressive Strength (in Day) = $a_1 + a_2$* Cement + a_3*MK + a_4*W/C

Where a_1, a_2, a_3& a_4 are coefficients belongs to real numbers regression equation for comp. strength (CS) at 3 days

= 4.0681x10-2x C – 0.11452xMK-127.82XW/C+129.412

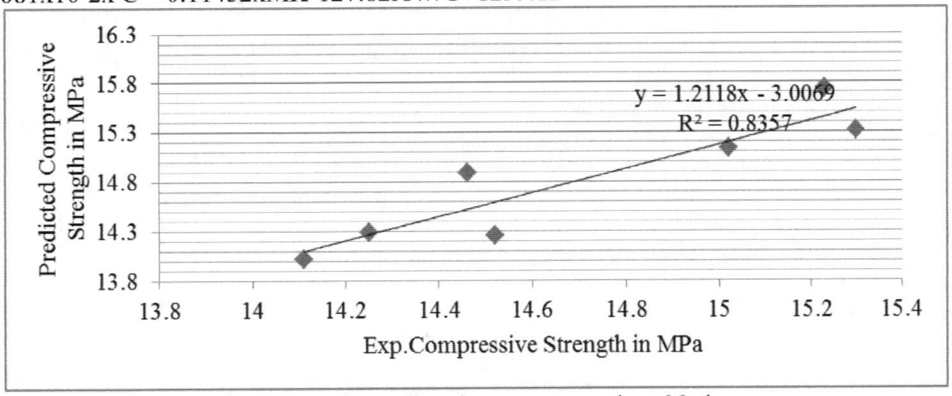

Fig. 3 Exp. & predicted comp. strength at 28-days

Fig. 3 represent the correlation curve among exp. and predicted comp. strength (CS) at age of 3 days for concrete specimen with/without MK addition.

Regression equivalence for comp. strength (CS) at 7 days

= -4.652x10-2xC-0.192xMK-321.6845xw/c+0.019453

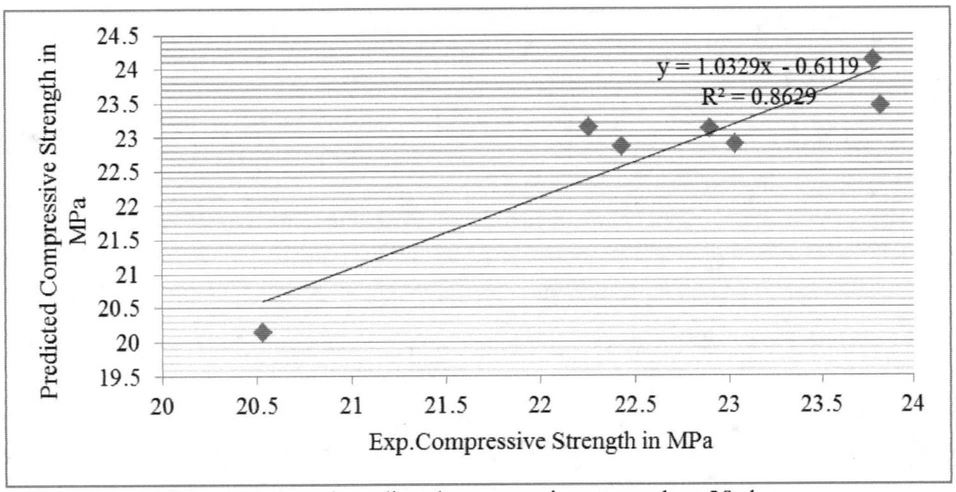

Fig. 4 Exp. and predicted compressive strength at 28-days

Regression equivalence of comp. strength at age of 28 days

= 8.681x10-2xC-0.08251xMK-304.261xW/C+174.8451

3.3.2 Regression model for total charge at age of 28 days

In this regression analysis function is consider of cement, MK and water cement ratio as independent variable and charge as descriptive variable for all specimens of concrete cured at the age of 28 days.

Eq. of total charge = f (cement, MK, ratio of water cement)
i.e. Q = f (C, MK, W/C)

Q at 28 Day = $a_1 + a_2$*Cement + a_3*MK + a_4*W/C
Where a_1, a_2, a_3& $a_4$are coefficients belong to numbers regression equation used at 28 days charge = 4.491x10-4xC-18.6728xMK+4281.3645xW/C+1241.124

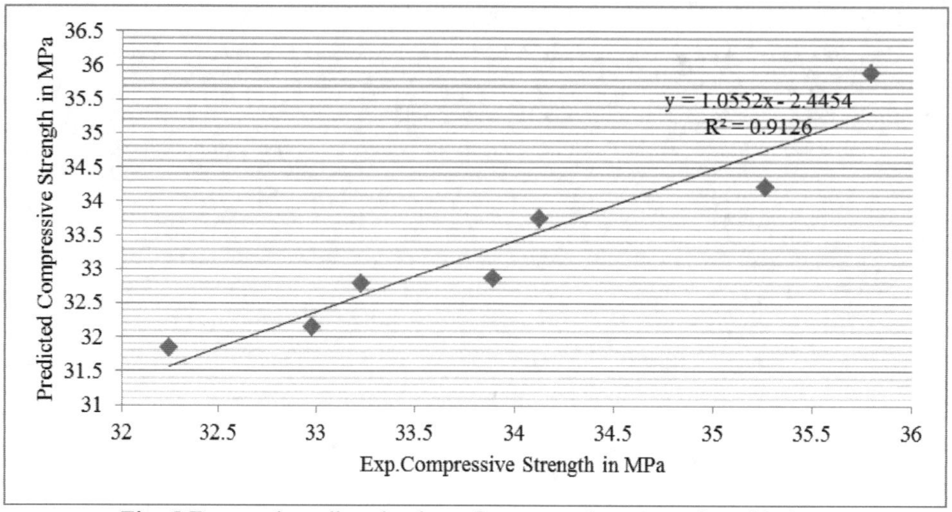

Fig. 5 Exp. and predicted value of compressive strength at 28-days

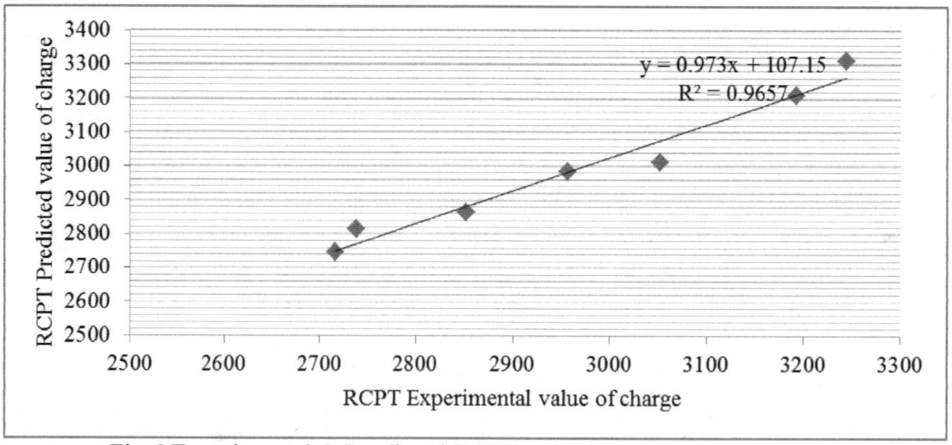

Fig.6 Experimental & Predicted RCPT values total charge at 28 days

4. CONCLUSIONS

The many closing comments of this research study are given below.

The buildup up to 20 percent of MK as replacement of cement for M-25 grade that is enlarging in compressive strength of concrete since compare to RCA concrete case of no substitution. The suggested model has accuracy that can be used in different engineering fields. In the predictive model explain excellent correctness, fewer percentage variation & fine agreement through the investigational data effect strength of compressive and value of RCPT. In the test results of experimental, the concrete by MK display small value of chloride ion permeability that shows that show of durable concrete at age of 28 days. Other main comment is to facilitate accumulation of MK reduces the concrete voids. As outputs of the investigational, it is understood to Cl^- ions permeability is enormous by upper ratio of water. Concrete with/without MK can be optional supplementary for pre-cast structure work.

REFERENCE

[1] Jadhav, R. M at al. (2015). A Study on behavior of metakaolin base recycled aggregate concrete. *Int. J. Struct. Civ. Eng. Res*, *4*, 50-62.

[2] Kapoor, K.et al. (2017). Permeability of self-compacting concrete made with recycled concrete aggregates and metakaolin. *Journal of Sustainable Cement-Based Materials*, *6*(5), 293-313.

[3] Moon, H. Y. Et al. (2005). Mechanical Properties and Resistance to Freezing and Thawing of the Recycled Aggregate Concrete with Metakaolin. *Journal of the Korea institute for structural maintenance and inspection*, *9*(4), 270-278.

[4] Rezaifar, O. et al. (2016). Concrete made with hybrid blends of crumb rubber and metakaolin: Optimization using Response Surface Method. *Construction and building materials*, *123*, 59-68.

[5] Kumar, R. et al. (2014). Effect of Metakaolin and recycled fine aggregate on workability and compressive strength of concrete. *J. Environ. Nanotechnol*, *3*(3), 82-87.

[6] Silva, R. V. Et al. (2015). Prediction of chloride ion penetration of recycled aggregate concrete. *Materials Research*, *18*, 427-440.

Synthesis of An Amidoximated Acrylic Copolymer Membrane (AACM) Treated with Nano Silver Particles to Study the Antibacterial Efficiency of The Membrane

Bhoopesh Kumar Sharma[1], Megha Walia[2], Archana Chaudhary[3], Yogesh Sharma[4], Shaile Thakur[5]
[1,3]*Associate Professor, Faculty of Science, Shree Guru Gobind Singh Tricentenary University, Gurugram – 122505, India*
[2,4]*Assistant Professor, Faculty of Science, Shree Guru Gobind Singh Tricentenary University, Gurugram – 122505, India*
[5]*Assistant Professor, Department of Forensic Science, Vivekanand Global University, Jaipur, Rajasthan, India*

Abstract

Polymers with cationic functional groups are proven to be an effective antimicrobial barrier. Particles, organic debris, and microbes can be well removed from drinking water using polymer membrane filtration. A variety of copolymers are utilized in the synthesis of such membranes. In this study, a copolymer membrane was synthesized first, and then Amidoximation was used to convert a portion of the copolymer's nitrile groups to amidoxime groups. The membrane was treated with HA at 60-80°C in an aqueous solution to obtain an Amidoximated film. After being treated with Nano Silver particles, the efficiency of the prefabricated and particularly produced copolymer film as an antibacterial water disinfectant was analyzed. Various techniques were used to analyse the virgin and Amidoximated Acrylic Copolymer Membranes (AACM), including X-ray diffraction (EDX), Fourier Transform Infrared (FTIR), Scanning Electron Microscope, Transmission Electron Microscope, and Thermogravimetric analysis, Many substantial changes in the copolymer characteristics, including functionality, crystallinity, thermal activity, hydrophilicity, elemental composition, surface morphology, and antibacterial activity, were identified during the Amidoximation phase. It was revealed that an Amidoximated copolymer membrane outperforms an untreated membrane in terms of antibacterial activity.

Keywords: Copolymer membrane, Amidoximation, Nano Silver Particles, Functionality of Fabricated Copolymer Membrane, Antibacterial Activity

1. Introduction

Water is considered life on this planet and is essential to the Earth's climate and the evolution of life [1]. Costal seawaters are the critical source of stool bacteria, including several diseases [2,3]. The management of microorganisms such as bacteria, fungus, and yeast in nature is one of the key elements for the survival of higher species [4]. Membrane filtration is an efficient way to remove particulates, organic materials, and microbes from drinking water. Polymers with cationic functional groups are proven to be an effective antimicrobial barrier. Particles, organic debris, and microbes can be well removed from drinking water using polymer membrane filtration. Copolymers are commonly used to synthesize such membranes. A variety of copolymers are utilized in the synthesis of such membranes. The membrane technique is superior to other approaches for improving water quality [5]. Development and advancement of new long-acting antimicrobial polymeric surfaces is always necessitated for destroying harmful germs and preventing infections in drinking water. Low molecular weight antimicrobial compounds have been proven ineffective in applications such as surface coatings, blends, and composites [6]. Over the last two decades, several researchers have developed and extensively tested antimicrobial polymers against Gram-positive and Gram-negative bacteria, viruses, yeast, and fungus species [7]. Positively charged polymers attract negatively charged bacterial cell surfaces. The membrane-forming capabilities of polyacrylonitrile (PAN) and acrylonitrile-based copolymers are remarkable in this concern as they are chemically resistant to most solvents. They also have strong thermal and mechanical stability, making them more effective in membrane production [6, 7]. PAN membranes have also been employed as membrane bioreactors for protein filtration [8, 9]. PAN hollow fiber membranes are already used in dialysis machines that focuses on molecular protein removal and high-flux dialysis therapy. Surface area, zeta potential, pore diameter, surface roughness, and antibacterial agent inclusion are all elements that influence a membrane's antibacterial effectiveness [10, 21].

On the other hand, it has been demonstrated in several prior studies that antibacterial activity is considerably required in nanofiber membranes for biomedical filtration devices in order to minimize bacterial contamination under moist conditions while retaining biocompatibility [11]. The copolymerization of acrylonitrile (AN) with various acrylic and vinyl monomers during the polymerization process has been significantly described in the literature [11-13]. Methacrylic acid, acrylic acid (AA), itaconic acid, acrylamide, and styrene are the most common monomers used for such interventions. Researchers [14] have already discovered several natural and artificial nanomaterials as antimicrobial agents in water disinfection systems, including chitosan, photo catalytic titanium dioxide, fullerene, gold, silver nanoparticles [15]. Due to their electrical characteristics, optical properties, biological qualities, and thermal conductivity Nano-Silver (AgNPs) particles are considered as a best alternative for the purpose [16]. NS (Nano-Silver) particles can effectively destroy and inactivate a wide variety of microorganisms as described in the previous researches [17-19]. The produced AgNPs and functionalized PAN nanofibers (PAN-AgNPs) have significant antibacterial activity and long-term durability against *Escherichia coli* and *Staphylococcus aureus* bacteria. A similar approach was used in the current research to fabricate and synthesize a copolymer film after the effective treatment with Nano Silver particles to determine its effectiveness as a water disinfectant. Nanofiber membranes, in particular, are a promising filtration platform for vapors and gases, as well as excellent filtering efficiency, in a variety of applications ranging from interior air filters to personal protective equipment. Also, the major objectives of the present work were to perform the Amidoximation of acrylic copolymer membranes, then to characterize the virgin and the Amidoximated acrylic copolymer membranes by using various techniques to study the efficiency of the synthesized membrane in this process for its antimicrobial properties.

2. Materials and Methods

Toluene, Polyvinyl alcohol, Silver nitrate, Azobisisobutyronitrile, Acrylonitrile, Acrylonitrile, Acrylic acid, DMF (Dimethylformamide), NaOH (Sodium Hydroxide), HCl (Hydrochloric Acid), and other compounds were utilised in this study.

All of the chemicals were AnalaR reagent grade, and they were purchased from Fluka and Sigma Aldrich. Silver nitrate (AgNO3), fructose, and polyvinyl alcohol were also given by Sigma Aldrich and Fluka (PVA).

All of the compounds were utilised in their natural state, without being purified. Before being used further, all of these solvents were dried and distilled using a conventional process [20]. Polymerization synthesis, Molecular Weight Determination, Copolymer Fabrication, Hydroxylamine treatment, and Amidoximes content analyses have all been published in one of our prior works [13].

2.1 Bulge Behaviour Determination

The slow diffusion of solvents into polymer chains, results in a swollen polymeric membrane. Bulge behaviour of copolymer membranes having Amidoxime (AO) content 3.5 meq/g was carried out with distilled water in water bath as shown in Table 1. After bulging, the excess water was removed by filter paper and the puffy sample was again weighted. The degree of bulging was calculated as [19]:

$$\text{Bulge (\%)} = \frac{W_s - W_0}{W_0} \times 100$$

Where, W_0 and W_s are the weight of the dry and swollen membrane, respectively.

Table 1: Abbreviations and composition of samples

Abbreviations used	Chemical Details
$AO_{3.5}$	Acrylic copolymer membrane contains 3.5 meq/g amidoxime content.
NS_{R20}	Nano Silver (NS) synthesized by 20-minute reduction time.
NS_{R40}	NS synthesized by 40-minute reduction time.
NS_{R60}	NS synthesized by 60-minute reduction time.
$AO_{3.5}$-NS_{R20}	$AO_{3.5}$ membrane contains NS formed by 20-minute reduction time.
$AO_{3.5}$-NS_{R40}	$AO_{3.5}$ membrane contains NS formed by 40-minute reduction time.
$AO_{3.5}$-NS_{R60}	$AO_{3.5}$ membrane contains NS formed by 60-minute reduction time.

2.2 Preparation of Nano Silver Particle (AgNPs)

Wet chemical method was used for the preparation of Nano silver (NS) particles. Silver nitrate was first converted to NS in an aqueous solution by reducing it with fructose. The solution of silver ions was prepared by dissolving 30 mg silver nitrate in 100 ml of distilled water under constant stirring condition at 25-30°C of temperature. Then fructose and polyvinyl alcohol were added in it as a stabilizing agent. Colourless solution changed into light yellow colour indicated the formation of NS particles. This solution was then kept in refrigerator at 4°C. Reduction reaction mechanism due to fructose can be given as [20].

$$2Ag^+ + H_2O + CH_2OH\text{-}(CHOH)_4\text{-}(CHO) \longrightarrow 2Ag^o + 2H^+ + CH_2OH\text{-}(CHOH)_4\text{-}COOH$$

2.3 Immobilization of Nano Silver Particles

To immobilize NS into the copolymer film, a fix weight of AO 3.5 meq/g copolymer membrane was first swelled into distilled water at 90°C in fixed temperature water bath. The swollen films were then instantly transferred into NS solution carrying vessel along with moderate shaking for few minutes. The manufactured films were then dried at 95-100°C for 120 minutes in vacuum oven and washed in distilled water to remove excess NS particles and other chemical impurities. The synthesized films were then once again vacuumed dried at 95-100°C for 240 minutes for further use.

2.4 Instrumentation Used

Synthesis of NS was confirmed by UV-visible Spectroscopy. UV-visible spectroscopy was carried out by Perkin Elmer Lambda E Z 201 spectrophotometer which was used for the determination of Plasmon peak of NS. Energy Dispersive X-Ray (EDX) studies of samples were carried out using STEREOSCAN 360 scanning electron microscope for the study of surface as well as fracture of NS immobilized copolymer. The morphological characteristics of NS particles were analysed using a PHILIPS CM -12 method, which was used to analyse the morphological characteristics of NS particles such as size and shape.

2.5 Antimicrobial Studies

All the samples were sterilized by γ-rays, so that pre-existing microbes can be eliminated. Viable cell count method was acquired to examine antimicrobial nature of samples. The American Association of Textile Chemists and Colourists (AATCC) 100 method was followed for this purpose. As this method is considered to be known to evaluate the antibacterial activity over the textiles. The antimicrobial activity was performed against model bacteria E. *coli* and S. *aureus* bacteria [23].

2.6 Method of counting colonies

Fresh colonies of E. *coli* or S. *aureus* were used to create a suspension in Muller Hinton Broth (MHB). Each of the samples (0.05 g) were inoculated with a 5 ml bacterial suspension in MHB containing 10^6 CFU/ml. Both of the samples were well shaken and incubated for 24 hours at 35-37°C. After 24 hours, the suspensions were shaken again, and dilute solutions were made, with the remaining bacteria enumerated using the spread plate method. 200 μl of the inoculum were uniformly spread on nutrient agar plate. The plates were incubated at 35-37°C for 24 hours, and the colonies were counted again. All of these tests were carried out in a sterile setting to avoid any contamination. Antimicrobial efficiency was expressed according to AATCC 100 and calculated as:

$$R(\%) = \frac{A-B}{A} \times 100$$

Where, A is the number of bacteria recovered from an inoculated test specimen after 24 hours of incubation with an untreated sample, B is the number of bacteria recovered with NS modified samples under "A" conditions, and R (percent) is the percent reduction ratio that indicates antimicrobial efficiency.

3. Results and Discussion

Bulge behaviour of membranes was carried out at different temperatures as shown in Table-2 to study the hydrophilicity in aqueous system (Fig. 1). From figure 1 it is clear that swelling of the membrane increases slowly up to 60°C and it increases rapidly beyond 60° and up to 70°C and once it passes 70°C the swelling enhanced quickly up to about 80°C and after it almost it is at the level off. This behaviour indicates that there is a sudden change in copolymer structure at about 70°-80°C.

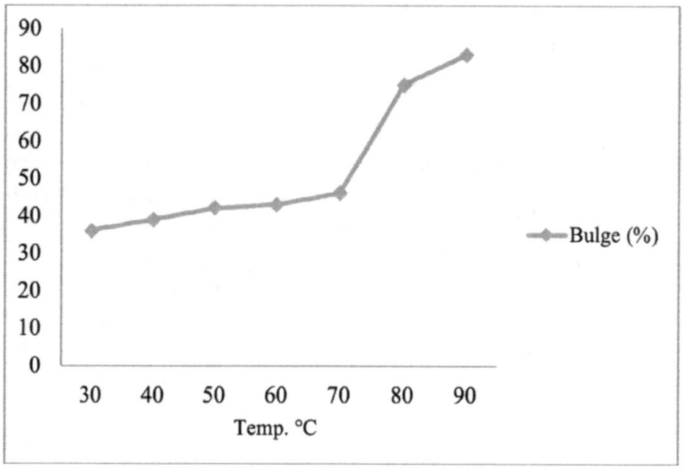

Figure 1: Bulge behaviour of membrane AO 3.5 meq/g at different temperatures

Table 2: Bulge behaviour of the membrane AO 3.5 meq/g at different temperatures

Temperature (°C)	Bulge (%)
30	36
40	39
50	42
60	43
70	46
80	75
90	83

Bulge measurement was carried out at 90°C temperature with respect to time and is given in "Fig. 2".

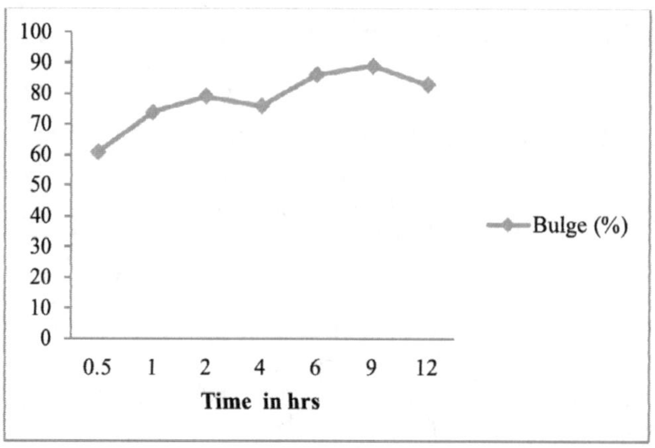

Figure 2: Bulge behaviour of membrane AO 3.5 meq/g at 90°C with respect to time

The bulge linearly increases up to 2 hours and then tends to almost saturate (Table 3). The results indicate that almost maximum bulge is achieved only in 2 hours.

Table 3: Equilibrium swelling behavior of membrane AO 3.5 at 90°C with respect to time

Time (min.)	Swelling (%)
30	59
60	72
120	77
240	75
360	87
540	90
720	81

3.1 UV- Visible spectroscopy

Results show the transformation of (silver) Ag^{+} ions in to NS (Ag) particles. A colourless silver nitrate solution turned in to light yellow solution and the colour shifted from light yellow to light green with the passage of time, which confirms the formation of NS particles in solution. The UV-visible absorption spectra of solutions with different reduction time were recorded in the range of 300-600 nm which confirmed it quantitatively and represented in Figure 3. The plasmon peak for NS with λ_{max} values were obtained at 383, 395 and 404 nm for 20 min, 40 min and 60 min reduction time, respectively.

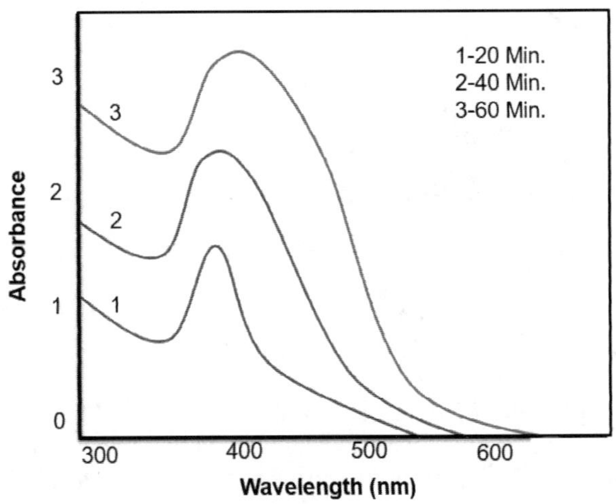

Figure 3: UV-visible absorption spectra of NS solutions.

During study it was found that with the progresses in reduction time, the absorption bands for NS broadened and shifted continuously towards larger wavelength which indicates the formation of larger NS particles.

3.2 Energy Dispersive X-ray Spectroscopy (EDX):

Fig. 4a and 4b shows the Energy Dispersive X-ray Spectroscopic (EDX) Results, which explain the qualitative elemental composition of virgin and Amidoximated samples.

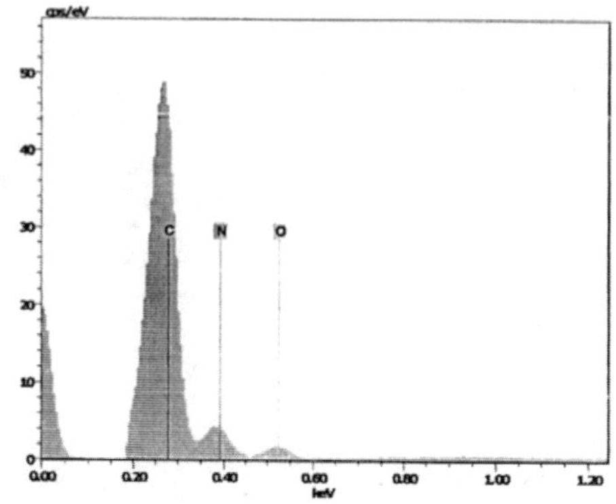

Figure. 4a: EDX of Acrylic Copolymer Virgin And Membrane

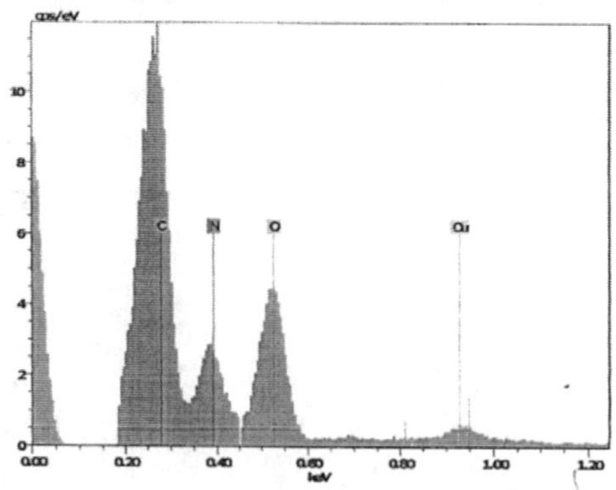

Figure 4b: EDX Of Acrylic Copolymer A.O Content 3.3 Meq/G

3.3 Transmission Electron Microscopy

During thorough observation for all the TEM images, NS particles appeared to be undispersed and spherical. The histograms for particle size distribution were obtained by counting 50 NS particles from TEM images for each sample as shown in Fig. 5. The analysis id represented in Fig. 6. The histogram shows the particle size ranges from 2.7-13.0, 2.4-14.7 and 5.4 -16.1 nm respectively with mean diameter of 6.1, 8.7 and 11.2 nm with standard deviation of 1.8, 2.6 and 2.5 nm for NS_{R20}, NS_{R40} and NS_{R60}, respectively.

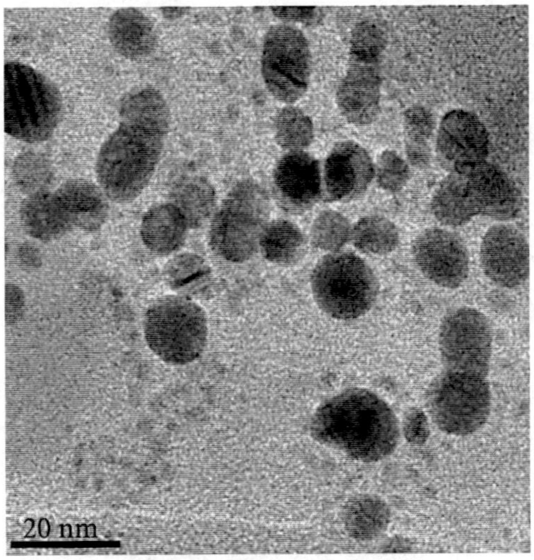

Figure 5: TEM image of NS_{R60}

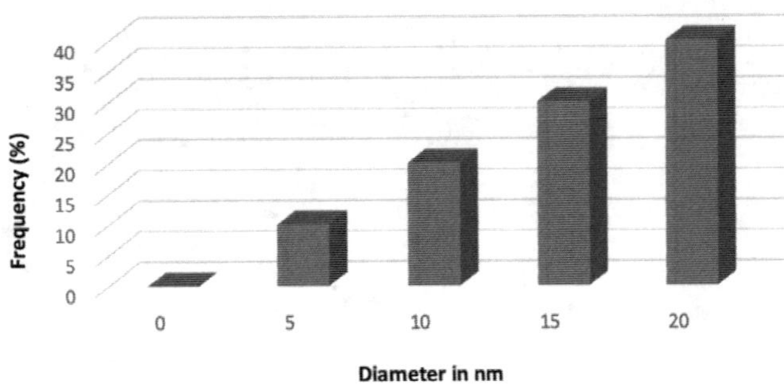

Figure 6: Particle size distribution histogram of NS_{R60} evaluated from TEM images

The variation in particle mean diameter with reduction time is presented in Fig. 7. It is interesting to see that with the increase in reduction time the mean particle diameter increases.

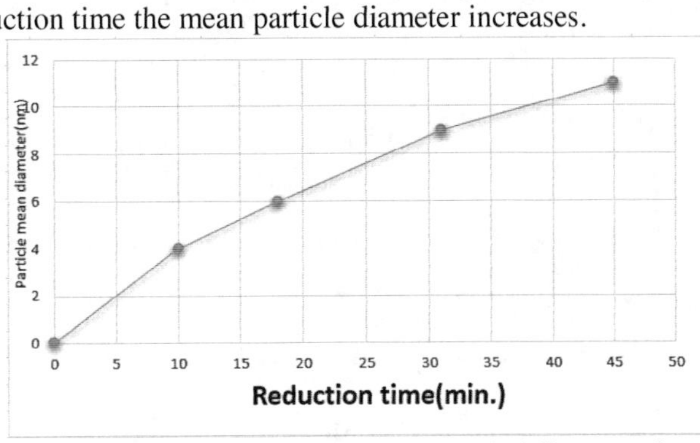

Figure 7: NS particle mean diameter (nm) with different reduction time

TEM and histogram show that as the reduction time increases, the NS particle intensity decreases, whereas particle mean diameter increases. Different studies suggested that nucleation leads to an increase in the number of scattering centers or the number of particles for a given system; therefore, it provides an increase in the scattered intensity. It is consistent with the mechanism of reducing Ag+ ions and the association of Ag° atoms to produce metallic Ag particles [21-22]. NS particle becomes more stable with the increased reduction time as the particles become more homogenous in size allocation.

3.4 Antimicrobial activity:

The antimicrobial activity of virgin as well as various NS containing samples were examined against model bacteria S. *aureus* (gram +ve) and bacteria E. *coli* (gram-ve) by an assessment of the number of viable colonies after being in contact with different samples for a period of 24 hours and are shown in "Fig 8, 9". It can be seen that an average number of viable *S. aureus and E. coli* colonies decreased by 95% in both cases as compared to virgin AO $_{35}$ membrane.

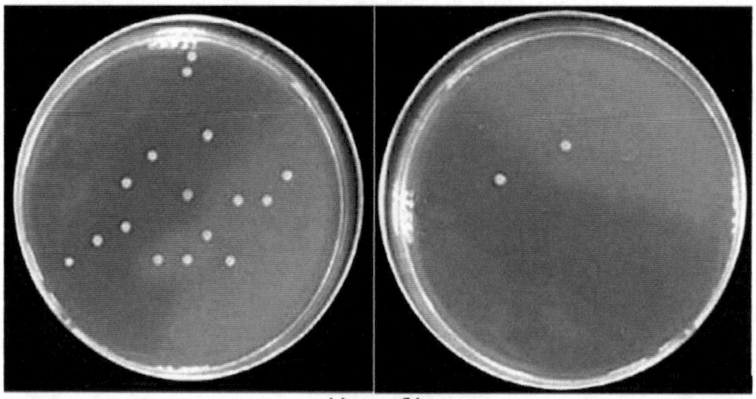

(a) (b)

Figure 8: Antimicrobial activity of membranes against S. aureus (a) Control and (b) AO$_{35}$ - NS$_{R60}$

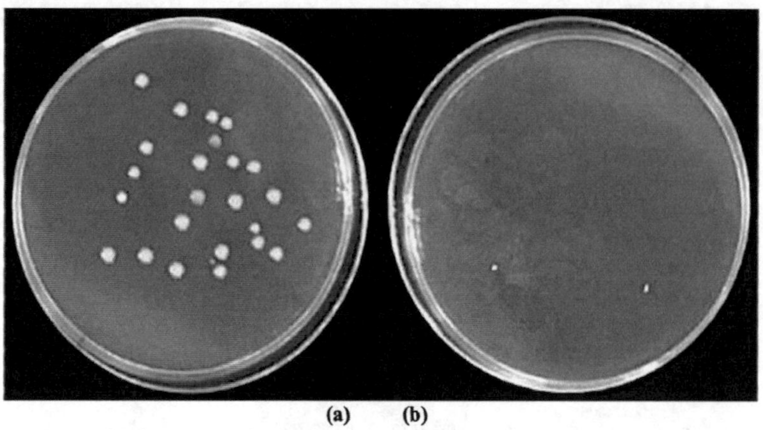

(a) (b)

Figure 9: Antimicrobial activity of membranes against E. coli (a) Control and (b) AO$_{35}$ - NS$_{R60}$)

4. Conclusion

- Copolymer membranes having AO (Amidoxime) content with immobilized with NS particles were found to be very effective in the removal of microorganism impurities from drinking water.
- The EDX studies showed that the presence of NS in the membrane surface as well as in the inner side of polymer matrix increases the efficiency of the membrane further.
- TEM measurement of NS particles was carried out to analyse the morphological characteristics such as size and shape of NS and were successfully studied.
- As the antimicrobial activities of virgin as well as various NS containing samples were examined against gram positive bacteria S. *aureus* and gram-negative bacteria E. *coli* by the estimation of the number of viable colonies that shows the reduction of these bacterial colonies by the application of Amidoximated copolymer membrane.
- The process indicated that the NS penetrates within the swollen AO membranes and stays back. It can be seen that on an average number of viable *S. aureus and E. coli* colonies decreased by 95% for both cases as compared to virgin AO 3.5 membrane.
- The reaction factors, such as temperature, time, and HA concentration, controls the amidoxime content.
- The FTIR analysis revealed that the intensity of virgin and amidoximated acrylic copolymer declines with increasing AO concentration, confirming that as amidoximation occurs, the nitrile content of the copolymer reduces on a regular basis due to the conversion of nitrile groups to amidoxime groups. Also, in comparison to pure PAN, the presence of AA decreases the degradation temperature of virgin copolymer.

Acknowledgements

The authors thanks to Dr. Sumit Kumar, and Dr. Pradeep Kumar (Associate Professor Shri Venkateshwara University Gajraula, UP, India) for their kind help during this work. The authors also help SGT University Gurugram for providing ample spacing and facilities for this study.

References

[1]. N.A.M. Zahri et al., Improved Method for Preparation of Amidoxime Modified Poly (acrylonitrile-co-acrylic acid): Characterizations and Adsorption Case Study, Polymers, 7, 2015, 1205-1220.
[2]. A. Mishra, S. Sharma and B. Gupta, Studies on the amidoximation of polyacrylonitrile films: Influence of synthesis conditions. J. Appl. Polym. Sci., 121, 2011, 2705–2709.
[3]. S.N.A.M. Jamil, R. Daik, I. Ahmad, Redox copolymerization of acrylonitrile with fumaronitrile as a precursor for carbon fibre. J. Polym. Res., 14, 2007, 379–385.
[4]. B.S. Karnik, S.H. Davies, M.J. Baumann and S.J. Masten, The effects of combined ozonation and filtration on disinfection by-product formation. Water Res., 39, 2005, 2839–2850.
[5]. P. Nicolas, A. Mor, Peptides as Weapons against Microorganisms in the Chemical Defense System of Vertebrates. Annu. Rev. Microbiol., 49, 1995, 277-304.

[6]. Shadmon, H., Basu, A., Eckhard, L., Domb, A. and Beyth, N., 2018. Synthesis, characterization and antibacterial activity of heterocyclic quaternary ammonium polymers. *New Journal of Chemistry*, 42(18), pp.15427-15435.
[7]. Kharaghani, D., Kee Jo, Y., Khan, M., Jeong, Y., Cha, H. and Kim, I., 2018. Electrospun antibacterial polyacrylonitrile nanofiber membranes functionalized with silver nanoparticles by a facile wetting method. *European Polymer Journal*, 108, pp.69-75.
[8]. M. Lwata, T. Adachi, M. Tomidokoro, M. Ohta, T. Kobayashi, Hybrid sol–gel membranes of polyacrylonitrile–tetraethoxysilane composites for gas permselectivity, J. Appi. Polym. Sci., 88, 2003, 1752.
[9]. H.A. Tasi, Y.S. Ciou, C.C. Hu, K.R. Lee, D.G Yu, Y.U. Lai, Heat-treatment effect on the morphology and pervaporation performances of asymmetric PAN hollow fiber membranes, J. Membr. Sci., 255, 2005, 33.
[10]. S. Gurunathan., J.H. Park, J.W. Han, J.H. Kim, Comparative assessment of the apoptotic potential of silver nanoparticles synthesized by Bacillus tequilensis and Calocybe indica in MDA-MB-231 human breast cancer cells, Targeting p53 for anticancer therapy. Int. J. Nanomed, 10, 2015, 4203–4222.
[11]. Timofeeva, L. and Kleshcheva, N., 2010. Antimicrobial polymers: mechanism of action, factors of activity, and applications. *Applied Microbiology and Biotechnology*, 89(3), pp.475-492.
[12]. J. R. Morones, J. L. Elechiguerra, A. Camacho, K. Holt, J. kouri, J. T. Ramirez, M. J. Yacaman, Nanotechnology, 16, 2005, 2346.
[13]. S. Rani, G. Praveen and G. Kumar, synthesis and characterization of copolymer film of acrylic acid with acrylonitrile and hydroxylamine treatment of the acrylic copolymer, VSRD International Journal of Technical & Non-Technical Research, Vol. IX (II), 2018, 83-88.
[14]. B Gupta, N. Muzyyan, S. Saxena, N. Grover, S. Alam, Radiat Phys Chem.,77, 2008, 42.
[15]. S. K. Mehta, S. Chaudhary, M. Gradzielski, J. Coll. Interf. Sci., 343. 2010, 447.
[16]. C. E. Hoppe, M. Lazzari, B. I. Pardinas, M. A. L. Quintela, Langmuir, 22, 2006, 7027.
[17]. K. Shimizu, M. Tsuzuki, K. Kato, S. Yokota, K. Okumura, A. Satsuma, J. Phys. Chem., 111, 2007, 950.
[18]. K. Shimizu, K. Sugino, K. Kato, S. Yokota, K. Okumura, A. Satsuma, J. Phys. Chem., 111, 2007, 1683.
[19]. Sharma, B. and Kumar, P., 2019. Method for Preparation of Amidoximated Acrylic Copolymer: Characterizations and other study. *IOSR Journal of Applied Chemistry (IOSR-JAC)*, [online] 12(3), pp.66-72. Available at: <http://iosrjournals.org/iosr-jac/papers/vol12-issue3/Series-1/H1203016672.pdf> [Accessed 16 March 2021].
[20]. Anjaneyulu, B., Dharma Rao, G. and Bajaj, T., 2021. Click chemistry: In vitro evaluation of glycosyl hybrid phosphorylated/thiophosphorylated 1,2,3-triazole derivatives as irreversible acetyl cholinesterase (AChE) inhibitors. *Results in Chemistry*, 3, p.100093.
[21]. Mukherjee, M. and De, S., 2015. Reduction of microbial contamination from drinking water using an iron oxide nanoparticle-impregnated ultrafiltration mixed matrix membrane: preparation, characterization and antimicrobial properties. *Environmental Science: Water Research & Technology*, 1(2), pp.204-217.
[22]. A. Madhavan and B. Sharma, 2019. Latent fingerprint development with biosynthesized Nano rust. *Advances in Science and Engineering Technology International Conferences (ASET)*, United Arab Emirates, pp. 1-4.
[23]. Govind, V., Bharadwaj, S., Sai Ganesh, M. R., Vishnu, J., Shankar, K. V., Shankar, B., & Rajesh, R. (2021). Antiviral properties of copper and its alloys to inactivate covid-19 virus: A review. BioMetals, doi:10.1007/s10534-021-00339-4

Fabrication and Estimation of Rice Husk Green Composite

Mohammad Nasir Khan, Usman Ahmad Ansari

Department of Mechanical Engineering, Aligarh Muslim University, Aligarh, Uttar Pradesh, India

Mohammad Nasir Khan (nasir4k@gmail.com)

Abstract. Composite materials form an important part of technical materials, from everyday products such as door frames to aircraft and space applications. The availability of natural fibres and the ease of production have prompted researchers to test inexpensive local fibres and examine their feasibility for reinforcement purposes and the extent to which these meet the specifications required for strong reinforced polymer composites. A lot of rice husk (RH) is believed to contain high silica fibre content. Depending on the size, the silica content and all impurities in the RH will vary. RHs mainly contain 15-20 wt. percent silica and various organic components are produced by thermal degradation. The low density and diversity of raw materials make silicon carbide easy to produce. In the current work, two different materials are considered for the production of composite materials, such as rice husk and polypropylene. These two available materials are considered as different components, and for each component, various parameters are evaluated to detect the usability of composite materials. In the research work, the proposed composite GUI production method was created using MATLAB, and different results were evaluated for different variables. According to the results, the RH mixture in PP has caused some changes in the various properties of the material, such as tensile strength, tensile modulus, fixture strength, porosity, and for different components, varied outcomes were obtained. According to application requirements, the composition of the material can be further considered. If good tensile strength is required, the minimum RH in the composition is considered. If a higher tensile modulus is required, the high RH composition of PP can be used for further consideration.

Keywords: Composite, rice husk, polypropylene, tensile strength, tensile modulus, fixture strength, porosity.

1. Introduction

The most important advantage of composite material production is that the specific characteristics of the component materials can be used to meet different requirements. Business and materials researchers are considering a more specific elaboration when defining a composition as a substance, which is

composed of content created by the physical combination of pre-existing constituent materials to create a novel material that has integral characteristics contrasted to the monolithic characteristics. This description distinguishes composite materials from different multiphase materials produced by batch processing, in which one or more phases are the product of phase change [1, 2]. Josh et al. [3] investigated the environmental efficiency of natural fiber composites and glass fiber reinforced mixtures during the life cycle, and noted that natural fiber composites have environmental advantages in the specific applications studied. Rana et al. [4] found that considering compatibilizer in jute fiber can improve its mechanical properties. The use of compatibilizers at 60% (weight) of the fiber load can increase the flexural strength to 100%, the flexural strength to 120%, and the impact strength to 175%. Shah and Lakkad [5] tried to match the mechanical properties of jute reinforcement and glass reinforcement, and the results showed that when jute fiber is used as a reinforcing material on the resin matrix, the mechanical properties are widely changed, and the reforming is relatively small instead of considering glass and other materials. Therefore, jute fibers can be considered as a reinforcing material, they have medium strength and modulus.
Kenechi et al. [6] focused on using rice husk as a plastic composite reinforcement material. Synthetic plastic mixtures are becoming popular at a higher rate because the same factors are taken into account in all areas of life and business, which leads to cost inflation, environmental degradation and mainly petroleum by-products, which are non-renewable materials. Mahzan et al. [7] explores the potential characteristics of considering RH waste as the sound absorption content of rice husk-reinforced composite materials. At room temperature (25-30 °C), the RH is first washed and dried. Then dry RH is used as a binder and mixed with polyurethane (PU) foam to make a mixture reinforced with rice husks. Six individual samples were produced based on the percentage of rice husk. The resulting sound absorption coefficient indicates that the mixture produces a large output in the low frequency region. The research results also show that the composite material reinforced by rice husk has a better sound coefficient than other natural materials, which indicates a good marketing opportunity. In another study, Habib et al. [8] investigated the properties of rice husk ash (RHA) taking into account furnaces made of iron cement. The impact of milling is first checked on particle size and surface area, and then XRD interpretation is performed to verify the availability of amorphous silica in the ash. In addition, the effects of average particle size and RHA percentage on concrete workability, fresh density, and water-reducing agent content and compressive strength were also investigated. Although grinding RHA will reduce its average particle size (APS), it is not a key factor in surface conditioning. The solution of the above-mentioned problems provided the motivation for this research; therefore, the current work attempts to consider the rice husk as a possible reinforcement material for the manufacture of plastic composite materials by examining its characteristics, uses, applications and operations.

2. Research Methodology

Rice husk obtained locally. In order to recapture the dirt or dust and other foreign matter on the fibre, it was washed with ordinary water many times and exposed to the sun for three days to absorb all the surface moisture. The RH samples are connected to fibres of different lengths. The scale separation is achieved by sieving in a vibrating sieve system with 4, 3.35, 2.8, 2.36 and 2 mm sieves. Most RH content ranges from +2 to -2.36. For experiments, rice husk fibres of this size were chosen. The fibre length and diameter have an effect on the composite output. RH is a short fibre, and fibre length standardization is assumed to minimize the impact of asset composite variability because of the fibre length.

The rice husk is finely grinded and collected. The crushed rice husk produces various impurities such as dust, tiny particles and fine particles of grain. To get pure rice husk, it needs to be washed. After the water purification, the rice husks are dried directly in the sun for 10 hours. Then it is weighted according to the appropriate percentages (10, 20, 30 weight percentages). Then mix the polypropylene in the tube and stir well for five minutes. RH is slowly poured, stirring to make the components fully mixed. First apply a release agent (Vaseline) on the mould until the mixture is added to the mould to block the composite material from a viscous state into the mould when it is taken out. Finally, after the mixture

IOP Conf. Series: Materials Science and Engineering 1224 (2022) 012021 doi:10.1088/1757-899X/1224/1/012021

was inserted into the mould, it was left for 1 hour in the oven at 200 °C and then it is slowly left to cool down (Figure 1).

Figure 1 Collection of the RH

2.1. *Design of Experiment*

Taguchi techniques are mathematical techniques designed to retouch the efficiency of the produced products, and most presently also considered to engineering, biotechnology, marketing and advertisement. The Taguchi approach is used whenever interest parameter settings are appropriate, not just for the manufacturing processes. Table 1 shows the levels of process parameters. Experimental layout using L9 (32) orthogonal array is shown in Table 2. In addition, Sequence of experiments with respect to % of rice husk and Polypropylene is shown in Table 3. Step by step processing of the sample preparation is shown in Figure 2.

Table 1 Levels of process parameters

Process Parameters	Level 1	Level 2	Level 3
% of Rice Husk	10	20	30
Sieve Size	250	355	500

Table 2 Experimental layout using L9 (32) orthogonal array

L9 (3^2) Experimental Trail	% RH	Sieve Size
1	1	1
2	1	2
3	1	3
4	2	1
5	2	2
6	2	3
7	3	1
8	3	2
9	3	3

2.2. *Sample Preparation*

Composite in pieces of 300 mm X 200 mm X 5 mm were make-believe as per the ASTM standards (D790) for mechanical tests.

Die size = 300 x 200 x 5 mm = 300000 mm^3

Volume of Die = 300cm^3

Calculation of weights for 10 % Rice Husk and 90 % Polypropylene

90% of Polypropylene = 270000 mm^3

Density of PP = 0.946 g/cc

Weight of PP = 270x0.946

Weight of PP =255.42 grams
10 % of Rice Husk= 30 cm^3

Table 3 Sequence of experiments with respect to % of rice husk and Polypropylene

Experiment Number	% of Rice Husk	% of Polypropylene	Size of RH in microns	wt. of RH (grams)	Wt. of PP (grams)
1	10	90	250	10.2	255.42
2	10	90	355	10.2	255.42
3	10	90	500	10.2	255.42
4	20	80	250	20.4	227.02
5	20	80	355	20.4	227.02
6	20	80	500	20.4	227.02
7	30	70	250	30.6	198.66
8	30	70	355	30.6	198.66
9	30	70	500	30.6	198.66

Figure 2 Step by step processing of the sample preparation

Volume of RH = 30 cm^3
Density of RH = 0.34 g/cc
Weight of RH=10.2 Grams
Calculation of weights for 20% Rice Husk and 80% Polypropylene
80 % of Polypropylene = 240000 mm^3
Density of PP = 0.946 g/cc
Weight of PP = 240x0.946
Weight of PP =227.02 grams
20 % of Rice Husk= 30 cm^3
Volume of RH = 60 cm^3
Density of RH = 0.34 g/cc
Weight of RH =20.4 grams.

3. Results and Discussion

3.1. GUI Interface for proposed methodology

The proposed methodology is better validated using the simulation of the same for which MATLAB 15a is being considered and the GUI based snaps are shown as under (Table 4): In the show results the research methodology is better evaluated using the various parameters like Tensile strength, Flexure Strength, Tensile modulus, Porosity and also the variables like weights of PP and RH are considered as dynamic parameters which changes with the change in the composition of the PP and RH.

Table 4 Input parameters

Size of Rice Husk (µm)	% of PP	% RH	Curing Temp. (C)	Compaction Pressure (ton)
250	90	10	100	10
250	80	20	100	10
300	70	30	110	12
300	60	40	110	12
300	50	50	110	12

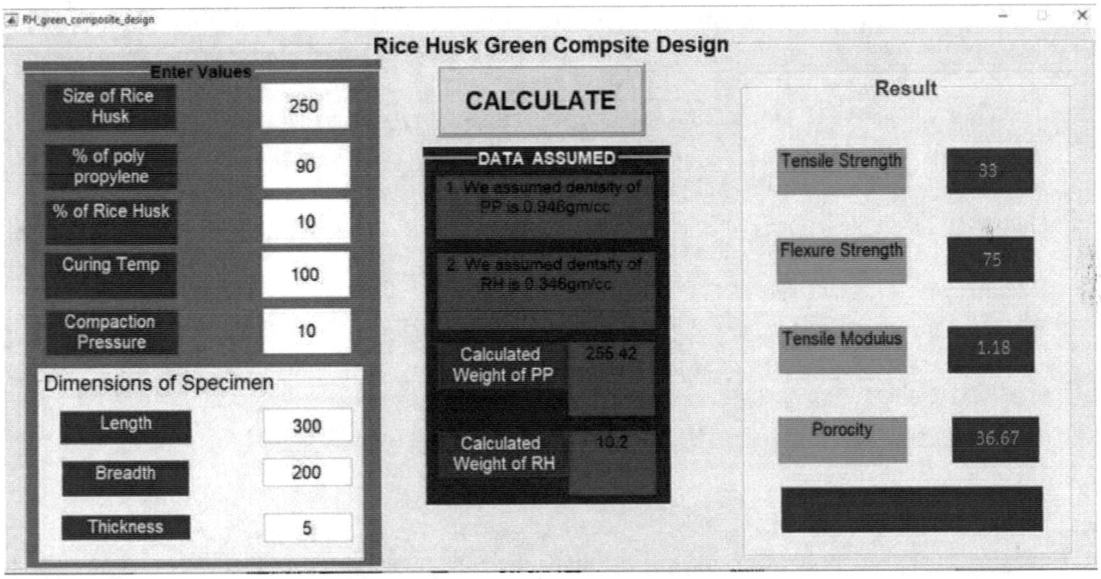

Figure 3 GUI for proposed methodology

In Figure 3, inputs are entered for the various parameters considered, and the GUI created on this basis estimates the results in various forms. Here the percentage of rice husk and polypropylene is 10/90, and the result showed tensile strength of 33 MPa, a flexural strength of 75 MPa, a tensile modulus of 1.18 GPa and a porosity of 36.6667%. In addition, when the percentage of rice husk and polypropylene is 20/80, the result is described as a tensile strength of 31 MPa, a flexural strength of 67 MPa, a tensile modulus of 1.30 GPa, and a porosity of 40%. Further, when the percentage of rice husk and polypropylene is 30/70, and the result is described as a tensile strength of 29 MPa, a flexural strength of 59 MPa, a tensile modulus of 1.58 GPa, and a porosity of 43.333. These values are further calculated for different ratios of RH and PP, and these values are displayed with the help of graphics generated using the GUI. According to the estimated results, it is obvious that as the RH content in the composite material increases, the tensile strength, tensile modulus, flexural strength and porosity increase. As the RH content in the composite material increases, the tensile strength decreases, as shown in the figure below.

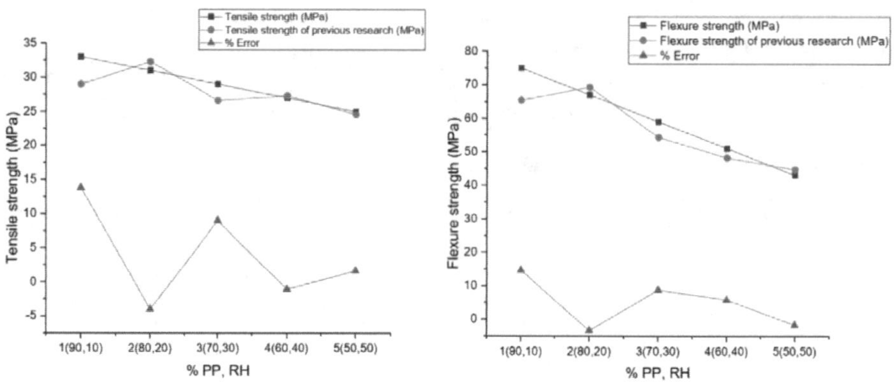

Figure 4 Comparison graph for Tensile and Flexure strength for all compositions created.

The graph shown above (Figure 4) is the depiction of the tensile strength for the different compositions considered for the polypropylene and rice husk. The PP and RH are considered in three different compositions as 90/10, 80/20, 70/30, 60/40 and 50/50 where the 90, 80, 70, 60 and 50 shows the constituent of PP and rest is of the RH. In addition, figure 5 showed that highest porosity was attained by 50/50 (PP/RH) composition followed by 60/40 composition.

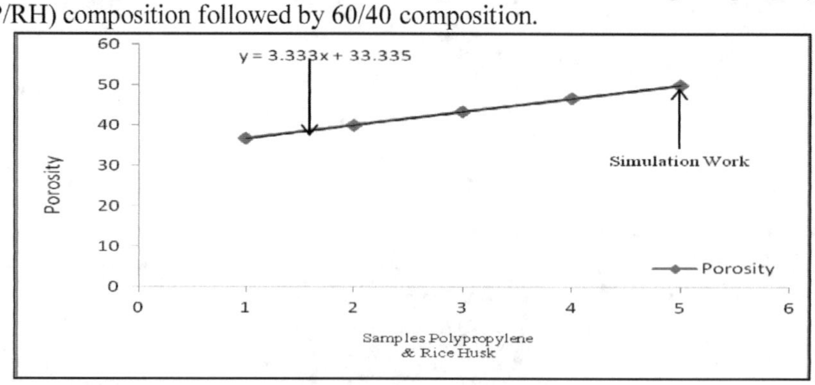

Figure 5 Comparison graph for Porosity for all compositions created.

In the current work, two different materials are considered for the production of composite materials, such as rice husk and polypropylene. These two available materials are considered as different composition, and for each composition, various parameters are evaluated to detect usability of the composite. In the research work, the proposed composite GUI production method was created using MATLAB, and different results were evaluated for different variables. According to the results, it is shown that mixing RH in PP will cause some changes in the various properties of the material, such as tensile strength, tensile modulus, flexural strength, porosity, and for different composition. According to the application requirements, the composition of the material can be further considered. If good tensile strength is required, the minimum RH of the composition should be considered. In the case of higher tensile modulus requirements, the high RH composition of PP can be used for further consideration. The composition discussed in the work is made of PP and RH with different compositions, which have properties such as tensile strength, tensile modulus, flexural strength, and porosity.

Table 5 Summarization table for output variables

% of PP	% of RH	Wt. PP (g)	Wt. RH (g)	Tensile Strength (MPa)	Flexure Strength (MPa)	Tensile Modulus (GPa)	Porosity (%)	Result	Ref.
90	10	255.42	10.2	33 (29)	75 (65.4)	1.18	36.67	Tensile Strength increases by 8.7%., F.S increases by12.80% (Best Sample)	[9, 10]
80	20	227.04	20.4	31 (32.3) @(85,15)	67 (69.3)	1.30	40	Tensile Strength decreases by 4.19%. F.S decreases by 3.43%	[9, 11]
70	30	198.66	30.6	29 (26.6)	59 (54.3)	1.58	43.33	Tensile Strength increases by 8.2%, F.S Increases by 7.9%	[9, 10]
60	40	170.28	40.8	27 (27.3) @(65,35)	51 (48.2)	1.84	46.67	Tensile Strength decreases by 1.17%. F.S increases by 5.4%	[9, 11]
50	50	141.9	51	25 (24.6)	43 (44.7)	2.10	50	Tensile Strength decreases by 1.5%. F.S decreases by 4.0%	[9, 10]
90	10	255.42	10.2	33 (29)	75 (65.4)	1.18	36.67	Tensile Strength increases by 8.7%. F.S Increases by12.80% (Best Sample)	[9, 10]

With the increase of the RH composition in the composite material, the tensile strength shows a moderate degradation change. Therefore, the tensile strength is the reason for the hardness of the composite material. Therefore, the composite material that requires hardness should maintain a lower RH content. Similarly, the flexural strength increases with the increase of the RH content in the composition, which is responsible for the deformation resistance of the composite material and has its own application fields (Table 5). The tensile modulus and porosity increase as the RH content in the composition that contributes to elasticity and absorption increases. The composite material can be well used in applications that require highly elastic materials and have good absorption capacity.

4. Conclusion

In the current work, two different materials are considered for the production of the composite material such as rice husk and polypropylene, both available materials are considered in different composition and for each of the compositions several parameters were evaluated to detect the usability of the compound. The best samples were obtained with samples of 90% PP and 10% rice husk. The tensile strength was found to be increased by 8.7% compared to the previous study, and the flexural strength increased by 12.80%. In the 90% PP and 10% rice husk samples, the porosity and tensile modulus are the smallest.

5. Future Scope

In the current work, the main consideration is to evaluate and determine various parameters. These parameters define the strength of materials considering PP and RH in various compositions.

- The work done can be further considered for real-time application in industry.
- Other tests can also be performed by XRD and SEM.
- Other thermal properties can also be tested.
- On the basis of the application the material can be selected for further research.
- The resulting composite material can be further applied to practical applications and can be verified through various tests to identify the quality of the composite material.

Conflict of interest

None.

References

[1]. Surappa, M. K., "Aluminium matrix composites: challenges and opportunities, Sadhana India", 2003, 28(1 & 2): 319-334.

[2] Zuhailawati, H., Samayamutthirian, P., Mohd Haizu, C. H., Fabrication of low cost of aluminium matrix composite reinforced with silica sand, Journal of physical science, 2007, 18(1): 47-55.

[3] S.V. Joshi, L.T. Drzal, A.K. Mohanty, S.Arora., "Are natural fiber composites environmentally superior to glass fiber reinforced composites?", Composites Part A: Applied Science and Manufacturing, 2004,35: 371-376.

[4] A. K. Rana, A. Mandal, B.C.Mitra, R. Jacobson, R. Rowell, A. N. Banerjee., "Short Jute Fiber Reinforced Polypropylene Composites: Effect of Compatibilizer" Journal of Applied Polymer Science, 1998, 69: 329-338.

[5] A.N. Shah, S.C. Lakkad, "Mechanical properties of jute-reinforced plastics", Fibre Science and Technology, 1982, 16: 309-312

[6] Nwosu-Obieogu Kenechi, Chiemenem Linus, Adekunle Kayode. "Utilization of rice husk as reinforcement in plastic composites fabrication- a review", American journal of materials synthesis and processing. 2016, 1(3): 32-36.

[7] S. Mahzan, A.M. Ahmad Zaidi, M. I. Ghazali M.N. Yahya M. Ismail, Investigation on Sound Absorption of Rice-Husk Reinforced Composite, Proceedings of MUCEET 2009, Malaysian Technical Universities Conference on Engineering and Technology, 19-22.

[8] Ghassan Abood Habeeb, Hilmi Bin Mahmud, "Study on Properties of Rice Husk Ash and Its Use as Cement Replacement Material", Malaysia Materials Research. 2010; 13(2): 185-190

[9] Yogita Sharma, Mohpreet Singh, Manoj Mittal, "Study of Mechanical properties of Rice Husk and PP based composites using Injection Moulding Process", IOSR Journal of Mechanical and Civil Engineering (IOSR-JMCE), 2017, 14(4): 19-23.

[10] Wang, Y. Wu, H. Zhang, C. Ren, L. Yu, H. Galland, M.A. Ichchou, M. "Acoustic characteristics parameters of polyurethane/rice husk composites", *Polym. Compos*. 2018, *40*: 2653-2661.

[11] S. D. Prasad, R. A. Krishna, "Production and mechanical Properties of A356.2/RHA composites", Int. Journal Adv. Sci. Technology, 2011, 33: 51-58.s

Evaluation of Antibiotic Resistant and Metal Tolerances Capability of Petroleum (hydrocarbon) Degrading Bacteria

Umema Mohsin[1], Madan Lowry[2], Pooja Devi[1], Himanshu Dixit[2], Shaile Thakur[1]

[1]Department of Forensic Science, Vivekananda Global University, Jaipur 302012 India, [2]Department of Zoology, University of Rajasthan, Jaipur 302004 India, [1]Faculty of Basic & Applied Sciences, Vivekananda Global University, Jaipur 302012 India

shelly.exams@gmail.com

Abstract. Diesel oil is considered one of the important products of crude oil which constitute a vital source of pollution in the environment. Biodegradation of complex hydrocarbon requires cooperation of more than a single species. This is true in case of pollutants made up of crude oil or petroleum. Soil contains varieties of microorganism that can be established in any natural environment. The bacterial isolates were characterized biochemically by indole test, methyl red test, citrate utilization Test, voges proskauer test, starch hydrolysis, catalase production test, nitrate reduction test, antagonism test, oligodynamic susceptibility and antibiotic sensitivity tests.

Keywords: Hydrocarbon, Biodegradation, susceptibility, oligodynamic, Antagonism

1. Introduction

Petroleum products are worldwide use as a major automobile fuel of them diesel oil is considered one of the crucial products of crude oil which constitute a vital source of pollution in the environment. Diesel oil can go into the environment via wrecks of oil tankers carrying diesel oil, cleaning of diesel tanks by merchants, war ships carrying diesel oil and motor mechanics. Diesel oil spills on agricultural land inhibits plant growth. Major factors responsible for the reduced plant growth in diesel oil contaminated soils range from direct toxic effect on plants. Reduced germination (Udo and Fayemi, 1975) due to inappropriate soil condition and insufficient aeration of the soil because diesel oil displaces air from the space among the soil particles. Degradation progression facilitated by microorganisms helps in removal of spilled oil from the environment by various physical and chemical methods (Ijah and Okang, 1993). Microorganisms have enzyme systems to degrade and utilize diesel oil as a source of carbon and energy (Ezeji et al. 2005). Monooxygenases play crucial role in microbial degradation of oil, chlorinated hydrocarbons and other compounds. Biochemical tests are used to differentiate such bacteria. The characteristics showed the physiology of the test organisms. This study was designed to isolate, characterize and identify diesel oil-degrading microorganisms.

2. Material and Methods

Biochemical test such as indole test, methyl red test, citrate utilization Test, voges proskauer test, starch hydrolysis, catalase production test, nitrate reduction test, antagonism test, oligodynamic susceptibilty and antibiotic sensitivity tests were carried out to find the enzymatic activity of isolated organism. Protocol used was followed as given by Dubey and Maheshwari, 2004.

2.1. *Antibiotic sensitivity*

Isolated halophilic bacteria were spread over nutrient agar plates. Wells were made in the nutrient agar plates for antibiotic sensitivity testing. Different concentrations of streptomycin (HiMedia, Mumbai, India) and tetracycline (HiMedia, Mumbai, India) (5 mg/ml, 10 mg/ml, 15 mg/ml, 20 mg/ml, and 30 mg/ml) were prepared and the wells were flooded with these antibiotics. Later, plates were incubated at 37oC for 24-48 hours. This method is designed to determine the smallest amount of the antibiotics needed to inhibit the growth of a microorganism. The concentration of antibody required for effective inhibition of microbial growth is called minimal inhibitory concentration (MIC), which was determined by measuring the diameter of growth inhibition (clear) zone surrounding the antibiotics containing wells.

2.2. *Oligodynamic Susceptibility*

Isolated bacteria were spread over the nutrient agar plates (HiMedia, Mumbai, India), various metals (Fe, Cu, Zn, Al, Hg) were used to evaluate its effect on microbial growth. All plates were incubated at 37oC for 24-48 hours. Effects of metals on the growth of microorganisms can be both stimulatory and inhibitory. Synergistic or antagonistic effects were observed on the growth of organisms in the plates around the metal ion.

2.3. *Antagonism*

In nature microbes grow in various associations ranging from antibiosis, symbiosis, symbions, and synergism to commensalisms. This association may exist within or between different groups of microbes. In procedure, the bottoms of nutrient agar plates (HiMedia, Mumbai, India) were divided into sectors. Loop full cultures of all organisms were streaked perpendicular to each other without touching. Plates were incubated at 37□C for 24 hrs.

3. Results and Discussion

3.1. *Biochemical analysis*

Isolates responded variably through the various parameters of biochemical analysis. Table1 shows the results of test such as; gelatin hydrolysis, urea hydrolysis, citric acid utilization, starch hydrolysis, MR, VP, nitrate reduction, indole, casein hydrolysis, phenyl deaminase test and carbohydrate fermentation. For tests such as urea hydrolysis, citric acid utilization, starch hydrolysis, MR, VP, indole production, casein hydrolysis and phenyl deaminase test, all isolates (i.e. I1 (Pseudomonas rhodesiae), I2 (Caulobacter henricii), I3 (Cupriavidus gilardii) and I4 (Phenylobacterium koreense) responded equally (Table 1)

Table 1: Biochemical identification of isolates

Biochemical Tests	Bacterial isolates			
	T1	T2	T3	T4

Tests	I1 *(Pseudomonas rhodesiae)*	I2*(Caulobacter henricii)*	I3*(Cupriavidus gilardii)*	I4*(Phenylobacterium koreense)*
Gelatin Hydrolysis	Positive	Positive	Negative	Positive
Urea Hydrolysis	Negative	Negative	Negative	Negative
Citric Acid Utilization	Positive	Positive	Positive	Positive
Starch Hydrolysis	Positive	Positive	Positive	Positive
MR	Negative	Negative	Negative	Negative
VP	Positive	Positive	Positive	Positive
Nitrate Reduction	Negative	Negative	Positive	Negative
Indole Production	Negative	Negative	Negative	Negative
Casein Hydrolysis	Positive	Positive	Positive	Positive
Phenyl Deaminase Test	Positive	Positive	Positive	Positive
Carbohydrate Fermentation	Negative	Negative	Positive	Negative

3.2. Oligodynamic Susceptibility

Table 1.2 shows susceptibility for the isolates to various metals such as; Cu, Zn, Al and Fe. Result showed that besides Zn, all other metals such as Cu, Al and Fe were not susceptible to isolates (I1, I2, I3 and I4). Isolates such as; I1, I3 and I4 were found susceptible to Zn. The only isolate I2 was not susceptible to any of the metal used in this study (Table 2).

Table 2: Oligodynamic susceptibility test of isolates

	Different Metals	**T1**	**T2**	**T3**	**T4**
Oligodynamic susceptibility Test	Cu	NS	NS	NS	NS
	Zn	S	NS	S	S
	Al	NS	NS	NS	NS
	Fe	NS	NS	NS	NS

Note: NS = Not Susceptible, S = Susceptible

3.3. Antibiotic sensitivity

All four isolates were examined for antibiotic sensitivity against streptomycin, ampicillin and tetracycline. For streptomycin isolates i.e. I1 and I4 were found susceptible except for I2 and I3. Antibiotic effect of ampicillin was mostly resistant in isolates I2, I3 and I4, however, for I1 it showed susceptibility. Furthermore, for tetracycline I2 and I4 showed susceptibility whereas, I1 and I3 noted with limited susceptibility (Figure 1, Table 3).

Table 3: Antibiotic susceptibility test of isolates

Antibiotic sensitivity and	Concentration of streptomycin					
		5ug/ml	10ug/ml	20ug/ml	30ug/ml	40ug/ml

IOP Conf. Series: Materials Science and Engineering 1224 (2022) 012022 doi:10.1088/1757-899X/1224/1/012022

evaluation of minimum inhibitory concentration	T1	0.4 cm	0.6 cm	1.0 cm	1.2 cm	Susceptible
	T2	0.7 cm	0.7 cm	1.4 cm	1.4 cm	1.9 cm
	T3	0.7 cm	0.9 cm	1.4 cm	1.8 cm	1.8 cm
	T4	0.5 cm	1.0 cm	1.6 cm	Susceptible	Susceptible
	Concentration of ampicillin					
		5ug/ml	10ug/ml	20ug/ml	30ug/ml	40ug/ml
	T1	0.4 cm	0.9 cm	1.2 cm	1.5 cm	Susceptible
	T2	0.4 cm	resistant	resistant	resistant	resistant
	T3	0.4 cm	resistant	resistant	resistant	resistant
	T4	0.3 cm	resistant	resistant	resistant	Resistant
	Concentration of tetracycline					
		5ug/ml	10ug/ml	20ug/ml	30ug/ml	40ug/ml
	T1	0.1 cm	1.5 cm	1 cm	1.2 cm	0.7 cm
	T2	0.3 cm	0.7 cm	1.2 cm	susceptible	susceptible
	T3	0.9 cm	0.8 cm	1.3 cm	0.5 cm	1.1 cm
	T4	Susceptible	Susceptible	Susceptible	Susceptible	Susceptible

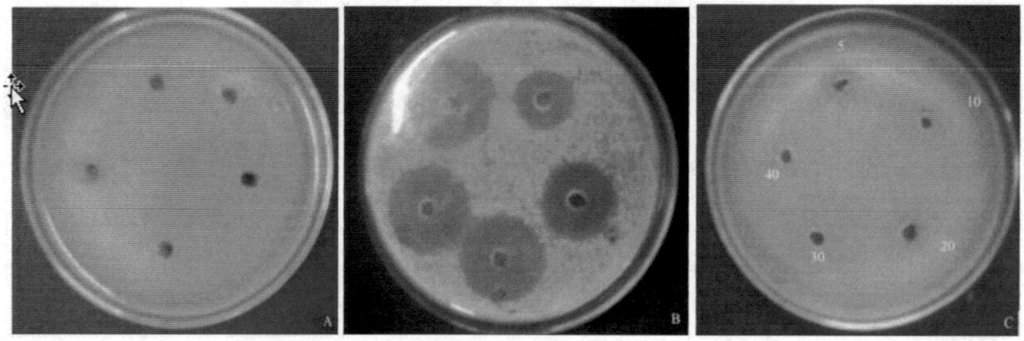

Figure 1: Antibiotic susceptibity (A) T-4 against tetracycline (B) T-2 agsinst streptomycine (C) T- 3 aginst ampicillin

3.4. Antagonism

All isolates i.e. I1, I2, I3 and I4 were found negative for examination of antagonism. No associations such as; antibiosis, symbiosis, symbions, and synergism to commensalisms were observed (Figure 2).

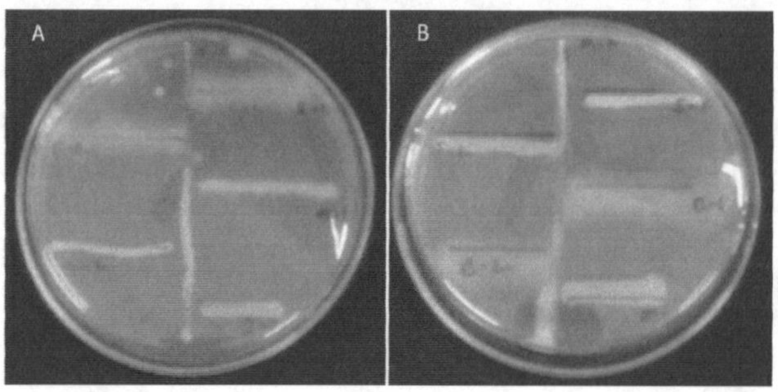

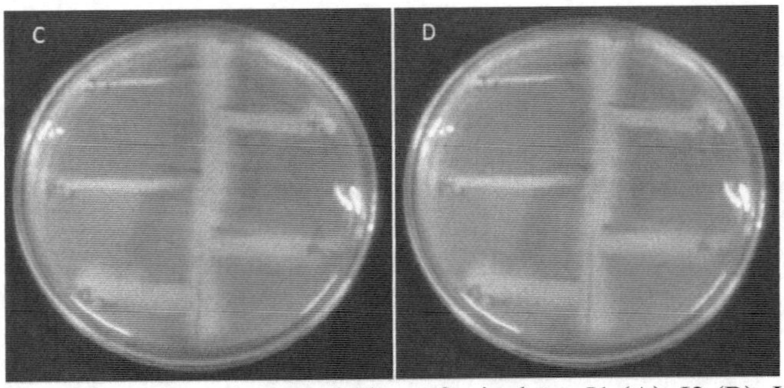

Figure 2: Antagonism test on nutrient agar plates for isolates I1 (A), I2 (B), I3 (C) and I4 (D)

4. Discussion

In this study, biochemical analysis showed positive results for gelatin hydrolysis, citric acid utilization, starch hydrolysis, casein hydrolysis and phenyl deaminase test. One of the first studies on Caulobacter sp. reports that when isolated for the first time organism failed to grow on gelatin, but once adapted to culture media, it produced surface growth and filiform growth along the line of puncture in gelatin stabs (Bowers et al., 1954). Same study further conducted on Caulobacter sp explains about colony morphology which mentions that surface agar colonies are up to 5 mm in diameter after 96 hrs of incubation, round, smooth, slightly raised, glistening, finely granular in the center, and grayish white with the center and the reverse side becoming brownish-yellow. Subsurface colonies are dense, brownish-yellow, and lenticular, up to 0.5 mm in diameter and one mm in length (Bowers et al., 1954). The above observation on colony morphology went partially in accordance with our study which noted light yellowish colony in color, irregular form, elevation umbonate, margins are irregular and undulate. For other biochemical markers a parallel study was missing or not found, therefore, present study was conducted exclusively for this investigation.

Antibiotic susceptibility is another important marker in distinguishing between species and strains. The first report of antibiotic sensitivity was published in 1980 by String fellow et al., who studied the protein synthesis elongation factors Tu and Tu-Ts from Caulobacter crescentus in relation to its sensitivity to kirromycin and activity in Qβ replicase (Stringfellow et al., 1980).

Through biochemistry we found that Phenylobacterium koreense were positive for citric acid utilization, starch hydrolysis, nitrate reduction, casein hydrolysis and carbohydrate fermentation. Aslam et al., (2005), also reported few of the above parameters which went in agreement with our result such as, positive result for nitrate reduction and negative for gelatin hydrolysis. It was however also important to notice that for VP our result differed as Aslam et al., reported negative for VP and our result turned out to be positive (Aslam et al., 2005). We also observed positive result for carbohydrate fermentation which confirms presence of acetoin in Phenylobacterium koreense. Various fermentive bacteria use Acetoin which utilizes natural four carbon molecule to store external energy. Acetoin is produced by decarboxilation of alpha-acetolactate, which is a common precursor in metabolism of branched amino acids.

We observed quite interesting results in antibiotic sensitivity tests, which indicated that Phenylobacterium koreense is ampicillin resistant and partial susceptibility to streptomycin and tetracycline. Nevertheless, it was not completely susceptible to any of the antibiotics used in this investigation. There were no parallel results available in any of the Phenylobacterium sp., to confirm our results. As limited susceptibility observed by streptomycin and tetracycline, it is possible that other antibiotic might have greater susceptibility. However it can be concluded that among all four isolate investigated in this study Phenylobacterium koreense tends to differentiate itself from other isolate both biochemically and metabolically.

We observed forming smooth colonies on blood agar, positive for oxidase and catalase, non-users of glucose, incapable of reducing nitrate or nitrite, and susceptible to colistin (Vaneechouttee et al., 2004).

Our results for biochemistry showed accordance to the above reported results, we also observed positive test for gelatin hydrolysis, citric acid utilization, starch hydrolysis, VP, casein hydrolysis and phenyl deaminase test, similarly, negative for nitrate reduction and carbohydrate fermentation.

Through antibiotic susceptibility test it was observed that Cupriavidus gilardii was susceptible to tetracycline and streptomycin, however, resistant to ampicillin. There were not much studied about the particular strain, nevertheless, one reported case of infection by Cupriavidus gilardii was investigated for antimicrobial sensitivity. Limited reports are available on antibiotic sensitivity of Cupriavidus sp., require more work to establish its antibiotic susceptibility/resistant. Oligodynamic action is the tendency of small amounts of heavy metals to exert a lethal effect on bacterial cells. Some metals have the ability to kill bacteria by entering into cells and act as a toxicant in characteristic way. Metals such as silver and copper have been used as a disinfectant for many generations. Oligodynamic metals kills almost all non-spore forming bacteria and viruses (Thurman & Gerba, 1988). According to a study, 1000 time lower concentration than a toxic level, silver can still serve as a disinfectant (Warrington, 1996). Our result for oligodynemic susceptibility test showed that all isolates were not susceptible to copper, aluminum, and iron, however, susceptible to zinc. Zinc is used as a weak antiseptic, and in paints as a white pigment and mold-growth inhibitor (Leikin and Paloucek, 2008). Previous studies reported that zinc chloride is a common ingredient in mouthwashes and deodorants, and zinc pyrithione is an ingredient in antidandruff shampoos. Zinc coated fittings on house hold items obstruct growth of algae. There are many shingles available that are made of Copper and Zinc (Tortora et al., 2010). Likewise, Zinc iodide and zinc sulfate are used as antiseptic (Rohe and Wolf, 2007). The mechanism of action of zinc and copper is still unknown, however, some studies suggest that metal ion denature vital proteins by tagging to reactive groups resulting in precipitation and inactivation of proteins. Therefore, high affinity of cellular protein for the respective metal ions causes' death of the cells, imbalance of ions are the reason of cellular deaths (Benson, 2002). Since these isolates were exclusively studied for oligodynamic tests, no parallel studies were observed to confirm our results. Why copper, aluminum and iron were not susceptible to isolates was not understood, therefore more studies are required to ascertain any conclusion.

Microbial antagonism is a method in which two different strain compete for survival with each other. It is a self-defense mechanism of host culture to protect against foreign strains. In this phenomenon the foreign strain could not thrive in the pressure of existing host and may extinct completely. For example, mouth bacteria which have this in built resistance against other bacteria coming in from inhalation, mouth to mouth contact or food consumption. During infection the act of antagonism can be used to eliminate harmful infectious bacteria (Ledermann, 2013). Ledermann in his study reported three types of antagonism i.e. bacterial killing by other bacteria, virus against bacteria and blockade of cellular receptors by bacterial filtrates. In the first type, the piocianase from Pseudomonas aeruginosa and the activity of Bacillus subtilis over Mycobacterium tuberculosis were the better examples; in the second, the French D'Herelle was a pioneer using bacteriophages against Shigella dysenteriae; and another French, Besredka, headed the third line with his "antivirus thérapie" on Staphylococcus aureus (Ledermann, 2013). Our result for antagonism between these four isolates i.e. Pseudomonas rhodesiae, Caulobacter henricii, Phenylobacterium koreense and Cupriavidus gilardii were found partially endosymbiotic. Growth of any of the bacteria versus each other seem to unaltered in nutrient media and grew uninterrupted. This was obvious because most of the investigated isolates were surviving together at the site of examination. The theory of endosymbiosis was first articulated in 1905 and 1910 by the Russian botanist Konstantin Mereschkowski, and later advanced by microbiologist Lynn Margulis in a 1967 (Mereschkowsky, 1910; Mereschkowski, 1905; William et al., 2012; Sagan, 1967). According to the theory, endosymbiosis is a theory of selection relying on the phenomenon of symbiosis No parallel study was found to compare these results as these four isolates were exclusively examined for antagonism in this study.

5. Conclusion

If the bacteria are sensitive to the antibiotic, a clear ring, or zone of inhibition, is seen around the wafer indicating poor growth. Antimicrobial susceptibility was tested using Strokes method, antibiotic diffusion tests, Agar and broth dilution methods and minimum inhibitory concentration methods. Through antibiotic susceptibility test is remarkable to notice that Pseudomonas rhodesiae was susceptible for all three antibiotics used in this investigation i.e. Streptomycin, ampicillin and tetracycline. Although this result was exclusive for Pseudomonas rhodesiae it was compared with other Pseudomonas sp. to understand its variation for antibiotic susceptibility whereas, our result for Pseudomonas rhodesiae showed susceptibility to ampicillin at 40 μg/ml.
In microbiology, the fast and simple biochemical tests can identify bacteria. Such as metabolic or enzymatic characteristics, for example the ability to ferment carbohydrate and the pattern in which it is metabolized can dictate the genus and species of bacteria. The other technique is to analyze acids, alcohols and gases when a bacteria is grown on to a selective media. Through biochemical analysis for the first time we came to know that Pseudomonas rhodesiae is positive for starch hydrolysis, gelatin hydrolysis, citric acid utilization, casein hydrolysis and, phenyl deaminase test. It was however, not uniquely present in Pseudomonas rhodesiae but was also found in other Pseudomonas sp. We observed quite interesting results in antibiotic sensitivity tests, which indicated that Phenylobacterium koreense is ampicillin resistant and partial susceptibility to streptomycin and tetracycline it was not completely susceptible to any of the antibiotics used in this investigation. There were no parallel results available in any of the Phenylobacterium sp., to confirm our results. As limited susceptibility observed by streptomycin and tetracycline, it is possible that other antibiotic might have greater susceptibility. However, it can be concluded that among all four isolate investigated in this study Phenylobacterium koreense tends to differentiate itself from other isolate both biochemically and metabolically. Through antibiotic susceptibility test it was observed that Cupriavidus gilardii was susceptible to tetracycline and streptomycin, however, resistant to ampicillin. There were not much studied about the particular strain, nevertheless, one reported case of infection by Cupriavidus gilardii was investigated for antimicrobial sensitivity.

Conflict of interest
None.

References

REFERENCES
1. Aslam, Z., Im, W. T., Ten, L. N., & Lee, S. T. (2005). Phenylobacterium koreense sp. nov., isolated from South Korea. International Journal of Systematic and Evolutionary microbiology, 55(5), 2001-2005
2. Bowers, L. E., Weaver, R. H., Grula, E. A., & Edwards, O. F. (1954). Studies on a strain of Caulobacter from water I.: isolation and identification as Caulobacter vibrioides Henrici and Johnson with Emended Description. Journal of bacteriology, 68(2), 194
3. Dubey RC, Maheshwari DK (2004): Practical Microbiology, S. Chand and Company Ltd., New Delhi
4. Ezeji, E. U., Anyanwu, B. N., Ibekwe, V. I., & Onyeze, G. O. C. (2005). Studies on the utilization of petroleum hydrocarbon by microorganisms isolated from oil-contaminated soil. International Journal of Natural and Applied Sciences, 1(2):1-6
5. George M. Garrity: Bergey's Manual of Systematic Bacteriology. 2. Auflage. Springer, New York, 2005, Vol. 2: The Proteobacteria Part C: The Alpha-, Beta-, Delta-, and Epsilonproteabacteria. ISBN 0-387-24145-0
6. Hill, G. B., & Moxey, J. G. (1980). Gasoline and diesel oil 4: 1-4. Petroleum Product Handbook Mc-Grew Hill,(NY)

7. Ijah UJJ, Okang CN (1993). Petroleum degrading capabilities of bacteria isolated from soil, W.A.J. Biol. Appl. Chem. 38(1-4): 9-15
8. Justesen, U. S., Holt, H. M., Thiesson, H. C., Blom, J., Nielsen, X. C., Dargis, R., ... & Christensen, J. J. (2007). Report of the first human case of Caulobacter sp. infection. Journal of clinical microbiology, 45(4), 1366-1369
9. Karafin, M., Romagnoli, M., Fink, D. L., Howard, T., Rau, R., Milstone, A. M., & Carroll, K. C. (2010). Fatal infection caused by Cupriavidus gilardii in a child with aplastic anemia. Journal of clinical microbiology, 48(3), 1005-1007
10. Kobayashi, T., Nakamura, I., Fujita, H., Tsukimori, A., Sato, A., Fukushima, S., & Matsumoto, T. (2016). First case report of infection due to Cupriavidus gilardii in a patient without immunodeficiency: a case report. BMC infectious diseases, 16(1), 1-4
11. Leikin, J. B., & Paloucek, F. P. (2008). Poisoning and toxicology handbook. Informa Healthcare
12. Ledermann, W. (2013). Antagonismo microbiano en la terapia de las enfermedades infecciosas. Revista chilena de infectología, 30(4), 446-450
13. Mereschkowsky, C. (1905). Uber natur und ursprung der chromatophoren im pflanzenreiche. Biologisches Centralblatt, 25, 293-604
14. Mereschkowsky, C. (1910). Theorie der zwei Plasmaarten als Grundlage der Symbiogenesis, einer neuen Lehre von der Entstehung der Organismen. Biologisches Centralblatt, 30, 278-288
15. Thurman, R. B., Gerba, C. P., & Bitton, G. (1989). The molecular mechanisms of copper and silver ion disinfection of bacteria and viruses. Critical reviews in environmental science and technology, 18(4), 295-315
16. Tortora, G. J., Funke, B. R., & Case, C. L. (2010). Microbiology an Introduction 10th Edition. USA: Benjamin Cummings
17. Vaneechoutte, M., Kämpfer, P., De Baere, T., Falsen, E., & Verschraegen, G. (2004). Wautersia gen. nov., a novel genus accommodating the phylogenetic lineage including Ralstonia eutropha and related species, and proposal of Ralstonia [Pseudomonas] syzygii (Roberts et al. 1990) comb. nov. International journal of systematic and evolutionary microbiology, 54(2), 317-327
18. Warrington PD (1996): Application of criteria for aquatic life. Water Quality Branch, British Columbia Ministry of Environment, Lands and Parks, Environmental Protection Department. Retrived from http://wlapwww.gov.bc.ca/wat/wq/ BCguidelines/ silver/bcsilver-04.htm
19. Martin, W., Roettger, M., Kloesges, T., Thiergart, T., Woehle, C., Gould, S., & Dagan, T. (2012). Modern endosymbiotic theory: Getting lateral gene transfer into the equation. Endocytobiosis & Cell Research, 23

"Experimental Studied of Density and Water absorption of Recycled Coarse Aggregate Concrete Incorporating GGBFS and FA"

Om Prakash Singh[1], Ravindra Kumar[2], Mahendra Kr. Singar[3]
Department of Civil Engineering, Vivekananda Global University, Jaipur, Rajasthan

Corresponding author email id: om.prakash.singh@vgu.ac.in

Abstract. This investigates work the density and water absorption of concrete specimens cured at the age of 28 days. Natural coarse aggregate (NCA) has been replace with 60 percent, and 100 percent by RCA of construction & destruction (C&D) waste of 30-year-old building and Ordinary Portland cement (OPC) has been replaced with 10 GGBFS and FA. To understand the physical properties of concrete of normal weight test was perform and to find suitable results. To examine the water absorption properties of GGBFS and FA with recycled coarse aggregate (RCA) stand mix concrete at the age of 28 days. Based on the experiment's study, it can be brief that the 40 percent replacement of NCA with RCA carried excellent results with 20 percent GGBFS. This research presents the outcome of GGBFS and FA as a cement replacement on the physical properties of RCA.

Keywords: NCA, RCA, GGBFS, FA, Construction & Demolition, Density, Water Absorption

1. Introduction

Concrete is the most widely used civil engineering material and the foundation of most facilities. Concrete likely to be the only construction material which will be available soon. Although concrete's strength is the most essential factor, it must also be durable, workable, and have a long service life. The development and expansion of the human race have produced severe damage on the environment and may endanger its sustainability (1). The exploitation of natural resources for construction purposes that of non-renewable resources, has led to millions of tons of C&D waste being produced every year. Since most countries have no specific treatment plan for these materials they are sent to landfill instead of being reused (2). The C&D waste is usually rejected on the side of the way, creating traffic and environmental issues (3). It enhances the administrative burden on the local government. NCA on the other hand, are in short supply and quickly depleting as construction activity increases. As a result, that critical problem must be resolved as soon. The issues can be solved by using C&D waste used to like RCA in concrete, which is now being used all around. Sustainable constructions, conservational issues of natural aggregate are all major concerns in the construction industry current scenario (4). As a result, this study aims to use as much C&D waste as possible in the manufacture of high-grade concrete. The usage of RCA in concrete would be a significant phase toward environmentally friendly building and natural resource protection (5). The essential elements of modern growth are environmental protection, natural resource management, and long-term construction. Sustainable development is now widely promoted across the globe. The building industry ingests natural resources and produces C&D waste (6). The concrete rubble has remained discovered to make up a significant ratio of C&D waste, accounting for about 40 percent of the total. Globalization and the expansion of the human race have resulted in environmental degradation. The use of natural resources for building purposes has resulted in the production of millions of tons of C&D waste per year.

The record countries do not have a single sustainability plan in place for these products, which are destined for the landfill rather than existence reused and recycled in innovative construction. The worldwide construction aggregates fair was forecast to grow at a rate of 5.2 percent per year until 2015, reaching 48.3 billion tons. Aggregates make up 70 percent to 80 percent of the capacity of concrete, 92 percent to 96 percent of tar paving, and nearly all of the volume of unbound and hydraulically bound materials (7). As a result, they show a significant part in the structure industry. Now contrast near the cement manufacturing, aggregate withdrawal and processing have a low environmental effect because it primarily consists of simple extraction without significant material alteration. The mandate for sand and gravel grows speedily in conjunction with innovative infrastructure projects, withdrawal of aggregates and rock stands attractive an ecological problem in many parts around us. This has resulted in an ecological imbalance in a variety of respects.

An increase in aggregate demand and a decrease in aggregate supply for concrete production necessitate the search for new aggregate sources. Environmental features are constantly being used to test building materials. Concrete recycling is becoming more common because it conserves ordinary resources and reduces through using readily available concrete as a source of aggregate for new concrete, the need for disposal is reduced. Via research and growth, the idea of sustainable construction growth necessitates the responsible use of regular resources and C&D waste (8). One such possible outlet is the use of RA in concrete construction. Construction products are increasingly being categorized based on their environmental attributes. Concrete recycling is important since it defends usual resources and removes the essential for removal by willingly obtainable concrete as a source of aggregate for innovative concrete. Supportable construction production necessitates judicious usage of usual resources and optimum recycling of construction waste, as well as extensive research and development. One such latent channel is the usage of RA in concrete construction.

1.1 The requirement of RCA in construction

The essential for RCA manufacture and usage is important presently. It's a really important topic, especially that now the situations are favorable for the construction of infrastructures.

- There is a material for the current protection of the country's natural resources, which stay neither immeasurable.
- There is a reduction in the quantity of extra waste material that is illegally deposited in uncontrolled regions.
- The goal of sustainable development is monitored, and as a condition, the framework for coexistence between humanity and beast is established.
- The presence of a national and policy of the treatment of C&D waste that should include solutions to key requests such as the regulator mechanism for C&D waste deposition with guidance, fines, and so on, is a prerequisite for the use of recycled concrete aggregate.
- Develop deposit areas for before and RCA treatment transport, especially near major cities, with the help of government and private team members to implement RCA.
- Encouragement of technical studies, and the implementation of the use of RCA in the presence of state development.

1.2 Need of alternate for Ordinary Portland cement

There is continuous technical progress and the incorporation of modern technology in the cement business. In India, only seven percent of the volume in the industry is produced from traditional wet and semi-dry course tech, and the remaining 93 percent of the volume is rooted in modern and eco-friendly dry process techniques (9). Cement production is not just one of the most emission building materials, and it is also a source of carbon-dioxide emissions, that account for about 65 percent of global warming among greenhouse gases and is second to only fossil fuels. As an outcome, the PC industry has become unsuitable for current industry thoughts and concepts. The incorporation of a high proportion of pozzolanic materials into the manufacturing of OPC has greatly improved the condition. It is critical to discover an alternative to OPC to assist the building sector in becoming more eco-sustainable. That should also have acceptable mechanical and durability properties that are similar to if it's not better than, typical cement-based concrete.

1.3 Necessity for the Present Work

In India, the Central Pollution Board (CPCB) takes projected solid waste group as approximately 48 million tons per annually of which 25 percent remain after the construction manufacturing. This generates noteworthy difficulties with solid waste management. The main influence toward the excess group is the C&D waste. The whole amount of excess is valued toward stand 12 to 14.7 million tons annually after the building manufacturing; obtainable of which individual three percent of the waste is cast-off for the bank. Projections of construction substantial requirements of accommodation subdivision direct a deficiency of aggregates toward the level of around 55,000 m^3. Recycling C&D waste may decrease the demand hole popular parts. Concrete and masonry scrap accounts for almost half of C&D wastes, and it is reportedly rarely recycled (10). As per the report submitted by Technology, Information, Forecasting and Assessment Council regarding the possibility of recycling of C&D waste from the building manufacturing the supreme main motive aimed at not accepting the recycling of waste since the building manufacturing remains "lack of awareness of the recycling techniques". Around 70 percent of participants stated lack of awareness among the reasons, and 30 percent of participants said they were aware of recyclable materials. The hence be industries responded that existing requirements may not provide for use of recyclable materials in construction. Around 67 percent of the workers replied that the non-availability of the recycled creation remained unique of the details aimed at not by it. Concrete is among the most important building materials. Concrete is moldable, adaptable, flame, widely available, and cost-effective. The properties of properly constructed and manufactured concrete include applying the knowledge, mechanical properties, and durability. Each year, the construction industry requires 10 billion tons of concrete. Coarse aggregate makes 70 to 80 percentage of the capacity of concrete and has a huge impact on its properties. As just a response, an alternate coarse aggregate product should be found to protect natural aggregate. In this background, the use of RA in concrete has become a popular method in recent times, with several countries encouraging this. Because of variations in RA quality between situations to a situation or the existence of adherent mortar on its periphery, many investigators used it for reduced concrete applications and insignificant activities. This project aims to M-25 grade of concrete through using of RCA found from C&D waste. In comparison, a multitude of application issues have been resolved and agreeable use was illustrated in several cement types. All things considered, the use of RA in cement is relatively rare. Overall Economic– straight expenses can likewise stand unfavorable .A constant supply of suitable aggregates is indeed a chicken-and-egg situation. Aggregate producers would not like to RA for concrete since there is no demand, and concrete producers cannot plan to make RCA without all the knowledge of high demand.

The main goal of such a thesis is to create a work that acts as a single world base of knowledge mostly on results of just using RCA on the concrete property by gathering sufficient data and information and quite well techniques, then objectively analyzing, evaluating, and republishing this. To a certain effect, the first objectives of this paper were to investigate the properties of RA obtained from C&D waste and compared them to NA. This is mostly natural, considering also that the main feature of RCA is the replacement of NA with RA, and thus a significant amount of attention must be given to the evaluation of such recyclable materials. This apart from thorough RA identification; one of the aims of such a work is to analyze the effect of using such materials on the most common engineering and durability properties of the concrete. Such experience must significantly assist future researchers, architects, and designers in understanding the role of RA in concrete structures. The density of concrete, water absorption of concrete, compressive strength, split tensile strength, flexural strength, and modulus of elasticity, chloride ion penetration, water permeability sulphate & acid attack effect are indeed the properties of concrete examined in this study. Lastly, it's also expected that several latest trends and similarities must develop from such an extensive numerical analysis of data gathered inside the systematic review on this subject, allowing a deeper understanding of the shortcomings of just using RCA to concrete. The data gathering on the object will then allow the development of simple and practical designs which can push the total use of RCA throughout the development of the RCC product.

2. Research Methodologies

The project goal is to improve the percentage of RA that is used to produce new concrete. To attain the key goal, the existing problems connected to RCA were methodically considered to understand the basic situation of RCA. To use a mixture of processing and mixing techniques, the numerous properties of RCA in both stages of fresh and hardened were inspected. Numerous influences affect the future performance of RA conclusion of the review of the literature, a treating technique known as "straight forward mechanical grinding" and "modified two-stage mixing methods" were selected for use in combination to solve the problems faced by RA and RCA in the construction. The adherent cement mortar and waste matter were removed from RA during the first phase that used a dry rolling method in instruction to progress the overall excellence of RCA. After adhered cement mortar was being eliminated, the RA supplements of elevated concrete are being improved. The effect of the processing technique on the several properties of RCA has been examined in this phase to choose the best RCA used. The different methodologies were adopted to improve the quality of RCA by removal of the adhered mortar content. Different test techniques and experimental measures were used to revision the probable conducts that the quality of the RCA is upgraded. A method is proposed in this work to particularly that is originated to be effective in removing a determined part of adhered mortar content with economically and improved the performance of the RCA.

Table 1 Mix of one cubic meter of concrete with RCA and GGBFS

Mix ID	RCA (%)	RCA (Kg/m³)	Cement (Kg/m³)	NCA (Kg/m³)	GGBFS (Kg/m³)	FA (Kg/m³)	Water (Kg/m³)	Plasticizer (Kg/m³)
RCA-1-60-GGBFS-20	60	747.70	268.8	498.46	67.2	659	148	3.36
RCA-2-60-GGBFS-20	60	747.70	268.8	498.46	67.2	659	148	5.04
RCA-3-60-GGBFS-20	60	747.70	268.8	498.46	67.2	659	148	5.04
RCA-4-60-GGBFS-20	60	747.70	268.8	498.46	67.2	659	148	5.04
RCA-5-60-GGBFS-20	60	747.70	268.8	498.46	67.2	659	148	5.04
RCA-1-100-GGBFS-30	100	1246.16	235.2	-	100.8	659	148	5.04
RCA-2-100-GGBFS-30	100	1246.16	235.2	-	100.8	659	148	6.72
RCA-3-100-GGBFS-30	100	1246.16	235.2	-	100.8	659	148	6.72

RCA-4-100-GGBFS-30	100	1246.16	235.2	-	100.8	659	148	6.72
RCA-5-100-GGBFS-30	100	1246.16	235.2	-	100.8	659	148	6.72

Table 2 Mix of one cubic meter of concrete with RCA and FA

Mix Id	RCA (%)	RCA (Kg/m³)	Cement (Kg/m³)	NCA (Kg/m³)	FA (Kg/m³)	NFA (Kg/m³)	Water (Kg/m³)	Plasticizer (Kg/m³)
RCA-1-60-FA-10	60	747.70	302.4	498.46	33.6	659	148	3.36
RCA-2-60-FA-10	60	747.70	302.4	498.46	33.6	659	148	5.04
RCA-3-60-FA-10	60	747.70	302.4	498.46	33.6	659	148	5.04
RCA-4-60-FA-10	60	747.70	302.4	498.46	33.6	659	148	5.04
RCA-5-60-FA-10	60	747.70	302.4	498.46	33.6	659	148	5.04
RCA-1-100-FA-20	100	1246.16	442.4	-	67.2	659	148	5.04
RCA-2-100-FA-20	100	1246.16	442.4	-	67.2	659	148	6.72
RCA-3-100-FA-20	100	1246.16	442.4	-	67.2	659	148	6.72
RCA-4-100-FA-20	100	1246.16	442.4	-	67.2	659	148	6.72
RCA-5-100-FA-20	100	1246.16	442.4	-	67.2	659	148	6.72

Nomenclature of mixes

In this study, one control concrete (natural aggregate) and 30 (recycled concrete aggregate) mixes are created. The nomenclature of the content is shown below.

NCA-1-60-0

Wherever NAC refers to NAC, first '1' refers to the source of aggregate and second '60' refers to aggregate % replacement of NA by RA, and third '0' refers to % of a mineral admixture.

NCA-2-60-0

Where NAC refers to NAC, first '2' refers to the source of aggregate and second '60' refers to aggregate % replacement of NA by RA, and third '0' refers to % of a mineral admixture.

NCA-3-60-0

Where NAC refers to NAC, first '3' refers to the source of aggregate and second '60' refers to aggregate % replacement of NA by RA, and third '0' refers to % of a mineral admixture.

NCA-4-60-0

Where NAC refers to NAC, first '4' refers to the source of aggregate and second '60' refers to aggregate % replacement of NA by RA, and third '0' refers to % of a mineral admixture.
NCA-5-60-0
Where NAC refers to NAC, first '5' refers to the source of aggregate and second '60' refers to aggregate % replacement of NA by RA, and third '0' refers to the replacement of cement % of a mineral admixture.
NCA-1-100-0
Where NAC refers to NAC, first '1' refers to the source of aggregate and second '100' refers to aggregate % replacement of NA by RA, and third '0' refers to % of a mineral admixture.
NCA-2-100-0
Where NAC refers to NAC, first '2' refers to the source of aggregate and second '100' refers to aggregate % replacement of natural aggregate by RA, and third '0' refers to % of a mineral admixture.
NCA-3-100-0
Where NAC refers to NAC, first '3' refers to the source of aggregate and second '100' refers to aggregate % replacement of natural aggregate by RA, and third '0' refers to % of a mineral admixture.
NCA-4-100-0
Where NAC refers to NAC, first '4' refers to the source of aggregate and second '100' refers to aggregate % replacement of natural aggregate by RA, and third '0' refers to % of a mineral admixture.
NCA-5-100-0
Where NAC refers to NAC, first '5' refers to the source of aggregate and second '100' refers to aggregate % replacement of NA by RA, and third '0' refers to the replacement of cement % of a mineral admixture.
NCA-1-60-GGBFS-20
Where NAC refers to NAC, first '1' refers to the source of aggregate and second '60' refers to aggregate % replacement of NA by RA, and third '20' refers to cement % of GGBFS.
NCA-2-60-GGBFS-20
Where NAC refers to NAC, first '2' refers to the source of aggregate and second '60' refers to aggregate % replacement of NA by RA, and third '20' refers to cement % of GGBFS.
NCA-3-60-GGBFS-20
Where NAC refers to NAC, first '3' refers to the source of aggregate and second '60' refers to aggregate % replacement of NA by RA, and third '20' refers to cement % of GGBFS.
NCA-4-60-GGBFS-20
Where NAC refers to NAC, first '4' refers to the source of aggregate and second '60' refers to aggregate % replacement of NA by RA, and third '20' refers to cement % of GGBFS.
NCA-5-60-GGBFS-20
Where NAC refers to NAC, first '5' refers to the source of aggregate and second '60' refers to aggregate % replacement of NA by RA, and third '20' refers to the replacement of cement % with GGBFS.
NCA-1-100-GGBFS-30
Where NAC refers to NAC, first '1' refers to the source of aggregate and second '100' refers to aggregate % replacement of NA by RA, and third '30' refers to the replacement of cement % with GGBFS.
NCA-2-100-GGBFS-30
Where NAC refers to NAC, first '2' refers to the source of aggregate, and second '100' refers to aggregate % replacement of NA by recycling RA
NCA-3-100-GGBFS-30
Where NAC refers to NAC, first '3' refers to the source of aggregate and second '100' refers to aggregate % replacement of NA by RA, and third '30' refers to cement % of GGBFS.

NCA-4-100-GGBFS-30
Where NAC refers to NAC, first '4' refers to the source of aggregate and second '100' refers to aggregate % replacement of NA by RA, and third '30' refers to cement % of GGBFS.
NCA-5-100-GGBFS-30
Where NAC refers to NAC, first '5' refers to the source of aggregate and second '100' refers to aggregate % replacement of NA by RA, and third '30' refers to cement % of GGBFS.
NCA-1-60-FA-10
Where NAC refers to NAC, first '1' refers to the source of aggregate and second '60' refers to aggregate % replacement of NA by RA, and third '30' refers to cement % of FA.
NCA-2-60-FA-10
Where NAC refers to NAC, first '2' refers to the source of aggregate and second '60' refers to aggregate % replacement of NA by RA, and third '10' refers to cement % of FA.
NCA-3-60-FA-10
Where NAC refers to NAC, first '3' refers to the source of aggregate and second '60' refers to aggregate % replacement of NA by RA, and third '10' refers to cement % of FA.
NCA-4-60-FA-10
Where NAC refers to NAC, first '4' refers to the source of aggregate and second '60' refers to aggregate % replacement of NA by RA, and third '10' refers to cement % of FA.
NCA-5-60-FA-10
Where NAC refers to NAC, first '5' refers to the source of aggregate and second '60' refers to aggregate % replacement of NA by RA, and third '10' refers to cement % of FA.
NCA-1-100-FA-20
Where NAC refers to NAC, first '1' refers to the source of aggregate and second '60' refers to aggregate % replacement of NA by RA, and third '30' refers to cement % of FA.
NCA-2-100-FA-20
Where NAC refers to NAC, first '2' refers to the source of aggregate and second '60' refers to aggregate % replacement of NA by RA, and third '10' refers to cement % of FA.
NCA-3-100-FA-20
Where NAC refers to NAC, first '3' refers to the source of aggregate and second '60' refers to aggregate % replacement of NA by RA, and third '10' refers to cement % of FA.
NCA-4-100-FA-20
Where NAC refers to NAC, first '4' refers to the source of aggregate and second '60' refers to aggregate % replacement of NA by RA, and third '10' refers to cement % of FA.
NCA-5-100-FA-20
Where NAC refers to NAC, first '5' refers to the source of aggregate and second '60' refers to aggregate percent replacement of NA by RA, and third '10' refers to cement % of FA.

3. Experimental

3.1 Density

Fig. shows that variation of density of concrete mixes containing percentages of RCA. The density of CC is 2496 kg/m^3, and it falls to R-1 to R-5 for RAC-60 percent (2456kg/m^3, 2443 kg/m^3, 2423 kg/m^3, 2415kg/m^3 and 2435kg/m^3), R-1 to R-5 for RAC-100 percent (2352kg/m^3, 2342 kg/m^3, 2358 kg/m^3, 2347 kg/m^3, 2336 kg/m^3) NCA is exchanged with RCA. The lighter weight and porous nature of the old mortar attached to RCA accounts for the reduction in density in both phase. The reduction in RAC density is 1.60 percent, 2.12 percent, 2.92 percent, 3.25 percent and 2.44 percent for 60 percent, and 5.77 percent, 6.17 percent, 5.52 percent,5.97 percent, 6.41 percent for 100 percent of RCA.

Table 3 Percentage decrease in density with RCA

Sr. No.	Mix ID	Density in kg/m³	% loss in density
1.	Control Concrete	2496	-
2.	RCA-1-60	2456	1.60
3.	RCA-2-60	2443	2.12
4.	RCA-3-60	2423	2.92
5.	RCA-4-60	2415	3.25
6.	RCA-5-60	2435	2.44
7.	RCA-1-100	2352	5.77
8.	RCA-2-100	2342	6.17
9.	RCA-3-100	2358	5.52
10.	RCA-4-100	2347	5.97
11.	RCA-5-100	2336	6.41

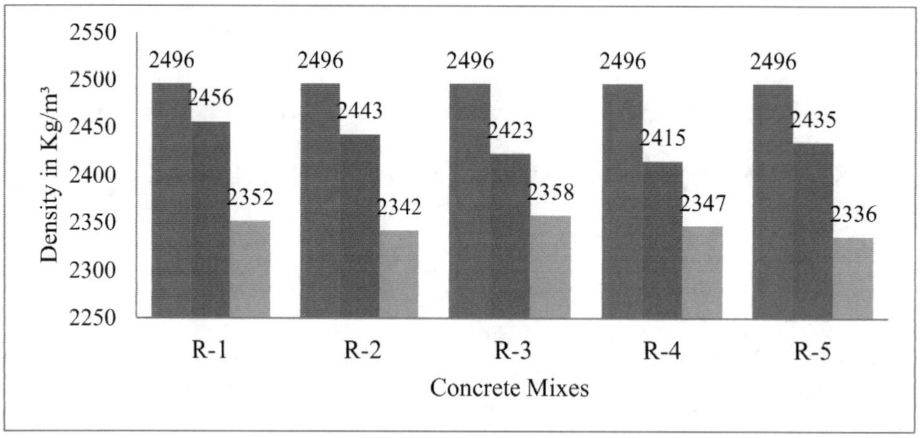

Fig.1 Variation in density of concrete mixtures with RCA-60% to 100%

The NCA replace with 60 percent RCA are substituted in (R-1 to R-5); the density of RAC with 20 percent of GGBFS decreases by 1.24 percent, 1.60 percent, 2.20 percent, 2.80 percent and 1.60 percent. The NCA replace with 100 percent RCA are substituted in (R-1 to R-5); the density of RAC with 30percent of GGBFS decreases by 5.52 percent, 4.80 percent, 4.61percent, 5.60 percent and 4.49 percent. This decrease in density is mainly due to the RCA lower specific gravity and higher water absorption ration compare to NCA and lower specific gravity of GGBFS comparison to that of OPC.

Table 4 Percentage Decrease in Density with RCA and GGBFS

Sr. No.	Mix ID	Density in kg/m³	% loss in density
1.	Control Concrete	2496	-
2.	RCA-1-60-GGBFS-20	2465	1.24
3.	RCA-2-60-GGBFS-20	2456	1.60

4.	RCA-3-60-GGBFS-20	2441	2.20
5.	RCA-4-60-GGBFS-20	2426	2.80
6.	RCA-5-60-GGBFS-20	2456	1.60
7.	RCA-1-100-GGBFS-30	2358	5.52
8.	RCA-2-100-GGBFS-30	2376	4.80
9.	RCA-3-100-GGBFS-30	2381	4.61
10.	RCA-4-100-GGBFS-30	2356	5.60
11.	RCA-5-100-GGBFS-30	2384	4.49

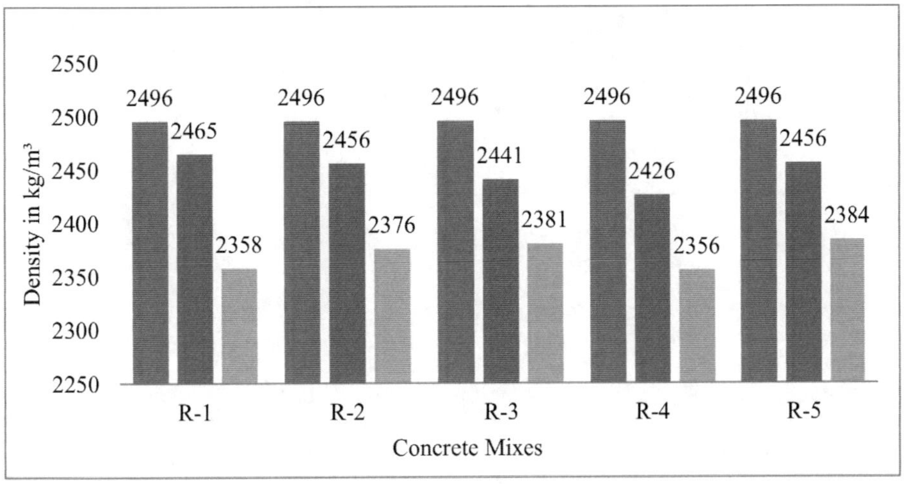

Fig.2 Variation in density of concrete mixtures RCA-60% with GGBFS-20% and RCA-100% with GGBFS-30%

If 10 percent (with RCA-60%), and 20 percent (with RAC-100%)of OPC becomes substituted by FA the density of RCA concrete decreases by 0.72percent, 1.60percent, 2.16percent, 1.00percent, 1.24percent and 4.40 percent, 4.73 percent, 5.29 percent, 4.81 percent, 6.05 percent. Finer FA, which acts as a filler material, can penetrate into the aggregates to make the concrete denser, the fineness of the FA has an effect on density. The FA in the first one is finer then that was used in the second, which is why the existing concrete is heavier.

Table 5 Percentage decrease in density with RCA and FA

Sr. No.	Mix ID	Density in kg/m³	% loss in density
1.	Control Concrete	2496	-
2.	RCA-1-60-FA-10	2478	0.72
3.	RCA-2-60-FA-10	2456	1.60
4.	RCA-3-60-FA-10	2442	2.16
5.	RCA-4-60-FA-10	2471	1.00
6.	RCA-5-60-FA-10	2465	1.24
7.	RCA-1-100-FA-20	2386	4.40
8.	RCA-2-100-FA-20	2378	4.73
9.	RCA-3-100-FA-20	2364	5.29
10.	RCA-4-100-FA-20	2376	4.81
11.	RCA-5-100-FA-20	2345	6.05

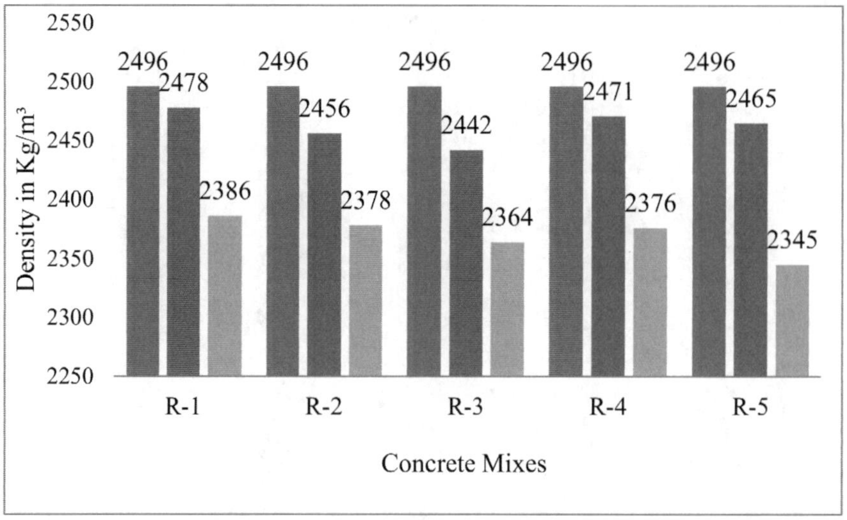

Fig.3 Variation in Density of Concrete mixtures RCA-60% with FA-10% and RCA-100% with FA-20%

3.2 Water Absorption

The effect of RAC with 60 percent and 100 percent on the water absorption quality for concrete mixes formulated for different RCA compositions during 28 days. Control concrete consumes 2.12 percent of water, rising (R-1 to R-5) to 9.43percent, 8.96percent, 11.32percent, 9.90percent and 10.38percent for 60 percent RCA. As the increase level of RCA than same scenario increase percentage of water absorption as RCA-100 water absorption of RCA like as (R-1 to R-5) 18.86 percent, 17.45 percent, 18.39 percent, 19.81 percent and 16.03 percent. This increase in water absorption in RAC is due to the fact that RCA will have a higher water absorption capacity that NCA.

Table 6 Percentage Increase in Water Absorption with RCA

Sr. No.	Mix ID	Absorption of water	% absorption of water
1.	Control Concrete	2.12	-
2.	RCA-1-60	2.32	9.43
3.	RCA-2-60	2.31	8.96
4.	RCA-3-60	2.36	11.32
5.	RCA-4-60	2.33	9.90
6.	RCA-5-60	2.34	10.38
7.	RCA-1-100	2.52	18.86
8.	RCA-2-100	2.49	17.45
9.	RCA-3-100	2.51	18.39
10.	RCA-4-100	2.54	19.81
11.	RCA-5-100	2.46	16.03

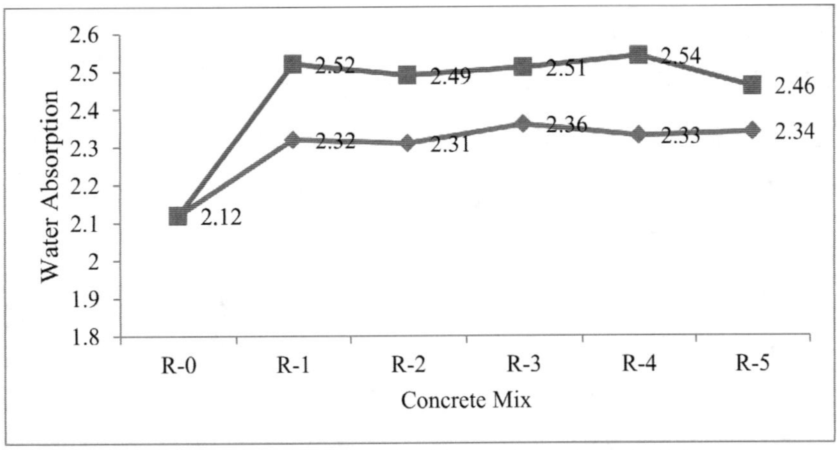

Fig.4 Variation in Water Absorption of RCA

So 20 percent GGBFS with 60 percent RCA is being used, the water absorption as towards 4.25 percent, 5.19 percent, 3.30 percent, 5.66 percent, 2.36 percent and as 30 percent GGBFS with 100 percent RCA is being used, the water absorption of increases towards 5.66 percent, 7.08 percent, 4.72 percent, 8.96 percent and 6.13 percent. It can be observed that the GGBFS seems to be effective in reducing the water absorption of RAC. It's obvious that the GGBFS is more effective at reducing RAC moisture content than at decreasing NAC water absorption. Because the finer GGBFS molecules that fill the pores and make the RCA denser, the pores in the RAC are distributed. The above analysis of results suggests which applying 20 to 30 percent GGBFS to NAC reduces water absorption, which would be primarily due to a start processing of fine GGBFS grains. Moreover, because the RCA is porous in nature, the blocking of pores in RCA due to the filling effect of GGBFS particles is more evident than in NCA. The stimulation of GGBFS in the absence with cement resulted in the formation of secondary compressive strength gel that covers any gaps throughout the concrete but enhances moisture content. Moreover, the finer particles of GGBFS are also having certain positive effect on blocking the pores and making the concrete microstructure denser due to which the water absorption is reduced. However, some amounts of GGBFS remain unused with increasing up-to 60 percent, after being expended through the silicate stage of GGBFS. These unused GGBFS leaches out with time leaving the empty pores. Hence, these additional pores increase the water absorption in the RAC for 30 percent GGBFS.

Table 7 Percentage Decrease in Water Absorption with RCA and GGBFS

Sr. No.	Mix ID	Absorption of water	% absorption of water
1.	Control Concrete	2.12	-
2.	RCA-1-60-GGBFS-20	2.21	4.25
3.	RCA-2-60-GGBFS-20	2.23	5.19
4.	RCA-3-60-GGBFS-20	2.19	3.30
5.	RCA-4-60-GGBFS-20	2.24	5.66
6.	RCA-5-60-GGBFS-20	2.17	2.36
7.	RCA-1-100-GGBFS-30	2.24	5.66

8.	RCA-2-100-GGBFS-30	2.27	7.08
9.	RCA-3-100-GGBFS-30	2.22	4.72
10.	RCA-4-100-GGBFS-30	2.31	8.96
11.	RCA-5-100-GGBFS-30	2.25	6.13

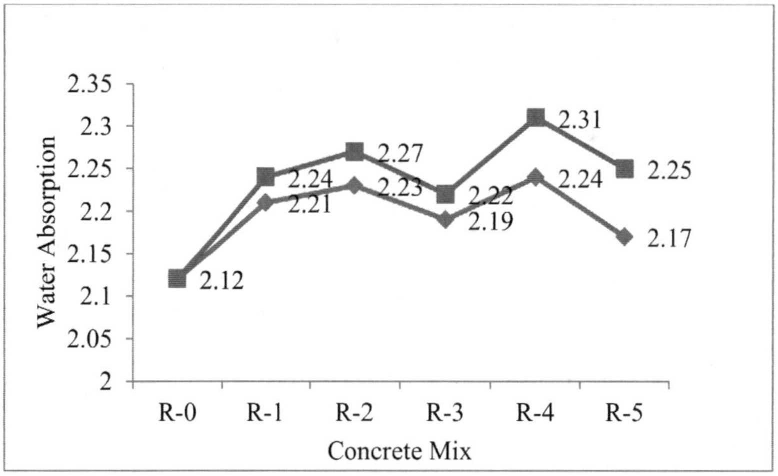

Fig.5 Variation in water absorption of RCA with GGBFS

FA remains presented up to 10 percent in adding to 60 percent RCA, it remains create that the water absorption engagement 5.19 percent, 6.60 percent, 4.72 percent, 5.19 percent and 3.77 percent content. The water absorption of concrete comprising 100 percent RCA with 20 percent FA develops 7.55 percent, 9.43 percent, 8.02 percent, 10.38 percent and 8.02 percent.

The above analysis of results proposes that applying 10 to 20 percent FA to NAC reduces water absorption, which would be mainly due to a start processing of FA grains. Furthermore, because the RCA is porous in nature, the blocking of pores in RCA due to the filling effect of FA particles is additional evident than in NCA. The presenting an appropriate of FA, like up to 20 percent in this case, the water absorption of RCA with 30 percent of FA can be reduced. The encouragement of FA in the absence with cement resulted in the creation of secondary compressive strength gel that covers several gaps through the concrete but improves moisture content. The amounts of FA continue unused with growing up-to 30 percent, and then being consumed finished the silicate stage of FA. Later, these extra pores increase the water absorption in the RAC for 20 percent with FA.

Table 8 Percentage Water Absorption with RCA and FA

Sr. No.	Mix ID	Water Absorption	Percentage Water Absorption
1.	Control Concrete	2.12	-
2.	RCA-1-60-FA-10	2.23	5.19
3.	RCA-2-60-FA-10	2.26	6.60
4.	RCA-3-60-FA-10	2.22	4.72
5.	RCA-4-60-FA-10	2.23	5.19
6.	RCA-5-60-FA-10	2.20	3.77
7.	RCA-1-100-FA-20	2.28	7.55
8.	RCA-2-100-FA-20	2.32	9.43
9.	RCA-3-100-FA-20	2.29	8.02
10.	RCA-4-100-FA-20	2.34	10.38

11.	RCA-5-100-FA-20	2.29	8.02

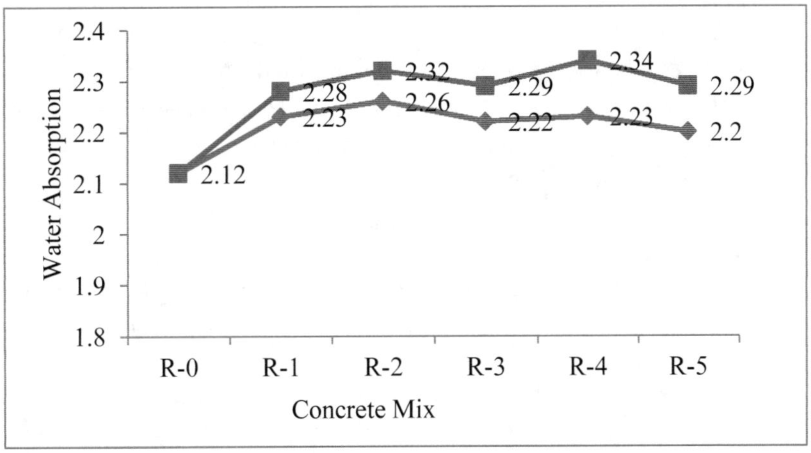

Fig.6 Variation in Water Absorption of RCA with FA

4. Conclusion

- All the control mix and mix with partial replacement of coarse aggregate by 60-100 percent with RCA creates practicable through an essential w/c ratio such as IS: 456: 2000 for grade of concrete.
- The workability in terms of compaction factor shows satisfactory results in the range of 0.85 to 0.90 on an average for concrete grade M-20 to M-40.This is as per the specified limit 0.8 to 0.92 in IS: 456: 2000.
- The workability in terms of slump shows satisfactory results in the range of 75mm to 89mm on an average for concrete grade M-20 to M-40.This is as per the specified limit 75 to 95mm in IS: 456: 2000.
- For all the design mix of NA replace by NCA gives the optimum value of a compaction factor in the range of 0.86 to 0.92 on an average as indicated in the graphical comparative presentation.
- Further replacement beyond 60 percent reduces the workability leading to bleeding, segregation and honeycombing.
- From the physical observation, w/c ratio and compaction factor results of RCA, the concrete can be utilized for ready mix applications.
- The density of dry concrete falls in the range of 60 percent of RCA lies 2415 to 2456 kN/m^3 and 100 percent of RCA lays 2336 and 2358 kN/m^3. It is as per the standard value 24.00 kN/m^3.
- The density of dry concrete falls in the range of 60 percent of RCA with 20 percent of GGBFS lies 2426 kN/m^3, 2465 $kN/m^3$and 100 percent of RCA with 30 percent of GGBFS lies 2356 kN/m^3,2384 kN/m^3,. It is as per the standard value 24.00 kN/m^3.
- The density of dry concrete falls in the range of 60 percent of RCA with 20 percent of FA lies 2426 kN/m^3, 2465 $kN/m^3$and 100 percent of RCA with 20 percent of FA 2442 to 2478 kN/m^3 and 2345 to 2386 kN/m^3,. It is as per the standard value 24.00 kN/m^3.
- The water absorption of dry concrete falls in the range of 60 percent of RCA lies 8.96 to

11.32 percent and 100 percent of RCA lays 16.03 to 18.86 percent of control concrete.

- The water absorption of dry concrete falls in the range of 60 percent of RCA with 20 percent of GGBFS lies 2.36 to 15.661.32 percent and 100 percent of RCA with 30 percent of GGBFS lays 4.72to 8.96 percent of control concrete.
- The water absorption of dry concrete falls in the range of 60 percent of RCA with 20 percent of FA lies 3.77 to 6.60 percent and 100 percent of RCA with 20 percent of FA lays 7.55 to 10.38 percent of control concrete.

Reference

[1] Lauritzen, E. K. (2004). International RILEM Conference on the Use of Recycled Materials in Buildings and Structures (Volume 1) (Vol. 1, p. 297). RILEM Publications.
[2] Kumar, V., & Jain, P. K. (2002). Commercializing new technologies in India: a perspective on policy initiatives. Technology in Society, 24(3), 285-298.
[3] Sridhar, K. S., & Kumar, S. (2013). India's urban environmental challenges: Land use, solid waste and sanitation. Yojana, 31.
[4] Faruqi, M. H. Z., & Siddiqui, F. Z. (2020). A mini review of construction and demolition waste management in India. Waste Management & Research, 38(7), 708-716.
[5] Tam, V. W., & Tam, C. M. (2007). Assessment of durability of recycled aggregate concrete produced by two-stage mixing approach. Journal of Materials Science, 42(10), 3592-3602.
[6] Shang, H. S., Zhao, T. J., & Cao, W. Q. (2015). Bond behavior between steel bar and recycled aggregate concrete after freeze–thaw cycles. Cold Regions Science and Technology, 118, 38-44.
[7] Yehia, S., Helal, K., Abusharkh, A., Zaher, A., & Istaitiyeh, H. (2015). Strength and durability evaluation of recycled aggregate concrete. International journal of concrete structures and materials, 9(2), 219-239.
[8] Tam, V. W., Gao, X. F., Tam, C. M., & Chan, C. H. (2008). New approach in measuring water absorption of recycled aggregates. Construction and building materials, 22(3), 364-369.
[9] Ismail, S., & Ramli, M. (2013). Engineering properties of treated recycled concrete aggregate (RCA) for structural applications. Construction and Building Materials, 44, 464-476.
[10] Al-Yaqout, A., El-Hawary, M., Nouh, K., & Khan, P. B. (2020). Corrosion Resistance of Recycled Aggregate Concrete Incorporating Slag. ACI Materials Journal, 117 (3).
[11] Ju, M., Jeong, J. G., Palou, M., & Park, K. (2020). Mechanical behavior of fine recycled concrete aggregate concrete with the mineral admixtures. Materials, 13 (10), 2264.

Gray Image and Colour Image Security using AES Algorithm

[1]Shaina Arora, [2]Nalin Chaudhary

[1]Department of Computer Science and Engineering, Vivekananda Global University, Rajasthan, Jaipur
[2]Department of Computer Science and Engineering, Raffles University, Rajasthan, Neemrana

shaina.arora22@gmail.com

Abstract. For picture security and confirmation we utilize cryptographic. In cryptographic we scramble the information. There are various symmetric and deviated encryption procedures, which are used for encryption and unscrambling reason. Symmetric encryption approach work on ordinary key (same key utilized for encryption and unscrambling reason), but unbalanced work on with one public key and private key. Public key utilized for encryption and private key for unscrambling. The security of the stowing away is measure utilizing the MSE, PSNR, Cross-Correlation and Entropy limits. For better stowing away get the low MSE and high PSNR respects. So to make the rapid reaction we need calculation which requires some speculation and give reaction quick. From this point forward additionally take apart the time reaction of purposed assessments to additionally foster security and expedient response.

Keywords: Public Key, Private Key, Symmetric Key, Asymmetric Key, MSE, PSNR

1. Introduction

Information (text, picture, sound, video and so forth) are the most generally utilized methods of correspondence in very field typically, like the examination, industry, clinical, military and so on critical picture moves happen over an unstable web organization. Accordingly, it is important to build up sufficient security so the advanced picture or computerized picture keeps from the unapproved people to getting to discharge data. Steganography and Cryptography are the most famous strategies for information security [1]. Information assurance has come to be a weighty advanced verbal trade downside through the web or the elective medium. Cryptography and transcription are the generally utilized procedure for information security. In cryptography information is transform one structure to another structure and in steganography restricted information is covered up into cover information [2]. Cryptography is a piece of CRYPTOLOGY where CRYPTOGRAPY and CRYPTOANALYSIS comes. Cryptography is a technique for Protecting or Secure Information. Cryptography implies how we secure our information by utilizing ENCRYPTION and DECRYPTION calculations [3]. Cryptography can offer the going with kinds of administrations: Confidentiality (mystery); Integrity (against altering); Authentication; Non-Repudiation

2. Types of Cryptography

1. Symmetric Key Cryptography:

It is additionally called Secret Key Cryptography or Conventional Cryptography, Symmetric Key Cryptography is an encryption system where the sender and gatherer of a message share a solitary, normal key that is used to encode and disentangle the message [4-5]. The Algorithm use is generally called a Secret key Algorithm or at times called a symmetric Algorithm.

A key is a piece of information (a limit) that chooses the valuable yield of a cryptographic calculation or code.

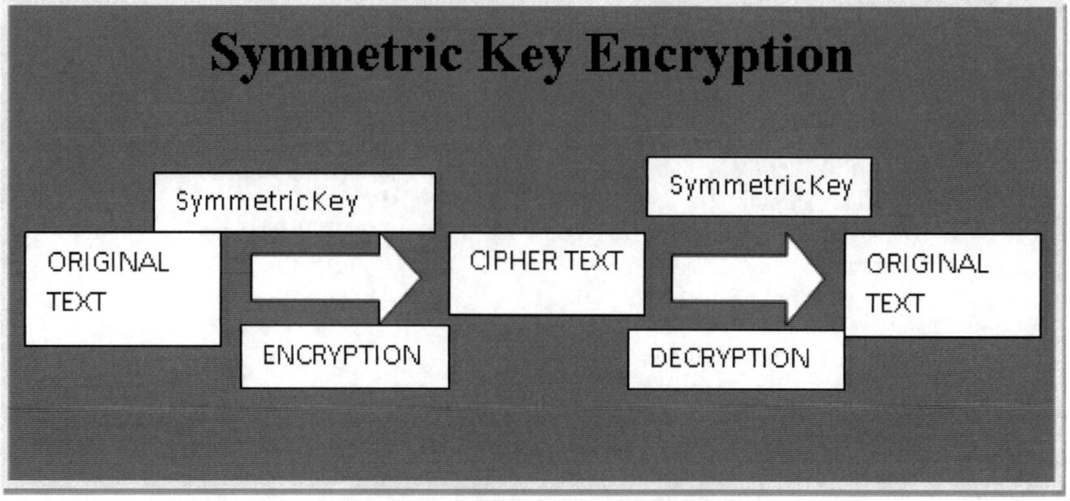

Fig. 1 Symmetric Key Encryption

Symmetric Key Following Algorithms include

- AES (Advanced Encryption Standard)
- DES (Data Encryption Standard)
- IDEA (International Data Encryption Algorithm)
- Blowfish (Drop-in replacement for DES or IDEA)
- RC4 (Rivest Cipher 4)
- RC5 (Rivest Cipher 5)
- RC6 (Rivest Cipher 6)

2. Asymmetric Key Cryptography:

Asymmetric cryptography, additionally referred to call as Public-key cryptography, insinuates a cryptographic calculation which requires two separate keys, one of which is private and one of which is public. The public key is used to scramble the message and the private one is used to unscramble the message.

Public Key Cryptography is an especially advanced kind of cryptography. Officially, it was created by Whitfield Daffier and Martin Hellman in 1975. The fundamental technique for public key cryptography was first found in 1973 by the British Clifford Cocks of Communications-Electronics Security Group (CESG) of (Government Communications Headquarters - GCHQ) but this was confidential until 1997 [6].

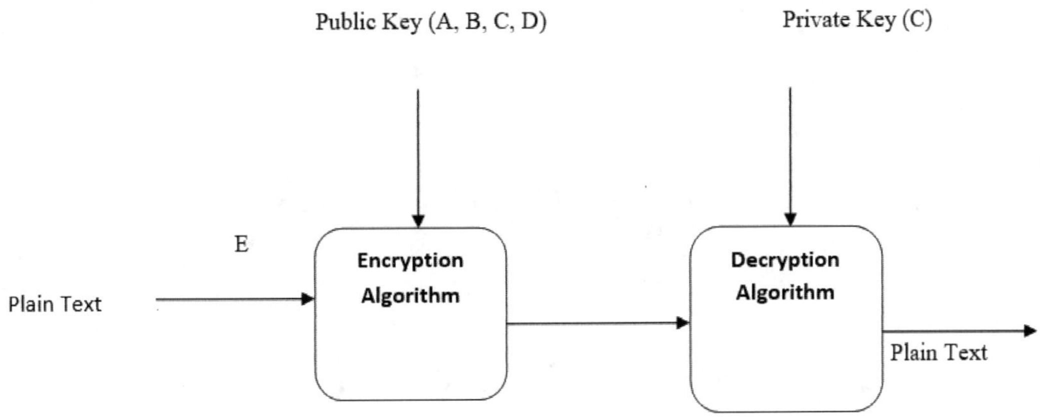

Fig. 2 Asymmetric Key Encryption

Asymmetric Key Following Algorithms Include
1. RSA Algorithm
2. ECC Algorithm
3. Diffie-Hellman Algorithm
4. DSA Algorithm

3. Symmetric Key Cryptography Algorithm

AES (Advance Encryption Standard):-
The Advanced Encryption Standard (AES) is a symmetric square code picked by the U.S. government to ensure arranged data. AES is executed in programming and equipment all through the world to scramble delicate information. It is fundamental for government PC security, network safety and electronic information assurance.
The National Institute of Standards and Technology (NIST) began improvement of AES in 1997 when it declared the requirement for an option in contrast to the Data Encryption Standard (DES), which was beginning to become defenseless against savage power assaults.
The Advanced Encryption Standard (AES) is the solitary freely available code endorsed by the US National Security Agency (NSA) for ensuring highly confidential data. AES was first called Rijndael after its two engineers, Belgian cryptographers Vincent Rijmen and Joan Daemen.

AES Features:

NIST indicated the new AES calculation should be a square code equipped for taking care of 128-bit blocks, utilizing keys estimated at 128, 192 and 256 pieces. Different measures for being picked as the following AES calculation incorporated the accompanying:

• Security: Competing calculations were to be decided on their capacity to oppose assault - when contrasted with other submitted figures. Security strength was to be viewed as the main factor in the opposition.

• Cost: Intended to be delivered on a worldwide, nonexclusive and sovereignty free premise, the competitor calculations were to be assessed on computational and memory proficiency.

• Implementation: Factors to be considered incorporated the calculation's adaptability, reasonableness for equipment or programming execution, and generally straight forwardness.

4. Gray Scale AES Image Algorithm

Original Image:

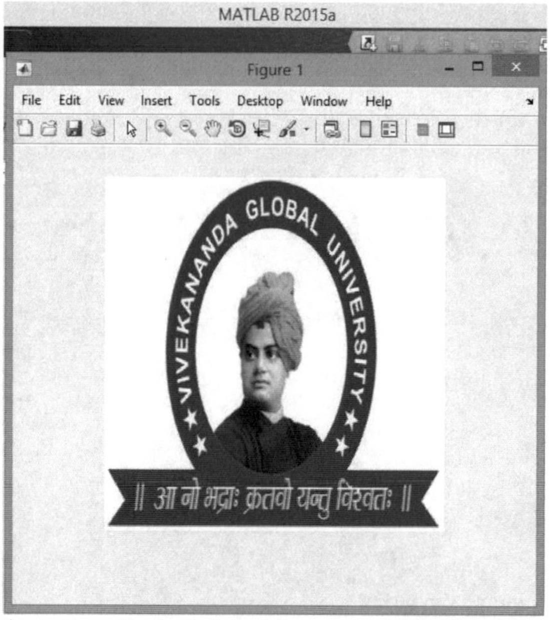

Fig 3 Gray Scale Original Image in AES Algorithm

Encrypted Image:

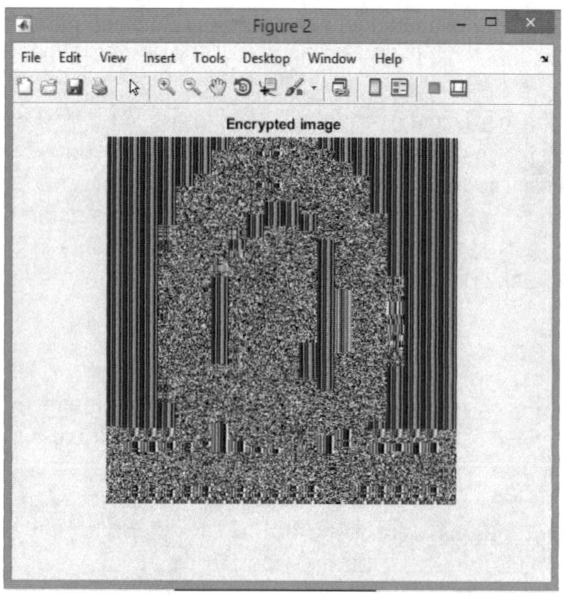

Fig 4 Gray Scale Encrypted Image in AES Algorithm

MSE Value:

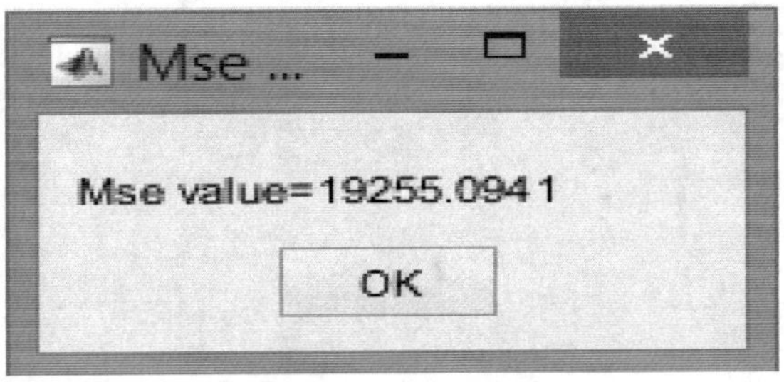

Fig 5 Gray Scale Image MSE Value in AES Algorithm

PSNR Value:

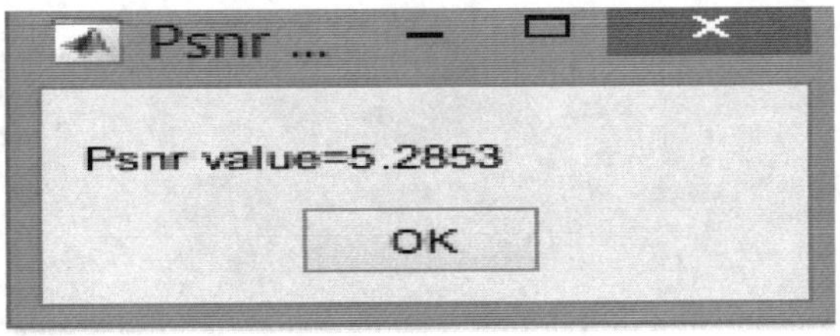

Fig 6 Gray Scale Image PSNR Value in AES Algorithm

Gray Scale AES Decrypted Image:

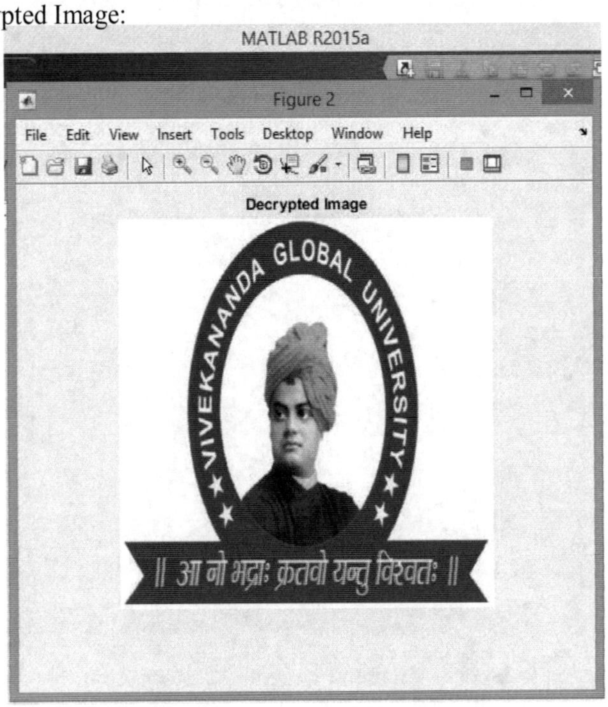

Fig 7 Gray Scale Decrypted Image in AES Algorithm

IOP Conf. Series: Materials Science and Engineering 1224 (2022) 012024 doi:10.1088/1757-899X/1224/1/012024

5. COLOR AES ENCRYPTED IMAGE

Original Image:

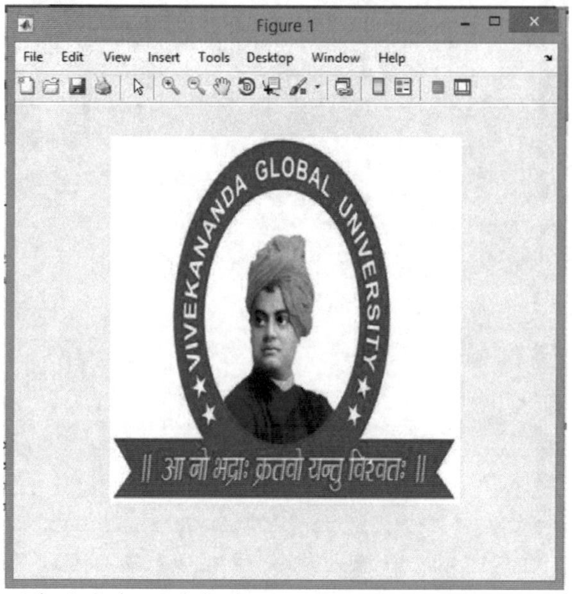

Fig 8 Color Original Image in AES Algorithm

Encrypted Image:

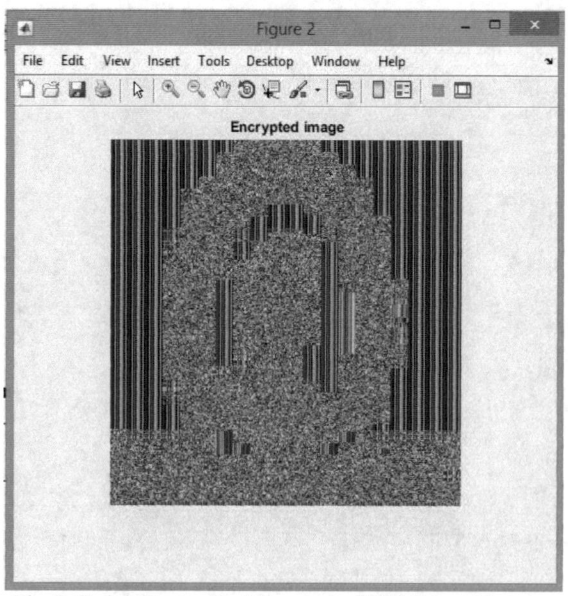

Fig 9 Color Encrypted Image in AES Algorithm

MSE Value:

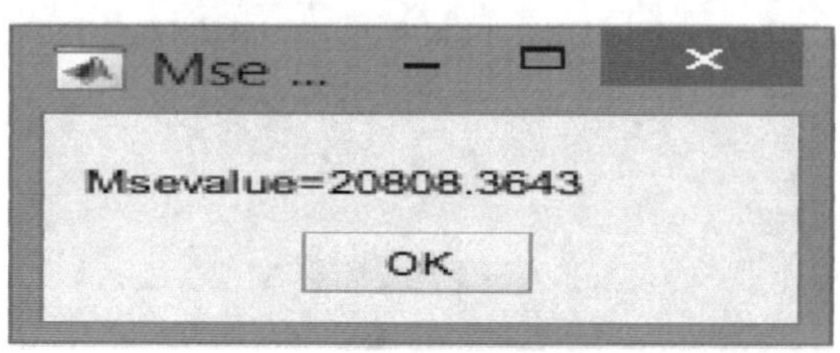

Fig 10 Color Image MSE Value in AES Algorithm

IOP Conf. Series: Materials Science and Engineering 1224 (2022) 012024 doi:10.1088/1757-899X/1224/1/012024

PSNR Value:

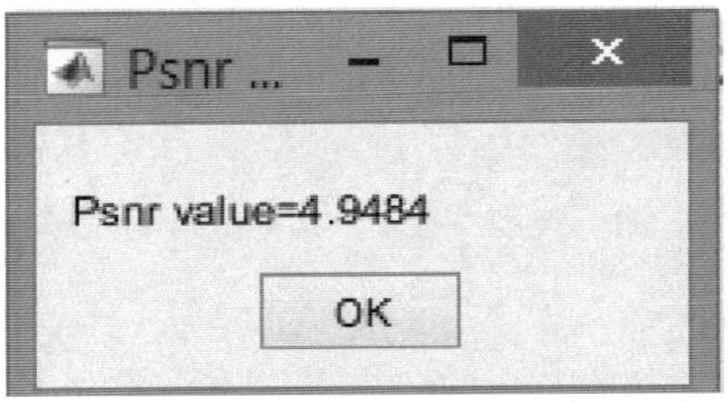

Fig 11 Color Image PSNR Value in AES Algorithm

Decrypted Image:

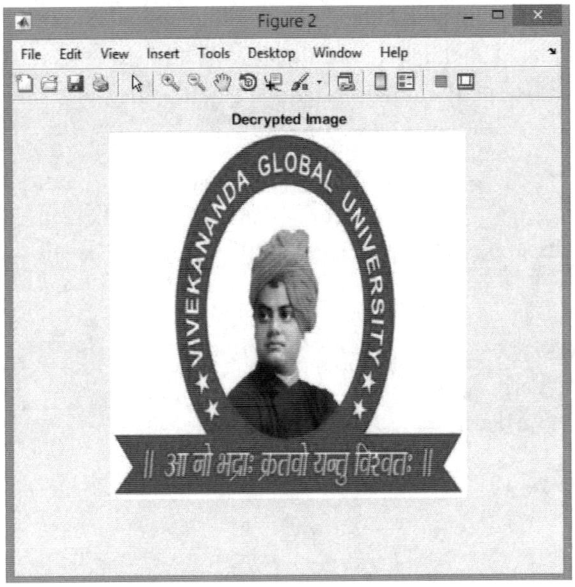

Fig12 Color Decrypted Image in AES Algorithm

6. AES Algorithm Gray Scale Image and Colour Image

Comparison:

AES Algorithm	MSE	PSNR
AES (Gray Scale)	19255.0941	5.2853
AES (Color Image)	20808.3643	4.9484

In Colour Image MSE three Planes Colour taken out, so it accompanies a solitary plane, so the MSE will get more from Gray Scale. PSNR conversely relative to MSE so if MSE Value Higher than PSNR Value is Low contrast with MSE.

Conversion Formula Colour to Gray Scale

0.289 R+0.5870 G+0.1140 B

7. CONCLUSION

In this Research Paper Study about Cryptography Symmetric and Asymmetric Key Cryptography and also calculated AES algorithm Gray Scale and Color Image Gray Scale difference between them. In this we calculated MSE and PSNR value. In future we will calculate other algorithms value and also research on Steganography.

Conflict of interest
None.

References

[1] Gangadari, Bhoopal Rao, and Shaik Rafi Ahamed. "FPGA implementation of compact S-Box for AES algorithm using composite field arithmetic." 2015 Annual IEEE India Conference (INDICON). IEEE, 2015.

[2] Kalaiselvi, K., and Anand Kumar. "Enhanced AES cryptosystem by using genetic algorithm and neural network in S-box." Current Trends in Advanced Computing (ICCTAC), IEEE International Conference on. IEEE, 2016.

[3] Pammu, Ali Akbar, et al. "High Secured Low Power Multiplexer-LUT Based AES S-Box Implementation." 2016 International Conference on Information Systems Engineering (ICISE). IEEE, 2016.

[4] Liao, Nan, et al. "A high-efficient fault attack on AES S-box." Information Science and Technology (ICIST), 2016 Sixth International Conference on. IEEE, 2016

[5] Wang, Yi, and Yajun Ha. "High throughput and resource efficient AES encryption/decryption for SANs." Circuits and Systems (ISCAS), 2016 IEEE International Symposium on. IEEE, 2016.

[6] Aryfandy Febryan, Tito Waluyo Purboyo and Randy Erfa Saputra, "Steganography Methods on Text, Audio, Image and Video: A Survey", International Journal of Applied Engineering Research, 12(21), pp-10485-10490, 2017.

Simulation and Optimization of an Independent Sustainable Micro Hybrid Electricity System for Rural Consumers

Satish Kumar Jangid, Mohammad Asif Iqbal, Rahul Mishra

[1,3]Assistant Professor , Department of Electrical Engineering, Vivekananda institute of Technology, Jaipur, India

[2]Associate Professor, Department of Electrical Engineering, Vivekananda Global University Jaipur, India

jangid.satish@vitj.ac.in, dr.mohammadasifiqbal@gmail.com, mishra.rahul@vitj.ac.in

Abstract. A fundamental necessity for sustainable development is to ensure reliable and cost-effective access to modern energy services, primarily for the lower income and impoverished sectors of the population. Successful rural electrification is vital to bridging the rural-urban gap, given that the vast majority of the energy-deprived population resides in rural regions of developing countries. if the existing system of energy services intervened it can influence the development process by livelihood activities, micro or small enterprises, lifestyle energy services, value-added activities, survival irrigation, and so on. Different micro hybrid systems preferably with combination of renewable energy resources will be a good option for addressing given barriers. India has a large Availability of Higher solar radiations, a society that depends on agriculture has enough biomass resources, plentiful cattle dung, forest conifers and agriculture waste. In this paper, we examine the hybrid system's techno economic feasibility and long-term viability. The system is made up of 60-kW solar photovoltaic panels (PV) and 40-kW biomass gasifier modules. Energy demand and resource availability are measured using data from comprehensive stakeholder discussions and field surveys, and they take into account daily and seasonal fluctuations in supply, end uses, and availability, as well as productive hours. The techno economic viability of electric renewable energy is evaluated using a hybrid optimization model (HOMER). Additionally, prospects for the growth of profitable uses and their extension into a long-term business model are investigated.

Keywords: Hybrid Energy, Solar Energy, Economic Model, sustainable development.

1. INTRODUCTION

We should ensure that every individual specially poor and deprived section of India is getting affordable energy services which are basic need for the sustainable development of any country. After the 73 years of independence a larger part of population is still residing in villages and electricity plays an important role in this rural urban gap.

Complete economic development of India is possible when its villages will be electrified. After the many years of efforts World Bank SE4ALL database says that still about 5% of population in India is living without electricity [1]. Grid extension alone cannot bear the burden of electrify rural and far away area because of uneconomical reasons unlike weak rural economy, terrain area, low purchasing capacity. Isolated areas like the proposed villages face some serious problems from the government side as lack of attention, distant market availability, and cost of installation infrastructure.

As per the recently available data in 2019 around 771 million people worldwide and 155 million people in developing Asia lacking access to electrical energy After so many efforts still around 6 million people still lacks access to electricity in India by the International Energy Agency (IEA). It automatically gives an idea that India is far behind the target of universal electricity access [2]. A large scarcity of energy access in remote areas, the expansion of energy networks to mitigating a gap in urban rural divide and climate change problem on global parameters these are some problems that exists in India's rural energy part. The best possible solution is to reach all without any lag [3]. Govt. data shows that all villages in rajasthan have been electrified by the various schemes like DDGVY, Saubhagya Yojna [4], but still a lot of households in some remote inaccessible areas still cannot afford traditionally joined electricity. Sawai Madhopur the proposed district of village has 27.7% forest cover and some partially electrified villages that are located far away. Distant villages will play an important role in the success of decentralized distributed technologies and global electrification efforts. It will be a most cost effective option for the rural areas till 2030 as the studies says [4].

This papers suggests a technically feasible and economically viable solution for rural distant households by installing a hybrid micro grid with a total 100kW capacity which consists of 60kW of solar PV and 40kW of biomass gasifier- based generation in a remote village of Bajoli in Rajasthan's sawai Madhopur district. Load characteristics is constructed by observing the consumption cycle of different load like irrigation, community, commercial and households. A software called HOMER which generated a hybrid optimization model for maintenance cost and various data of the system [5].

2. SITE SELECTION AND RESOURCE SURVEY

The proposed village which is selected for the installation of the hybrid system is a small village in the sawai Madhopur district of rajasthan at khandar tehsil. Longitude and latitudes of this location are 76.68, 26.09. It is located at a distance of 18 km from the sub-district headquarter khaddar and 58 km from the district head quarter. According to 2009 statistics bichpuri gujran is the gram panchayat of the village. It has a total 905 hectare area. 2496 inhabitants reside in the total 350 houses in the village. If we talk about the electricity availability some households are still without electricity and rely on kerosene for illumination and biomass for cooking and heating. An extensive survey was done to estimate the electricity demand. This survey consists conversation with villagers, interviews, data collections and resource availability calculation [6].

2.1. Economy of the proposed village and demand

Village consists family size of 4 to 5 in 350 households. One secondary and one primary school is situated in village. Various scattered kirana shops are available in village for fulfilling daily requirement of the people. Agriculture is the prime source of income of the people in village with people holding 2-3 acre of land. Paddy, Pearl Millet, Green Gram, Black Gram, Sesame, Ground Nut, Soyabean in monsoon season and wheat, Mustard, Taramira are sown in winter and harvested in the spring by farmers [7]. The village is located in the vicinity of Banas River, thus making it an ideal for site for agricultural practices. So, the setup of renewable energy powered irrigation pumps can be huge benefits to the farmers in terms of costs benefits and will also increase the yield of produce.

The economy of village is completely reliant on its agricultural base such as farming activities. Due to lack of efficient power and energy sources many families in village have to depend on firewood for cooking and kerosene lamps for lighting, which are not only highly unreliable and uneconomical but also poses a great hazard to human health and environment. Children require efficient, brighter and more reliable sources of lighting in schools and homes to be able to study effectively for longer durations. Lack of lighting on the streets also raises security concerns for women and children making it difficult for them to step out of homes after sunset. It also increases crime rates after dark.

2.2. Biomass Availability

Straw Goods:

It has been observed that total straw goods yield in reported 1050 acre land is around 760 ton yearly. The amount of grain obtained from the fields is calculated from the sacks of grain available in a year, which is estimated one hundred kilogram per sack. Straw good content from the grain is calculated from the ratio of straw to grain, which is measured to be two ratio one, means one kilogram of grain leaves two kilogram of straw. Actually it is very complex to observe availability of straw hence a non-direct method is used.

Woody Biomass:

The estimations were concluded after assorting through many questionnaires and weighing of some sample loads. On average, 2260 kg of wood is collected and used per year per household. [8] To estimate the potential for biomass resources, woody biomass (wastes generated from forests and plantations, straws, stems, stalks, leaves, husk, shells, peels, and whiskers) and soft biomass (agricultural, commercial, and animal waste) were surveyed for weekly consumption and extrapolated for the year. Through analysis it was found that around 570 ton of wood is collected by 250 households in a year, with a documented collection of 35 kg of wood at least two to three times a week.

Animal Dung/Manure:

Table 1 gives the information of survey done from village. Generally in the proposed village cows, buffaloes and goat are left freely to graze in the nearby forest or with an attendee. If a common centralized animal waste collection method is developed there can be a huge potential for biogas generation. With overall calculation and estimation it was observed a waste of seven kilogram per day per cattle, so total animal waste could be two thousand one hundred seventy tones annually.

3. POTENTIAL AND NEED FOR INSTALLATION WITH A BUSINESS MODEL

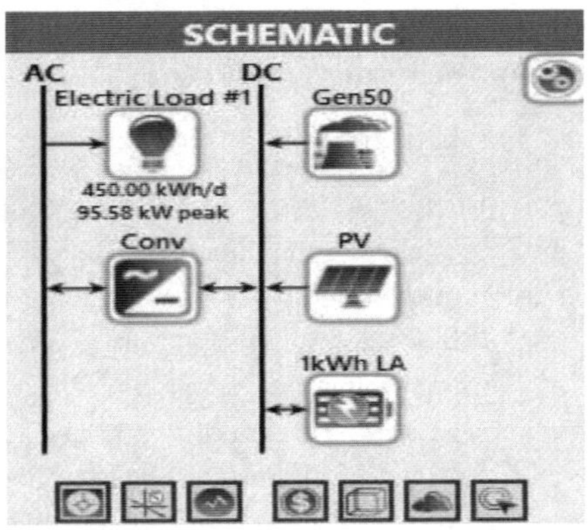

Fig. 1. Screenshot designed hybrid system in HOMER.

Due to its remoteness from the district headquarter; the people of Bajoli are dependent on the farming activity. People of village usually travel nearby towns for the employment and important things. At least one person from the family has gone to big cities for the employment due to lack of source of income in village. Most of the families rear Cattle for dairy only. Villagers travel to other towns and villages through privately owned buses which come two times in a day. People in villages share food, fuel, and firewood in times of requirements. Small shops also established by some of villagers to supply local needs and income. The selected village Bajoli satisfied most of the requirements for setting up a micro hybrid system like sufficient land availability, proper solar irradiance available, rich biomass and animals available for their dung. But as proper income model is not available in the village this system can help them.

4. HYBRID SYSTEMS AND METHODOLOGY

Here in this paper we have proposed a small independent micro hybrid system which can power a small village of a particular household capacity and can be connected or extended to grid later on which can increase the reliability of supply during outages and increase time of electricity. These technologies generally termed as distributed technologies.

4.1. Solar Photovoltaic Module

A 60kW solar photovoltaic array is used with a lifetime of 25 years and 80% derating factor. As we have sufficient waste fields available in village bajoli PV system can be installed easily. Location of village is in the dustier part of Rajasthan so proper and timely cleaning is required for the panels.

4.2. Biomass Gasifier Module

A 40kW small scale biomass system is proposed with a 15 years life time capacity and minimum load ratio of 50%. Bajoli have a lot of potential for the biomass fuel because people here have a lot of animals and cattle like goat buffalo, cows some so animal waste is available in sufficient quantity. Simultaneously wood and agriculture waste is also available as most of the villagers involved in farming activity.

4.3. Optimization of Proposed Hybrid System using HOMER

This software is developed by National Renewable Energy Labs, USA [8] and we have modeled and designed our proposed hybrid model through this package. In HOMER we can optimize, simulate, and can analyze technical and cost effectiveness by considering a particular load demand and supply every hour throughout the year. Several studies have been done [10] .

4.4. System Components and Cost

As discussed earlier the combined system consists of a 60kW solar and 40kW biomass gasifier modules. The system cost comprises combined cost of biomass resources, solar resources, PV panels and biomass gasifier, batteries, converters, connection costs, O&M Costs. Fig. 1 shows the screenshot of the designed model on HOMER software for the system. Component costs (Table 2) are estimated from the market surveys from independent manufacturers and wholesale markets. Project is estimated of total 25 years life time. Discount rate and inflation rate is 10% and 5% respectively. The fixed and variable costs are site-specific and would vary with geography of operations.

4.5. Development of Load Profile

Load profiles were developed in three steps: first, to form an initial skeleton of village's demand profile, studies of villages with similar socioeconomic, geographic, population density and electricity utilization were used for reference [9]. Second, information was collected from a range of field visits to schools, houses and local markets, and through consulting the established enterprises, and demand surveys. The demand surveys reflected on the current issues faced by the villagers in sectors and adopted plans to provide lighting in schools, street lighting, establishing paper bag making center, and provision of water heater and fridge for storing vaccines and other biomedical stuffs. Finally, with the use of conventional methods of bottom up analysis a time-dependent probabilistic distribution of loads was generated to determine the varying demand across the day to reflect the end uses. This was all done through analyzing various segments of customers with small populations typical of minigrid operations. This analysis was done by categorizing all the types of electric operations, devices and their typical loads with respect to the number of consumers and studying the probability of their usage at a given time of the day.

The loads are divided into domestic (households), commercial, community, and irrigation sectors. Similarly Commercial loads, Community loads and Irrigation loads are given in table. Monthly variation in load is shown by dividing time duration in three groups of months in summer, monsoon and winter. Loads are distributed uniformly for particular services of household, irrigation, community and commercial. In summer more cooling appliances are needed and in winter night load is decreased but water heating is used so variation in load is shown accordingly. During rainy days no water pumps are required. This variation in load can be revised according to time.

IOP Conf. Series: Materials Science and Engineering 1224 (2022) 012025 doi:10.1088/1757-899X/1224/1/012025

Table 1 Cost of System Components

Load Sector	Appliances	Watts	Quantity per Households	No of Households	Summer Usage Hr/Day	Winter usage Hr/Day	Monsoon Usage Hr/Day
Domestic							
Households	LED Bulbs	6	6	200	8	10	10
	TV	70	1	200	3	4	4
	Fan	50	2	200	8	0	8
	Radio	15	1	200	4	4	4
	Phone Charger	4	2	200	6	6	6
Commercial							
Shops	LED Bulbs	6	4	5	8	8	8
	Fan	50	2	5	10	0	8
Small Cold Storage Plant		6000	1	1	5	2	1
Mixer	Mixer	500	1	5	5	5	5
Commercial Flour Mill	15 HP Motor	12000	1	3	7	7	7
Community							
School	LED Bulbs	6	10	1	10	8	8
	Fan	50	6	1	8	0	4
Health Center	LED Bulbs	6	25	1	10	10	10
	Fan	50	10	1	8	0	5
	Fridge	475	1	1	6	6	6
	Water Heater	1000	1	1	6	6	6
Street Lights	LED Bulbs	10	40		8	8	8
Agriculture							
Fields	Water Pumps	3730	1	8	7	4	0

Table 2: Rating and Requirements of Appliances across Sectors in Different Season

System Component	Size	Number	Capital Cost (US$)	Replacement Cost (US$)	O&M Cost (US$)	Life Time (Years)
PV Array	1kW	60	950/kW	950/kW	25/kW	25
Biomass Gasifier	40kW	1	1495/kW	0/kW	0.20/kWh	25
Battery	200 Ah,12V	240	226/Battery	220/Battery	6.0/Batt/Year	5
Converter	1 kW	100	117/kW	117/kW	3.0/kW/year	15

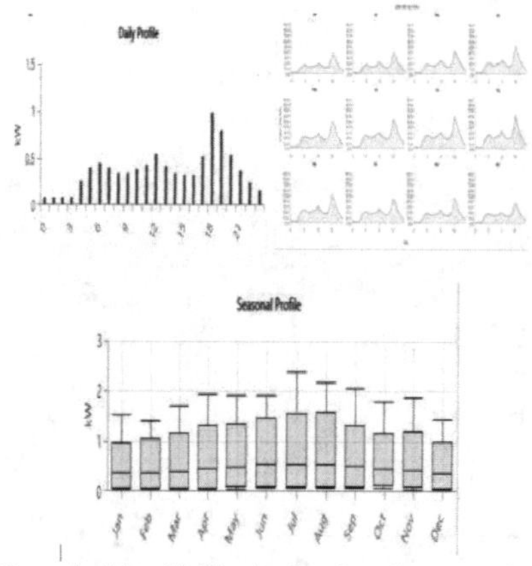

Fig. 2. (a) 30 days demand curves of Households. (b) Single day Load curves for Households Load.

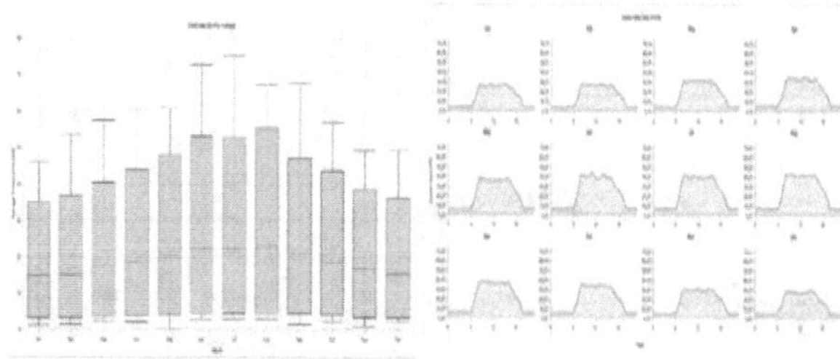

Fig. 3. (a) 30 days demand curves of commercial load. (b) Single day Load curves for commercial Load.

4.6. Observation of solar irradiance

The potential of solar energy for Bajoli village with the Co-ordinates of latitude 26.09, longitude 76.68 was studied. Through the findings annual average solar radiation was 5.46 kWh/m2/day and the average clearness factor was 0.612 (Fig. 6). Around the months of February to May are best to leverage potential of solar energy because of the highest clearness resulting from low fog and low cloud cover which lessens during late summer and monsoon season caused by cloud and rain activity. It is estimated through the surveys and field visits that an approximate of 2250 ton woody biomass is available which is variable according to the seasons.

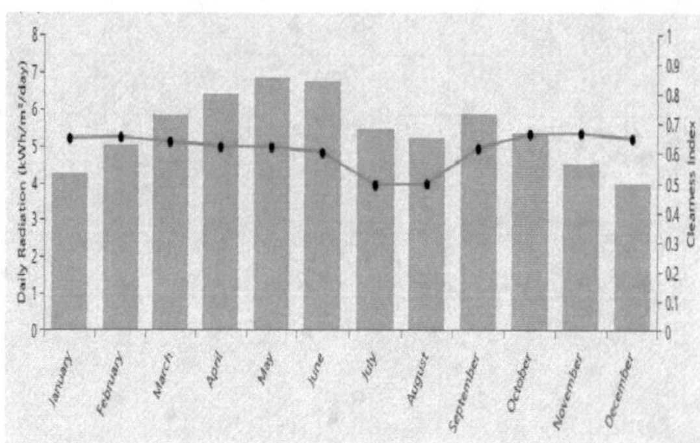

Fig. 4 Irradiation Data obtained from the NASA Database on February 21, 2021.

5. RESULTS AND DISCUSSION

5.1 Techno economic Feasibility Analysis Through HOMER

By using described software we have optimized perfectly this hybrid system with various load profiles of three types of loads. The conclusive system model comprises a 60kW solar photovoltaic array system, a 40 kW biomass gasifier arrangement which is a size your own component in the library of homer software, a generic large 100kW capacity converter connected between AC and DC Bus with 15 years life time, 240 strings of generic 12 volt lead acid battery with 1 kwh energy storage capacity of each. A total of 0.100 % which is 164kWh/year capacity shortage was achieved in the system. Cost summary report in US$ of the above-given architecture detailed in Tables 4 and 5. The hybrid system was optimized at a levelized cost of energy of $0.3206/kW using the software.

As fuel is required for biomass gasifier modules its cost which includes after paying to villagers or collectors or providers is projected to be $30 per tonne of biomass. This would be a source of income for people of village who currently collect firewood, wood chips and straw for their home needs and can collect the agriculture residual waste, forest foliage, and fire-wood for the gasifier. In addition, more biomass resource can be easily collected from the waste fields near the villages and animal dung [9].PV output peaks at 141 kW during the summer months as per software output, when solar irradiation is at its best, and decreases to 0 kW at night, explaining the low mean. Peak demand of electricity load is 95.58 kW, while the minimum of 0.063 kW is observed at night.

The model was able to reach zero power and energy shortages, which necessitated the use of a large battery backup numbers. The current trend of lower battery prices and higher long-term storage will help reduce battery storage costs & lower the renewable energy systems COE [10]. Notable electricity generation is useful in the case of appliance durations, day-to-day demand, and operating reserve capacity, and can be used again in the future if demand increases.

One of the three energy access goals, reducing climate change, is also a critical component of the research. Per capita greenhouse gas (GHG) emissions in rural India are estimated to be 26 kgCO2 e per year per person for homes that use kerosene and electricity for lighting purpose [11]. The entire device powering 350 houses will generate a total of 51.0 kg CO2 e per year in this case, which is a substantial reduction in emissions. Since biomass can be sourced sustainably, resulting in zero emissions, such low levels of emissions are feasible. Table 3 demonstrates the Cost of installed hybrid system whereas table 4 shown the lifetime costing of different components used.

To make the hybrid system financially viable, the capital investment must come from the government or a donor entity, and the operating costs must be recovered by the use of differential energy pricing for household and industrial end uses. It's also important to have a profit margin in the price structure to ensure the system's long-term viability. The parts that follow give an overview of some of the potential approaches to achieving these targets.

5.2 Economic Analysis of Hybrid System Feasibility

The key innovation challenge in HOMER studies is to develop the supporting business process and use the ecosystem which can facilitate the implementation of this outcome and ensure that it function successfully in future. The coming portion discusses three major types in this regard. With institutionalized partnerships, three-tiered developments in structure, funding, and final mile connectivity are proposed. Second one, a picture of business process or model is analyzed in order to perform a comparative and targeted study. Third, a business model for the Bajoli village site is suggested.

Table 3 Cost of installed hybrid system

Total Net Present Lifecycle Cost (US$)	631624
Levelized Cost of Electricity (US$/kWh)	0.3206
Operating Cost per year (US$)	36122.21

Table 4 Component wise Breakup of Life-Time Costs

Component	Capital ($)	Replacement ($)	O&M ($)	Fuel ($)	Total ($)
Biomass Gasifier	59800	0	650.94	192050	252500
240 kWh LA Battery	54240	52800	56982	0	164022
Flat Plat PV	57000	0	70706	0	127706
System Converter	11700	11700	4192	0	27592
Distribution Connection	51978	0	7886	0	59804
System	234658	64500	140416	198050	631624

5.3 Government Policies and Regulations:

These types of startups generally face difficulties from the side of governments while they start to working in rural energy supply. These may be political, policy related and financial risks [12]. Some risks like appropriation and thefts can be controlled through better governance, law and order and village level continuous check-ups. Policy and implementation threats, such as sudden policy changes or failure of implementation plans, require greater clarity across local, global, and international guidelines. Novel billing and metering models, such as village-level funds or pay-as-you-want subscriptions, can mitigate revenue risks caused by consumer non-payment or low demand. For the selected site optimized values of selected parameters have been presented in table 5. The key policy and regulatory barriers currently plaguing the Indian rural energy market are defined as a lack of effective policies and programme, a lack of simulated private sector, a poor institutional system, insufficient governance structure, misdirected focus and priorities, ineffective distribution mechanism Due to the resource, deployment, sector, and investment complexities of decentralized renewable systems, these issues are compounded, necessitating initial capital and revenue support. Rural Energy Access Authorities have been proposed as leadership agencies at both the federal and state levels. Furthermore, they will act as a focal point for bringing energy efficiency, repair, and maintenance of micro-grid energy systems to the public through the government's various skilling and apprenticeship programmes. They will also play an important role in disseminating accurate knowledge about distributed energy systems and emphasizing the benefits in terms of health, education, and opportunities.

Table 5 Technical and economical Parameters of the system

Configuration	Unit	Optimized Design
Solar PV	kW	60
Biomass Gasifier	kW	40
Battery Storage	kW hour	240
Converter	kW	100
Total Capital Cost	US$	234658
Total Replacement Cost	US$	64500
Total annual Capital Cost	US$	9386.32
Total Operating Cost	US$	140416
Unit Cost of Energy	US$	0.3206
Generation Shortage	%	0.01%
Unmet Load	kW	83.3
Excess Electricity	kWh/year	78363

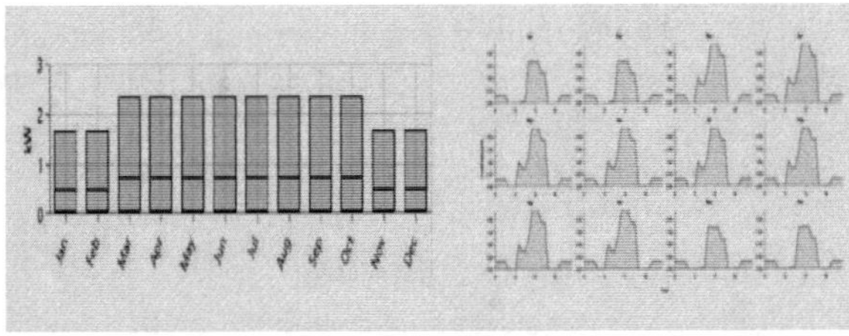

Fig. 5. (a) 30 days electricity use Graphs for community applications. (b) Daily Electricity Graphs for community applications.

5.4 ***Study of different available Economic Models:*** With using multiple types of revenue models from the wide classifications, a diversified financing, ownership, and maintaining schemes can be seen. The following is a list of a couple of these categories.

Pay-as-you-want (Prepaid Energy Access): PAYW contracts allow consumers to use energy for which consumer have to pay in advance according to their preferred schedule or limit, similar to how cellular phone talk-time and data are sold. This is best for consumers who have a fluctuating income and are at risk of missing payments for services. Households will have more leverage over their usage, expenditures, and savings if payment chunks are smaller. Companies who choose to grow the pace of collection of revenue (7 day, 15 day, or 30 day) or adopt different payment methods may have more flexibility.

Utility-controlled development:

Partnerships between energy companies and solar and off-grid companies are a common occurrence. The companies(off grid) which are benefitted from the power companies has mighty money flows and balance sheets, while latter gain access to new markets, goods, and retail customers in new environment [7].

Operation and maintenance given on contracts and installed by community: Community-owned and funded micro grids, which are funded at the level of village using country, regional, or local funds, may be a viable option. Specially allocated energy funds will be needed for the advance costs in high-impact countries where local funds are insufficient. These will be useful in low-income areas where skilled workers are scarce and skilled man forces and technically knowledgeable experts are in scarcity [10].

Single Company Method:

Another recent development is where a single company plans, draws, runs, have ownership, and maintain only integrated solution. A single company can showcase full financial, consulting, advance technology grid, generation resources, operations and maintenance, and a micro grid-sized technical controlling method i.e. SCADA. The strategy is restricted to companies that have a variety of electrical and software capabilities, despite the fact that it offers one-stop solutions.

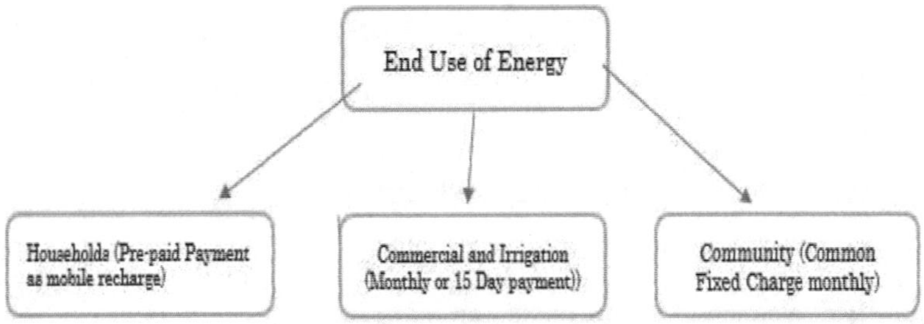

Fig. 6. Proposed business model.

5.5. Final Economic Model Proposed for the village:

We suggest the following simplified revenue and business models for the Bajoli project (Fig. 6). Its modalities operate on couple of stages: first at the last mile user level and another at the level of system. It has been divided into three groups at the last mile user level: families, industrial & irrigation, and society. Term "community" refers to all of the village's popular electricity use, such as street lights and panchayat illumination. Panchayat's big seminar rooms, educational building etc.

The household will be responsible for the upfront installation costs. A 30 days or 15 days electricity use tariff may be utilized for industrial and irrigation end use. Cross-subsidization may be explored by high billing of industrial and agriculture use. Commercial subscribers' payment schedules may be calculated depending on the essence of their market and payment versatility.

6. CONCLUSION

Above Research founded the technical, cost effective, & community viability of a combined biomass and solar photovoltaic system which is to be set up in a partially electrified village of Rajasthan named bajoli. This paper proceeds in three sections: first is estimation of available fuel and solar photovoltaic irradiance and capacity with load requirements, second is making of load changing table profile according to village household electricity consumption as well as community and agricultural load, this table is plotted after constant monitoring of load and taking questionnaire sessions with villagers of bajoli. Third and last section of flow of research paper is calculating the optimal operating feasibility and cost effectiveness of this proposed system using HOMER. HOMER also provides the data of emission. The output of this combined system showcases that installation is reliable, optimally designed and can meet the electricity demand of all types of loads in the village.

After considering worldwide results and existing system an economic model is also proposed which shows that it can be run as private enterprise. Study of various systems suggests that public and private partnership can be availed for installation cost sharing and for the successful operation of such systems recovery of running or operational cost plays a very important role.

We can also add some study results of impact of grid integration with present system during future high load conditions.

REFERENCES

[1] Mohammad Asif Iqbal, and Sunil Kumar Gupta (2020). Comparative analysis between numerical simulation of PPV/PCBM and InGaN based solar cells. Materials Today: Proceedings. 10.1016/j.matpr.2020.05.519.

[2] Mohammad Asif Iqbal, and Sunil Kumar Gupta. "TCAD Based Simulation and Performance Optimization of PPV/PCBM and Perovskite PV cells." International Journal of Computing and Digital Systems 10 (2020): 2-7.

[3] Mohammad Asif Iqbal, "Analysis & Comparison of Various Control Strategy of Hybrid Power Generation," IEEE, vol. 2, no. Iconce, pp. 184–189, 2014, doi: 10.1109/ICONCE.2014.6808717.

[4] Mohammad Asif Iqbal, Arun Dev Dhar Dwivedi "Modelling & efficiency analysis of ingap/gaas single junction PV cells with BSF," Int. J. Eng. Adv. Technol., vol. 8, no. 6, pp. 623–627, 2019, doi: 10.35940/ijeat.F8081.088619.

[5] Mohammad Asif Iqbal,Arun Dev Dhar Dwivedi "Efficiency Improvement Approach of InGaN Based PV cell by Investigating Different Optical & Electrical Properties," SSRN Electron. J., vol. 1, no. 1, pp. 1–9, 2019, doi: 10.2139/ssrn.3355989.

[6] Mohammad Asif Iqbal, Arun Dev Dhar Dwivedi "TCAD Based Simulation & Performance Optimization of InxGa(1-X)N based PV cell," Glob. J. Res. Eng. F Electr. Electron. Eng., vol. 19, no. 4, pp. 27–33, 2019.

[7] Mohammad Asif Iqbal, Gaurav Srivastava "A Review on Field Programmable Gate Arrays Control Based Photovoltaic Energy Management," Iconic Res. Eng. JOURNALS, vol. 1, no. 9, pp. 350–355, 2018.

[8] Mohammad Asif Iqbal, Gaurav Srivastava "A Review on Donor Material of Bulk Heterojunction Polymer(Donor) PV cells," Iconic Res. Eng. Journals, vol. 1, no. 9, pp. 30–33, 2018.

[9] Mohammad Asif Iqbal, Shilpa pachori "Solar Energy Programs for Rural Electrification: Experiences & Lessons from South Asian countries," Int. J. Trend Sci. Res. Dev., vol. Volume-2, no. Issue-3, pp. 860–872, Apr. 2018, doi: 10.31142/ijtsrd11049.

[10] Mohammad Asif Iqbal, "Study of Polymer(Donor) PV cell," ICONIC Res. Eng. Journals, vol. 1, no. 10, pp. 29–32, 2018.

[11] Mohammad Asif Iqbal, "Investigations of Indium Gallium Nitride Properties For Enhancement of Performance of PV cells," Int. J. Creat. Res. Thoughts, vol. 6, no. 1, pp. 605–608, 2018.

[12] M. A. Iqbal & S. Sharma, "Multi-Agent Model of Hybrid Energy System," Am. Int. J. Res. Sci. Technol. Eng. Math., vol. 5, no. 2, pp. 164–168, 2014.

Performance Analysis of High Speed and Low Power Binary Adders

Satyandra Bhati, Neeraj Sharma, Meetu Nag

Department of Electronics & Communication Engineering, Vivekananda Global University, Jaipur, Rajasthan,

neeraj.sharma@vgu.ac.in

Abstract. In all the arithmetic operations, addition is one of the most important and initial operations used in most of the mathematical equations. The operation is performed by many adders present in the digital world. These adders give us carries with preferred delay and power. The three main features like structure, logic, and compact circuit layout help design a better adder. This Paper aims to analyse and compare various additions for high-speed, low-power and fast calculation. The various adder designs seen in digital signal processing applications require computationally efficient adding and cumulative operations, so blocks with the required attributes must be carefully selected. Various techniques have been proposed to design efficient adders in terms of performance, low power consumption and area. This work focused on 16-bit implementations of highly optimized area-efficient Ripple Carry Adders (RCA) and Look Ahead Carry Adders (LAC), and carry select Adder. Finally, we can prove that the Carry look ahead Adder is very fast in all existing designs. These processes are simulated and synthesized in all ISE Xilinx 14.7 software.

Keywords: VLSI, Adder, Low Power, Xilinx

1. Introduction

Anyone can see that even some books on computer arithmetic contain many different circuit architectures that are widely used in practice and have different characteristics of interest. Many researchers have conducted research based on the proportional representation and analysis of data, even when they encounter the binary adder structure [1].

This treatise provides a qualitative assessment of classifying binary adder architecture. Among the vast members of the Adder, we wrote the VHDL code for Ripple carry, Carry select and Carry look to emphasize common recital attributes in the correct position in that class. The next section provides a brief description of the structural design of the Adder [2].

First-class based on very slow ripple carry Adder with minimal area and power. The second class has several levels of carry skip, the prerequisites for the carry select adder are small, and the computation time is reduced. The Carry look ahead Adder of the third class or parallel prefix adder of fourth class support fast addition methods with maximum area complexity [3].

1.1. Need for Low Power Design

There are various interpretations of Moore's Law, which predicts the growth rate of integrated circuits. According to one estimate, it doubles every 18 months. Others claim that device density increases 10-fold every seven years. I agree that all growth is fast, and there are no signs of a slowdown, regardless of the exact numbers. New generations of process technologies are being developed while current-generation devices are safe from underlying physical limitations. The need for low-power VLSI chips arose from such evolutionary forces in integrated circuits. The Intel 4004 microprocessor, developed in 1971, has 2300 transistors, consumes about 1 watt, and has a speed of 1 MHz [4].

Then came the Pentium in 2001, with 42 million transistors, about 65 watts of power, and clocked at 2.40 GHz [5]. As the power dissipation increases linearly over the years, the power density increases exponentially due to the size getting smaller and smaller out of Integrated Circuits. If this exponentially increasing power density were to continue to increase, a microprocessor designed a few years later would have a capacity equivalent to that of a nuclear reactor. Such high-power density leads to reliability problems such as electrical migration, thermal stress and device degradation caused by the hot carrier, resulting in performance loss. Another factor driving the demand for low-power chips is the increased market demand for battery-powered portable consumer electronics. The desire for smaller, lighter and more durable electronic products indirectly leads to lower power consumption. Battery life is emerging as a product differentiator in many portable systems. As the heaviest and largest component of many portable systems, batteries do not have the same rapid density growth rate compared to electronic circuits. The primary source of power dissipation in these high-performance battery-powered portable digital systems such as laptops, cell phones and personal digital assistants is gaining in importance. Low power consumption is a major concern for these systems, as it directly affects performance by affecting battery life [6]. In this situation, low power VLSI design has become important as an active and rapidly growing field. At the circuit design level, significant energy savings potential exists through the appropriate choice of type for the implementation of combinational circuits. This is because of all the important parameters governing power dissipation switching [7].

1.2. Logic Style Requirements for Low Power

Reducing each of these parameters results in a decrease in power dissipation. However, reducing the clock frequency is only possible at the architecture level, and at the circuit level, the frequency is generally considered constant in order to meet certain speed requirements. All other parameters are affected to some extent by the type of logic applied. Therefore, some general logic type requirements for implementing low power circuits can be stated at this point [8, 9].

- Switched Capacitance Reduction
- Supply Voltage Reduction
- Switching Activity Reduction
- Short-Circuit Current Reduction

2. Motivation

The need for low-power VLSI systems comes from two main factors. First, the stable expansion of operating frequency and processing power per chip, supply a bulky current, and appropriate cooling technology must remove the heat caused by large power consumption. Second, the battery life and low-power design of portable electronic devices frankly prolong the operating time of these convenient devices. [10]

This design implements additional logic to determine the Carry and increase additional speed without waiting for the total to complete. This is done by adding an intermediate signal called propagate creation. The Signal generate describes the conditions for digit addition to generating a carry, regardless of the Cin signal. The propagate describes the condition of single addition propagating the Carry only if the Cin signal is a one.

2.1. Objective

Ripple-and-look-ahead architectures can be designed using an iterative approach [11]. This paper shows two novel methods of binary adders. The first approach presents a new Ripple and look ahead architecture design that simultaneously uses the Ripple and Look Ahead architectures to reduce the critical path and area. The following objectives of this work are given below:

- To design Ripple adder, carry select and look ahead Adder is an efficient manner.
- To reduce the area, power and delay to meet the fast operation of the logic.

2.2. Overview of Various Adder Architectures

Adders are one of the key components of any data path. Since any component in VLSI digital design, the alternative of adder architecture is constrained by the imperative factors of area, speed, power and delay. Amongst the various architectures of adders available, some of them are provisionally described in this section. Figure 1 shows technology decreasing day by day [13].

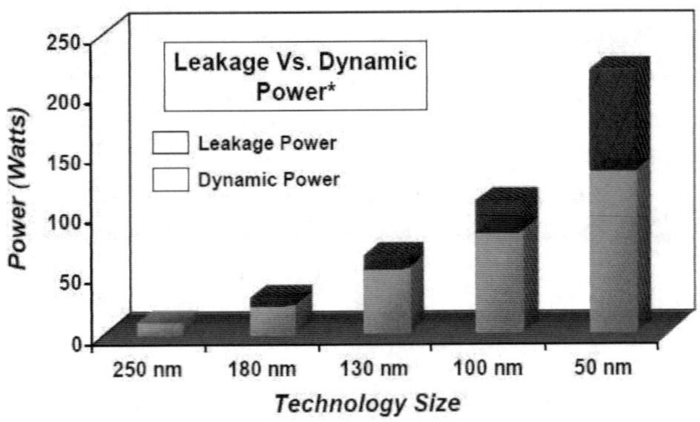

Figure 1: Overview of Various Adder Architectures

(a) Ripple Carry Adder

Ripple Carry Adder (RCA) is the simplest add-on tool for digital design. This Adder takes two binary inputs, N bit (where N is an integer) and produces (N + 1) the production bits (sum N bit and implementation 1bit. The RCA makes N full adders in cascade, and the FA bit connects to the next FA bit from analysis time. Figure 3.7 shows a representation of an N-bit RCA. [12] The input area or b, the execution of each FA is labeled counts as C_{out} (corresponding to the execution (C_{in}) of the next FA), or the sum bits are labelled as a sum [13]. Before counting, the operand and C_{in} need to be entered in each small amount. If we want to get close to the delay of the spread of this Adder, we need to study the delay of the worst case rather than the possible combination of double entry. This is also called the important path. The most significant amount can only be calculated when the previous FA administration is known. In the worst case (if the total lead bit is 1), this lead bit needs to be varied in the system from the smallest number of effective permutations to the most effective location Figure 3.8 has a black line representing the defined path. [14]

(i) Delay

Another important factor than the time between carry-in to carryout to sum is the turnaround time to implementation, as the carrier propagation chain will conclude the entire circuit delay until the Adder. Therefore, considering the worsening of the signal path above, we can write the following equation. For a k-bit RCA, the interrupted path in the worst case is:

TRCA-k bit = TFA(x0, y0, c0) + (k-2) * TFA (Cin Ci) + TFA (Cin Sk-1).

Complexity and delay for n-bit RCA structure logic equations

ARCA = O (n) = 7n……………..………1

TRCA = O (n) = 2 ……………….……2

Consequently, the time for this execution of the Adder where t RCA carry is the delay for the Carry out of a FA and t RCA sum is the delay for the sum of a FA. We can see that the delay is comparative to the length of the Adder [15]. An example of the worst-case propagation entry mode for a 4-bit addict is the conversion of inputs 1111 and 0000 to 1111 and 0001, causing the currency to change from 01111 to 10000. We have shown the results of the 1-bit ad. [16]. In terms of VLSI digital design, this is the easiest Adder to use. It only needs to design and order FA elements and then sequence those N elements to generate N-bit RCA. The concert in the FA band determines the overall RCA speed. Entering the pathway amplifying the FA (t RCA bearing) delay will reduce the RCA prompt. The application of the FA group can be selected to reduce the application delay [17].

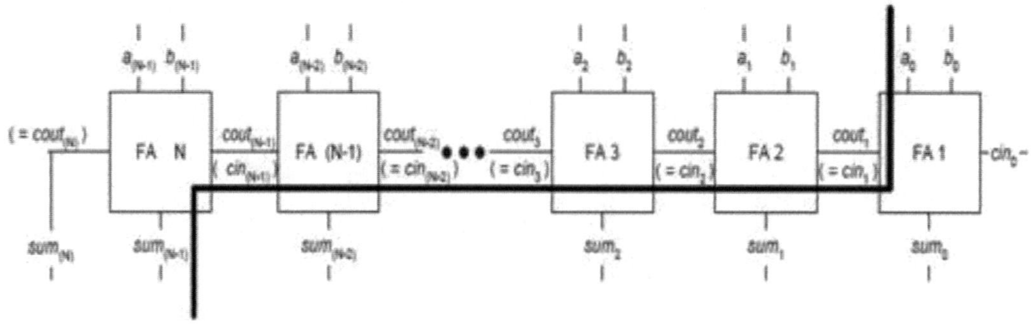

Figure 2: Critical Path for an N-bit Ripple Carry Adder

(b) Carry Look Ahead Adder

From the critical path from the equations, the delay is linearly dependent on N, i.e. the length of particular Adder.

Pi = Ai XOR Bi ………………………..….3
Gi = AiBi Carry generate………..…4
Si = Pi XOR Ci Sum…….…………..…….....5
Ci+1= Gi + PiCi Carry out ………………..….6

Equations 3 to 6 show that the carryout Signal may be thrown mainly on the delay. Algorithms that condense the time to compute carryout and condense N linear habits can significantly shorten the extra work. Equation 3 shows that the carry-in can be computed with g, p, and carry-in [18]. Signals g and p are not required for carry-in and can be computed when the two input operands are present. The CLA adder uses a partial full adder to calculate the resulting signal propagation signal required for the performance of the equation. Figure 2 shows the schematic of the added 4-bit CLA. The logic for each PFA block is shown in Figure 3. The CLA logic block implements the logic of equations 7 to 10, and the gate of this block is shown in Figure 3. For a 4-bit CLA adder, the 4th execution signal can also be considered the 5th total bit. Having single-level carry look adder logic for a long adder is impractical but can be explained by adding other levels of CLA logic [19]. To achieve this, each adder block requires two additional signals: group creation and group propagation.

C1 = G0 + P0C0 7
C2 = G1 + P1C1 = G1 + P1 (G0 + P0C0) = G1 + P1G0 + P1P0C0 8
C3 = G2 + P2C2 = G2 + P2G1 + P2P1G0 + P2P1P0C0 9
C4= G3 + P3C3 = G3 + P3G2 + P3P2G1 + P3P2P1G0 + P3P2P1P0C0 10

The equations for the two signals assuming a 4-bit adder block size are given in Equations 7 and 8. Group creation can occur when one Carry produces an adder block, or group propagation occurs when a carry-in to an adder block is propagated to a carryout. Multiple levels of CLA logic allow building carry prediction adders of arbitrary length. The size of the CLA adder block is 4 bits. This is a familiar factor for most word sizes, as there are practical limits to the gate size that can be executed. To show the use of other levels of CLA logic, Figure 4 has a schematic of a 16-bit CLA adder, creating a group of each 4-bit adder sub cell and acquiring and executing a group radio signal. This indicates that it is the second level of CLA logic that computes the Signal. If the Adder has several levels of CLA logic and not more than the final level, then a c4 signal should be generated. All other levels reinstate this c4 Signal with group creation or group radio waves.

ACLA = O (n) = 14n
TCLA = O (log n) = 4 log

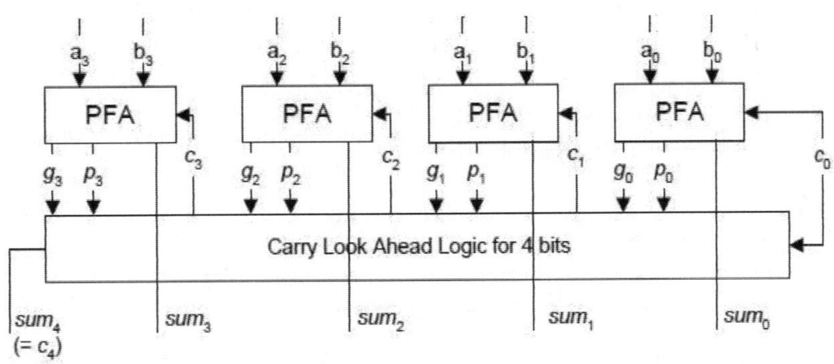

Figure 3: 4-bit Carry Look-Ahead Adder

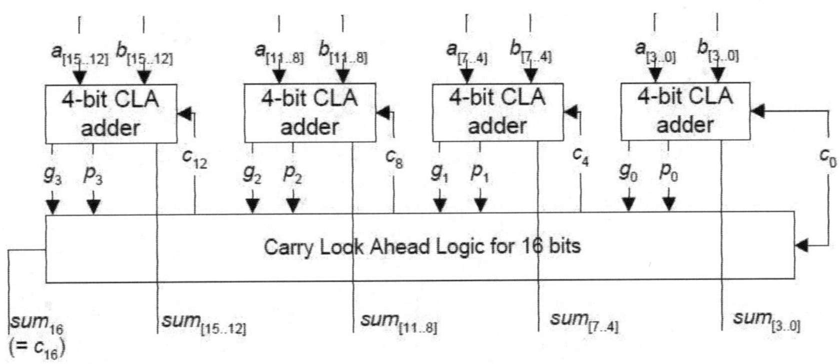

Figure 4: Schematic for a 16-bit CLA Adder

Complexity and Delay for n-bit CLA structure use third-level CLA logic and four blocks of 16 docs to create 64-bit docs. The CLA logic generates the c16, c32, and c 48 signals that are used as input to carry the 16-bit adder block and generates C64 to the sum64 Signal. If the format requires a length of 32, the creator uses two 16-bit blocks and two load-bearing signals (c16, c32) in the third phase of the CLA logic. Free. A similar device to the CLA. Logic simplifies building a long ad using this system, and the adder block can be installed via the subunit. Defining the critical path of the CLA adder is difficult because the leading path gates have different nibs. If you want to have a rough idea, first think that the delays of all gates are the same. The delay of the 4-bit CLA adder requires a gate delay to count the waveform and generate the Signal and a two-gate delay to count the carrying signal and calculate the total Signal. Therefore, it is necessary to postpone the gate. It represents the postponement of the four gates. For 16-bit CLA servers, a single gate delay (from PFA) is used to calculate the frequency and output signal and a two-gate delay used to calculate the group and generation propagation is the first step in the logic rule. The delay of two gates is a gate of logic carrying a first level and a second level carrying a logic signal. Compared to the four-bit CLA ad, the second-level logic in the 16-bit CLA address provides two other gate delays, resulting in six gate complaints. Continue as follows (64 bits extension requires eight gate delays; 256 bits extension requires ten gate delays) The CLA address delay is use redundancy to add two numbers to further increase the rate of addition. That is, we can add two of any number of total bits.

2.3. Ripple carry algorithm

It successively adds the two least significant bits to produce a total bit of equal importance and one Carry output bit of higher importance. Post results after hours. Bits continuously i = 1,2, . . . N 1 adds the two operand bits and the importance i carry bit to produce a total bit of equal importance and one Carry bit of greater importance. The Post result changes the Si time after switching the input operand that caused the change. Even if the Ripple carry algorithm references constant updates, the updates are rarely reflected in a source switch.

Figure 5 shows a prominent event designed to add two carry generations. The Carry propagates across the Adder, with the upper 4 bits executing the path through another path. More specifically, assuming that the adders all start with 0 carry, the first step in the algorithm is to generate the sum of heats 0 and 3 and the Carry in columns 1 and 4. The total calculation for the other columns will be recreated later. The second stage produces results such as columns 1 and 4. Three-step carry propagation is used to recalculate the Carry in column 3. This will have to start over because the calculations in the previous three columns will be unfounded. This will propagate the new Carry to column 4. Recalculating column 4 produces another

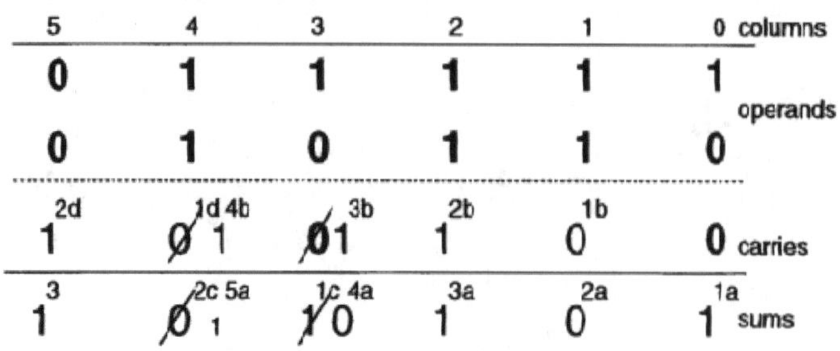

Figure 5: Ripple Carry Addition

The algorithm exits after step 5 because it totals a bit but does not generate another carry bit. Note that the correct most significant total bit was generated in step 3, but further two-step addition was not complete because the lower low bit was changed.

(c) Look Ahead Adder

Carry look ahead logic uses the concept of generating and propagating Carry. It is most natural to think of creation and propagation in the context of binary addition in Carry look ahead adders, but the concept can be used more commonly than this. Carry look precedence depends on two points.

• If coming in from the right, calculate whether the position propagates the Carry for each digit position.

• You can combine these calculated values to quickly infer whether that group propagates the Carry coming from the right for each group of numbers.

• For all 1-bit adders, calculate the result. At the same time, the predictor performs the calculation.

• Assuming the Carry occurs in a particular group, the Carry will appear from the left edge of the group within a maximum of 5 gate delays and will begin propagating to the left through the group.

• If the Carry is continuously communicated through the next group, the predictor has already inferred this. Therefore, before the Carry comes out in the next group, the look ahead device can immediately notify the next group on the left (with one gate delay) that it will receive the Carry, and at the same time, notify the next group. The next predictor is on the left side, where the Carry is in progress.

3. Simulation and Results

3.1. Adder Implementation

Execution of 16-bit Ripple carry Adder and look ahead adder RCA has been done using Xilinx 14.1 and simulator has mention out by I-Sim 14.1e tool. Ripple carry Adder and look ahead Adder. Optimized Adder is meaningful that the worst-case delay of Ripple Carry Adder or look Ahead Adder. Delay among Carry-in and carry-out each full Adder; I investigate another optimized design of full Adder to see if it will get a better presentation. The top level of the Ripple carry Adder is the same, though the full adder schematic is different, as Figure 6. The intuition of this full adder optimization is to minimize the delay of Carry in and carry out of the full Adder, which is the answer direct of Ripple carry adder

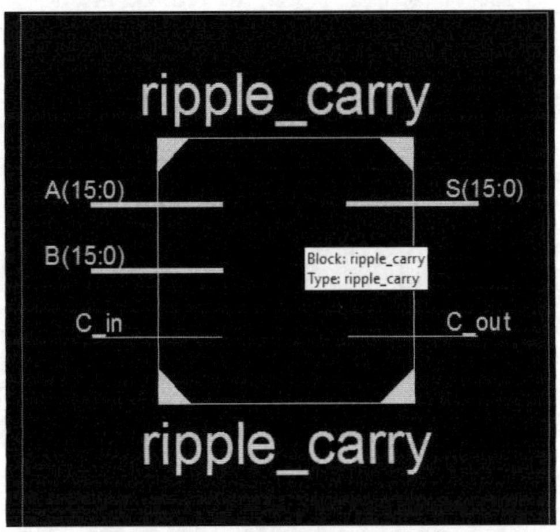

Figure 6: Ripple Carry Adder

Binary input, a(15:0) ,b(15:0) and c_in are the inputs which are going to (N-1 down to 0) s(15:0, are outputs shown in figure 6. The RTL view of the 16-bit binary coded decimal converter. RTL view of 16-bit Ripple carry Adder and look ahead adder input-output and RTL view.

RTL view of 16-bit Ripple carry Adder, RTL view of sequential output like counter and flip flop which count signal propagation and show I/O lines of mux and flip flops as shown in simulation. It generates indication bits for carry propagation and generation (P and G) of each full Adder and calculates all the carries simultaneously. The drawing of the spread full Adder and carry look ahead component is as Figure 7. Carry look ahead logic uses concepts of generating or propagating carries.

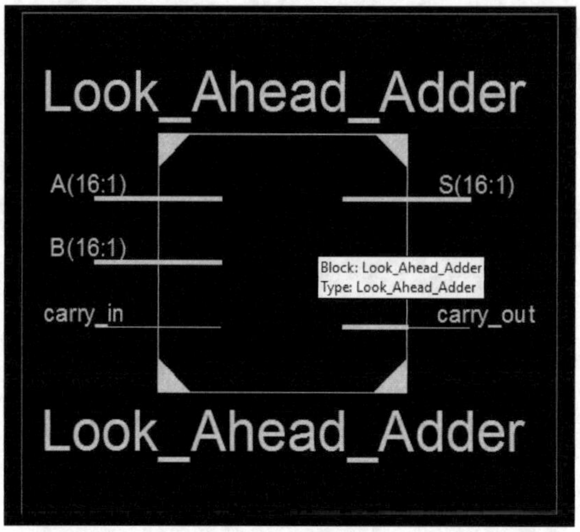

Figure 7: RTL View of 16-Bit Carry Look Ahead Adder

A Carry Look Ahead Adder (CLA), or Fast Adder, is a type of Adder used in digital logic. Carry look ahead adder recover speed by reducing time obligatory to settle on the carry bit. The Ripple carry Adder requires the Carry to propagate to the gate level before getting a gun at the output terminal. An n bit ripple carry Adder has 2n gate levels. Propagation time can be a factor that hinders the speed of the circuit. One of the most widely used processes to reduce latency is to use the Carry look ahead mechanism.

3.2. Simulation and Synthesis Results

Run simulations and synthesis and predict results of all adders. Functional verification (simulation) and synthesis of all adders (where high-level techniques are converted to RTL) are run, summarized the results. Here is a summary of the 16-bit Carry look build results. Design and implement a circuit to build a 16-bit carry look ahead adder (how the carry look-a-head functions and how to derive the final circuit layout), builds and simulates the circuit using Xilinx. Accurate display of all three simulation bags and REAL time to completion of XST: 21.00 seconds Total CPU time to completion of XST: 21.27 seconds Number of slices: 18/5472,4 Number of Input LUTs: 32/10944 Few IOs: 50 Number of combined IOBs: 50 240

➢ Design Statistics: IOs: 50

Cell Usage:

- BELS: 32
- LUT3: 32
- IO Buffers: 50
- IBUF: 33
- OBUF: 17

IOP Conf. Series: Materials Science and Engineering 1224 (2022) 012026 doi:10.1088/1757-899X/1224/1/012026

Every part of single-ended outputs in this plan are with slew rate imperfect productivity drivers. The hindrance on speed-critical single-ended outputs can be dramatically condensed by designating them as fast outputs.

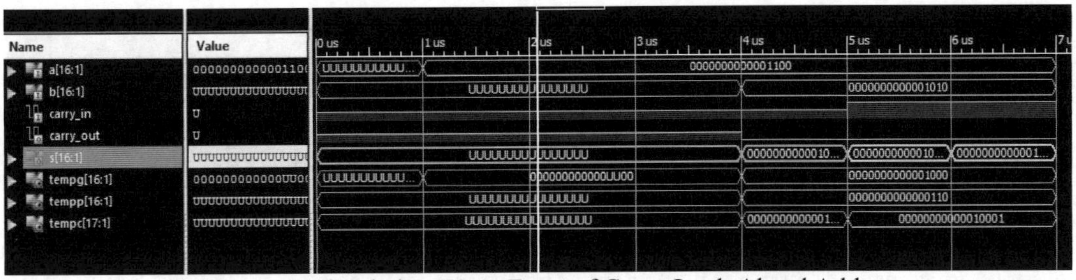

Figure 8: Simulation Wave Form of Carry-Look Ahead Adder

Figure 8 shows the state and state _next Signal of the Simulation Wave Form of Carry-look ahead Adder shown in figure 12 simulation waveform of Simulation Wave Form of Carry-look ahead Adder.

3.3. *Timing Analysis*

It is important to realize the importance of timing analysis before taking the design to the next step. Timing analysis allows you to set time limits on rigorous designs to improve performance ultimately. The Carry look pre-adder allows addicts to exploit space to determine the number of associated logic gates. Logic gate limits fix space utilization.

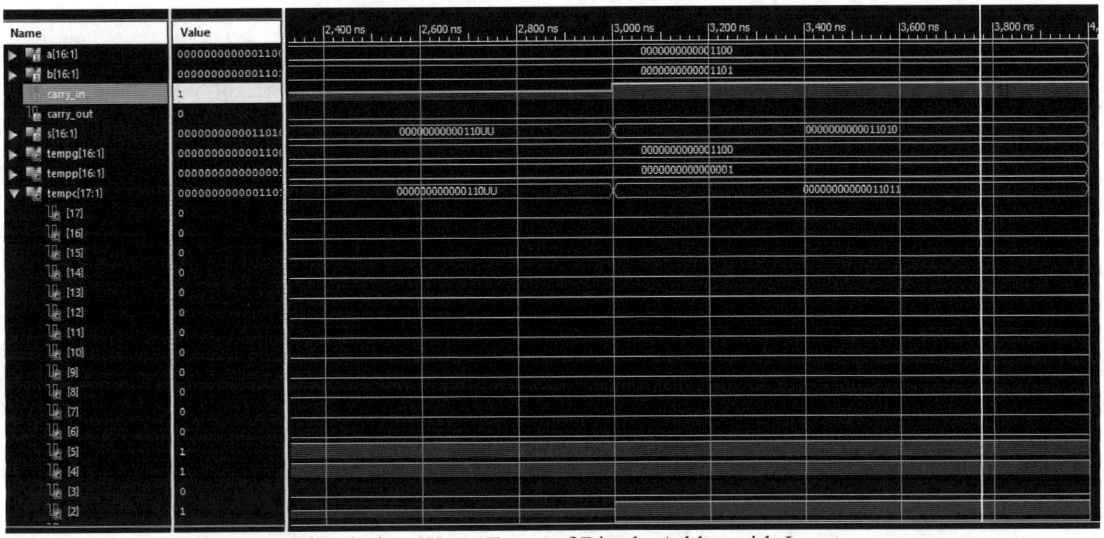

Figure 9: Simulation Wave Form of Ripple Adder with Input

Figure 9 gives an overview of the overall simulation of a 16-bit binary code decimal converter. Executed using Simulator (Isim) 14.7, described using a VHDL dataflow model that represents the delay and time of the configuration of all 16-bit Carry look pre-adders and ripple adders with the entire combinatorial logic of binary code decimal design. The 16-bit Carry look pre-adder, ripple adder, and multi-operand design are set so that all inputs have a 100% clock charge. Based on the proposed algorithm, a binary code decimal configuration was designed, and the 16-bit Carry looks pre-adder and ripple adder of the proposed algorithm were replaced with the architectural structure for a fair comparison. The composite results show that there is a diminution in impediment power and area. This results in reduced power delay product or smallest duration: 2.600ns (maximum frequency: 384.645MHz) & smallest input arrival time before clock: 2.765ns & maximum output duration after clock: 4.714ns greatest combined path delay.

All the 16 -bit Carry-look ahead Adder and Ripple Adder and adding and shifting arrangement were illustrated utilizing VHDL data flow modelling and imitation employing simulator (Isim) 14.7. The 16 -bit Carry-look ahead Adder or ripple adder are mapped on RTL Compiler on 14.7 (Isim). All inputs are set to have a clock rate of 100%.

4. Conclusions and Future Work

All three models of Ripple Adder, Look-Ahead Adder, and Carry Select Adder are designed, and the results are compared in terms of delay implemented in VHDL using the Xilinx 14.7 ISE tool. The performance of the three adders on the delay is evaluated by implementing the logic using the predictive Adder in the adder part. The planned intend is optimized in terms of latency or hardware complexity. First, since the gate used here is a carry look ahead adder parity storage gate, Adder maintains full parity. Therefore, if no errors are detected, no intermediate checks are required. Total REAL time to completion of XST: 21.00 seconds Total CPU time to the conclusion of XST: 21.27 seconds, number of 4 input LUTs 3210,944. The number of occupied slices 245,472 Contains only related logic Slice 24 uses of 24, in the world of low power VLSI design, anything is possible, and in the future, all adder designs may be more accurate than all previous designs. This ultimately improves area efficiency because it can be implemented with fewer gates in the computer hardware. The proposed ripple adder forms the centerpiece of a multi-operand binary adder. The simulation results show the capabilities of the proposed ripple adder, carry select Adder, look ahead Adder in addition to the multi-operand decimal adder for conventional designs.

The Carry Select Adder, which is designed in this paper, is very suitable for implementing the VLSI circuits. The optimized delay power among the adders, power consumption, and reducing the transistors is carried out. The overall objective of the project is to construct a static and compact carry select adder that can be implemented for further purposes

In an attempt to develop arithmetic algorithms and architecture-level optimization techniques for low-power, high-speed multiplier design, the method presented in this paper has achieved good results. However, there are limitations to this work, and several possible future research directions are as follows.

Possible directions are radix higher than base four recordings. radix4 recoding is being considered in the work of this paper because it is a simple and popular choice. Higher radix recoding allows power saving to reduce the number of PPs further. To improve the performance, you can use a higher-order deep-level pipeline architecture to improve speed. All designs can be identical at the circuit level where Vt can be changed for low power applications, made at the architectural level while keeping one common base for comparison. The pipeline technology is widely used to improve the performance of digital circuits. A simple calculation shows that a gate delay of around 33 is required to complete the multiplication of 32-bit operands without pipeline processing. In this number, nine gate delays belong to Wallace structure (4:2 compressor), ten gate delays belong to MBR, and 14 gate delays belong to 64-bit carry prediction adder. If you add three registers here, each will transmit one gate delay, resulting in a total of 36 gate delays. So, splitting this delay into three would require about 12 gate delays for each stage in the pipeline to complete.

5. References

1. J. L. Hennessey and D. A. Patterson and Morgan Kaufmann "Computer Architecture A Quantitative Approach" San Francisco:, 1990.

2. G. Ragunath, Viren Sugandh and R. Sakthivel, "Delay Optimized Binary to BCD Converter for Multi-operand Parallel Decimal Adder" International Conference on Vision Towards Emerging Trends in Communication and Networking (ViTECoN) Year: 2019

3. Bayadir A. Issa and Israa S. AL-Furati, "High Precision Binary Coded Decimal (BCD) unit for 128-bit addition" 2020 International Conference on Electrical, Communication, and Computer Engineering (ICECCE) Year: 2020

4. Radha N., M. Ajeeth, S. Akash and P. Muralikrishnan, "An Efficient Implementation of Decimal Adder Using Parallel Prefix Addition" 3rd International Conference on I-SMAC (IoT in Social, Mobile, Analytics and Cloud) (I-SMAC) Year: 2019

5. Amit Kumar Panda, Rakesh Palisetty and Kailash Chandra Ray "High-Speed Area-Efficient VLSI Architecture of Three-Operand Binary Adder" IEEE Transactions on Circuits and Systems I: Regular Papers Year: 2020

6. Amr Mohammad, Mohammed E. Fouda, Lobna A. Said and Ahmed G. Radwan, "Memristor-CNTFET based Ternary Full Adders" IEEE 63rd International Midwest Symposium on Circuits and Systems (MWSCAS) Year: 2020

7. Avinash Jain, Somya Bansal, Shamim Akhter and Shaheen Khan, "Vedic-Based Squaring Circuit Using Parallel Prefix Adders" 7th International Conference on Signal Processing and Integrated Networks (SPIN) Year: 2020

8. S. Sridevi Sathya Priya, Benita Poul Raju, B. Benita and Dharani, "A Design of low power Adders" 4th International Conference on Devices, Circuits and Systems (ICDCS) Year: 2018

9. Jinping Fan, Yujie Gu, Masahiro Hachimori and Ying Miao, "Signature Codes for Weighted Binary Adder Channel and Multimedia Fingerprinting" IEEE Transactions on Information Theory Year: 2021

10. Junqi Huang, T. Nandha Kumar and Haider Abbas, "General Expressions for Performance Evaluation of Binary Adders" 17th International Conference on Electrical Engineering / Electronics, Computer, Telecommunications and Information Technology (ECTI-CON) Year: 2020

11. Aung Myo San and Alexey N. Yakunin, "Reducing the hardware complexity of a parallel prefix adder" IEEE Conference of Russian Young Researchers in Electrical and Electronic Engineering (EIConRus) Year: 2018

12. Jaeyoon Park and Youngmin Kim, "Design and Implementation of Ternary Carry Lookahead Adder on FPGA" International Conference on Electronics, Information, and Communication (ICEIC) Year: 2021

13. Aung Myo Sana, lexey N. Yakunin, Han Myo Htun and Aung Thu, "Development and Result Verification of a Multi-Bit Binary Parallel-Prefix Adder-Subtractor" IEEE Conference of Russian Young Researchers in Electrical and Electronic Engineering (EIConRus) Year: 2020

14. D Yaswanth, S. Nagaraj and R.Vishnu Vijeth, "Design and analysis of high speed and low area Vedic multiplier using carry select adder" International Conference on Emerging Trends in Information Technology and Engineering (ic-ETITE) Year: 2020

15. Sudhanshu Janwadkar and Rasika Dhavse, "Implementation and Performance Evaluation of Novel Line Adder Architecture for Portable Systems: A Vedic Mathematics" Approach 2020 IEEE REGION 10 CONFERENCE (TENCON) Year: 2020

16. Nafiz Hossain, Nowrin Hossain, Zarrin Tasnim Sworna and Mubin Ul Haque, "A Fast and Compact Binary to BCD Converter Circuit" IEEE International WIE Conference on Electrical and Computer Engineering (WIECON-ECE) Year: 2019

17. Anurag Chauhan, Lokesh Mahor and Piyush Tiwari, "Low Power Quaternay Adder Using CNFET" IEEE VLSI DEVICE CIRCUIT AND SYSTEM (VLSI DCS) Year: 2020

18. Sai Krishna Anumula and Xingguo Xiong, "Design and Simulation of 4-bit QCA BCD Full-adder" IEEE Long Island Systems, Applications and Technology Conference (LISAT) Year: 2019

19. Diksha Manikkule and Pravin Jaronde, "Encapsulation of Full Adder Using 180nm CNTFET" 9th International Conference on Emerging Trends in Engineering and Technology - Signal and Information Processing (ICETET-SIP-19) Year: 2019

20. Binta Lamba and Anurag Sharma, "Improved Pipeline Inner Product Computation Unit for Binary Numbers" 4th International Conference for Convergence in Technology (I2CT) Year: 2018

21. Neelam Somani, Chitrita Chaudhary and Sharad Yadav, "Reversible Adder Design for Ripple Carry and Carry Look Ahead (4, 8, 16, 32-bit)" International Conference on Computing, Communication and Automation (ICCCA) - 2016

22. M.-E. Hwang and K. Roy, "ABRM: Adaptive-Ratio Modulation for Process-Tolerant Ultradynamic Voltage Scaling," IEEE Trans. VLSI Syst., vol. 18, pp. 281-290, 2010.

23. Chetan Vudadha, Sai Phaneendra, Syed Ershad Ahmed, Sreehari Veeramachaneni, Moorthy Muthukrishnan and Srinivas M.B "An Improved Sum Computation Block for with High Sparseness", in 20th International Workshop on Logic & Synthesis (IWLS 2011), June 2011, San Diego, CA, USA.

24. J.-F. Li, J.-D. Yu and Y.-J. Huang, "A design methodology for hybrid Carry-lookahead / Carry Select Adders with reconfigurability", in Proc. IEEE International Symp. on Circuits and Systems (ISCAS), pp.77-80, May 2005.

IOP Conf. Series: Materials Science and Engineering 1224 (2022) 012027 doi:10.1088/1757-899X/1224/1/012027

The Future of Machine Vision in Industries- A systematic review

Allashyam Charan, Chundu Karthik Chowdary, Peddagowda Komal

Department of Mechanical Engineering, Vivekananda Global University, Jaipur, Rajasthan, India

Corresponding Author: Allashyam Charan, 18tec2rb002@vgu.ac.in

Abstract. The Machine vision is the substitution of the human visual sense and judgment capabilities with a video camera and computer to perform an inspection task. It is the automatic acquisition and analysis of images to obtain desired data for controlling or evaluating a specific part or activity. Inspection of components using machine vision technologies provides solution for quality and process control. Now a days, Industry has huge demand for productivity improvement with the implementation of computer control automation. Machine vision reduces the scrap production due to non- conformity by controlling the manufacturing process through machine vision and to prevent the value addition for scrap product in subsequent stage of manufacturing process. The factory vision based industrial robot change the traditional way of mechanical assembly, quality control of product and rapid manufacturing. The various applications of machine vision technologies are automotive pharmaceutical, food and beverage, electronics, packages, and process control.

Keywords: Video measurement, design application industrial, image processing, computer vision, rapid manufacturing, Automatic Inspection, Machine vision, Distance measurement, stereo camera, disparity maps.

1. Introduction

According to the Automated Imaging Association (AIA), machine vision encompasses all industrial and non-industrial applications in which a combination of hardware and software provides operational guidance to devices in the execution of their functions based on the capture and processing of images. Though industrial computer vision uses many of the same algorithms and approaches as academic/educational and governmental/military applications of computer vision, constraints are different [1]. Machine Vision is the knowledge and approaches used to provide image-based automatic examination for quality control, process control, and robot guidance. Basically, machine vision is used for optical gauging, quality assurance, sorting, part assembly inspection, presence and absence and controlling the manufacturing process [2].Liquid automatic filling technology used in the coke factories, petroleum, etc. At the time of filling some dangerous liquids like acids, there will be a serious cause to humans. If we use machine vision instead of human then there will be having advantage to humans [3]. A machine vision system is an integration of optical technology, computer technology and it's a basic understanding of visual capability of human and analysis. It having remarkable advantages like accuracy, consistency. Most of the electronic companies uses machine vision technology to restrict the defect products ratio. Therefore there is no need to inspect the manufactured parts [4] as shown in Figure 1.

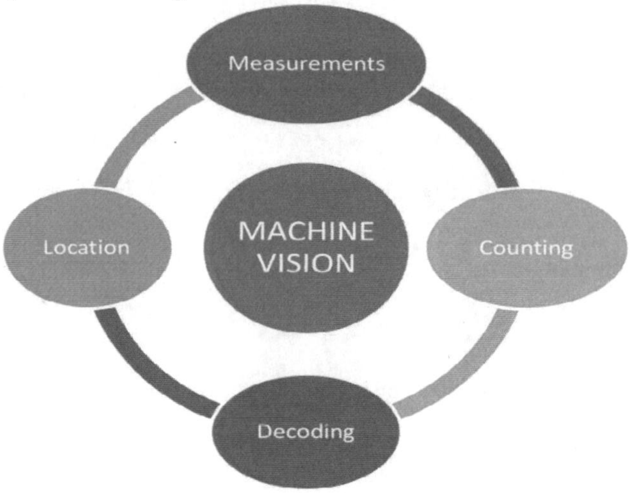

Figure 1. Flow chart of machine vision

2. Applications of Machine vision

2.1 Packaging inspection

It is very crucial for the pharmaceutical companies to count capsules before placing them into containers. The key feature for this solution involves using computer vision to check whether the capsules formed are broken or in good condition. The pc vision based system also implemented to a pc which performs the counting function if the capsule is seemed to be defective, this information then sends a signal to the counting functioning, and at the time of shipping it helps to remove the possibility of defective medical capsules [5].

2.2 Reading barcodes

Reading and identifying the hundreds or thousands of barcodes per day is not an easy task for humans and simply cannot do. For example, electronic devices like mobiles and cell phones requires very smaller and smaller printed circuit boards or PCBs. By doing this task manufacturers are pressured and they cannot produce higher volumes of PCBs forever growing tech market. For that the process known as "penalization". In this process the number of identical circuit boards are printed on a large panel and then each and every circuit is separated by machine for the final testing, in order to inspect every board, however, a machine vision based technology called Panels can was developed to read barcodes-These are known as unique identifiers of each circuit that is present on PCN panel [5].

2.3 3D Vision Inspection

A machine vision inspection system contains a Dalsa Genie Nano camera which is been used in an industries production line to undertake tasks, Humans can struggle sometimes. In this case, the system uses high-resolution images used to build up a full 3d model of components and connector pins. A product pass through the manufacturing plant, the machine vision system takes multiple scans of images of different angles to produce a 3D model, these images, This allow the system to identify if connector pins on circuity are faulty which could have disastrous effects later down the production line[5].

2.4 Object detection

The objective is to locate or detect whether an object of interest is present or absent in a given image. The example is whether an egg tray or an ice tray is placed in a refrigerator. The vision system simply identifies the parts by virtue of a "golden image" or a pattern that has been pertained which it uses to compare it with the real time images from the camera [6].

2.5 Measurement

As the name suggests involves an object dimensions to be accurately determined. This is done by locating certain points on an image and measuring geometrical dimensions like distance, radius, diameter, etc. Example of such applications are measuring the inner diameter of an engine cylinder bore [6].

2.6 Healthcare

Scanning, surgical navigation and skin cancer detection are some of the examples of Machine vision in healthcare industry [7].

2.7 Logistics

Automated Data Capture (ADC)or Automated Inspection(AI) are important process in logistics for verification and control of the product list of these terms refers to automatically identifying objects, collecting data about them, entering them directly into computer systems. QR codes, bar codes and RFIDs are some of the examples of ADC [7].

2.8 Automotive

Dimensional gazing is a method in production process in automotive and machine vision helps by calculating the distance between points or geometric positions on an object and determines whether these measurements meet specifications. Machine vision allows robotic guidance which is needed to place parts onto a vehicle during the body- in-white stage of the assembly process. Presence-absence checking is also assisted by machine vision solutions in automotive plants [7].

3. How machine vision can help humans

Machine vision is widely recognized as being used within automation and production lines. Using machine vision systems allows a system to reduce human involvement in a variety of processes [8]. Correct use of machine vision systems helps not only significantly improve productivity but also delivers greater accuracy of work output by identifying defects prior to customer receipt. Due to machine vision being implemented in industries such as conveyors, they can be integrated for use in a hazardous place or clean spaces that could cause a human to become contaminated or injured. It also improves product quality as well as saves the time and cost. This helps to stop faulty products ever making it to the end-user and causing negative publicity [8].

4. Image processing in machine vision

The digital data coming from the hardware can be analyzed by using image processing algorithms. There are three main steps in image processing in machine vision:

Pre-processing: preprocessing consists of noise removal and contrast enhancing.

Image recognition

- Segmentation: A threshold is applied and the edges of the image is determined in the process.

- Feature Extraction: Size, color, length, shape or combination of these features can be extracted in this process.

Imaging methods emerging methods are designed to provide fast, efficient image analysis under interpretations of images. Some of these are:

- Conventional 2D light imaging: this is the most common technique in machine vision to capture details of object under analysis. 2D light imaging is limited by the wavelength of light that can be detected, this technique is not suitable for some applications police stuff some of key differentials in the technique our frame rate, resolution, monochromatic color systems [9].
- Multispectral Imaging: in this image data that ranges across the electromagnetic spectrum is used to create images. This type of imaging provides visual information behind the limited range of and human eye under visible light that includes ultraviolet and infrared Spectra. This is mainly developed for military applications I'm at least based imaging systems were developed with multi spectral imaging to detect features for mapping such as coastal lines and vegetation. Multispectral imaging is employed in many different areas that includes artwork, documentation and mainly landmine detection [9].
- Hyperspectral Imaging: Hyperspectral Imaging provides an even more in depth analysis of an object spectral fingerprint to form a 3D hyperspectral data cube by utilizing the vast portion of the electromagnetic spectrum. There are four ways one of these systems can sample the hyperspectral cube: spectral scanning, spatial scanning, and snapshot amazing, spatial- spectral scanning. This imaging has been used in molecular biology, oil exploration, surveillance and physics [9].
- 3D Machine vision Techniques: The 3D machine vision techniques are Laster triangulation, time-of-flight, stereo vision, and structured light. This systems require a larger degree of complexity and processing power then their 2D counterparts, they help to provide more in -depth analytical data in areas where 2D imaging falls less.
  i. Dimension and space management
  ii. Detection of surface and assembly defects
  iii. Measuring features such as angles, shapes, holes, and curves
  iv. Object scanning/digitization
  v. Robot Guidance
  vi. Verification against 3D CAD models [9]
- Line Scan Imaging: Line scan imaging build up an image at a time using a linear array with a very high degree of precision. A line scanning camera uses a single light sensitive pixels that scan a moving object on a conveyor belt at a very high frequency, capturing multiple "slices" of an image. Line scanning is just used in product development and QA application. It is used to cylindrical objects, Distortion, which would be produced by area scanning techniques, is limited as the line scanning camera or sensor recorded the same optical position across the entire cylinder [10].

5. Why 3D machine vision is better than 2D?

2D machine vision is a great tool for many applications, but it has some fundamental problems. 2D cameras only see the world in the flat images as shown in figure 2. This means we can't inspect objects as they actually as they measure them as you could expect to be able to. It also means that you inspection software needs to apply complex algorithms to try and guess what an object may look like in a single image. For example, darker shiny surfaces will not show up clearly and so they will lack detail. Lack of contrast poses a problem.

The important point about 3D is that virtue of being able to capture the extra third dimension data reliable, because 3D Machine vision systems are immune to the environmental factors adversely affecting 2D systems the aspects of lighting, contrast, distance to the object is better than compared to two dimensional image processing, working in 3D is indeed more time, processor and the software intensive

IOP Conf. Series: Materials Science and Engineering 1224 (2022) 012027 doi:10.1088/1757-899X/1224/1/012027

please stop 3D algorithms and software tools mean that 3D Machine vision systems are now more than capable of keeping up with production line throughout the requirements [11] as shown in figure 3.

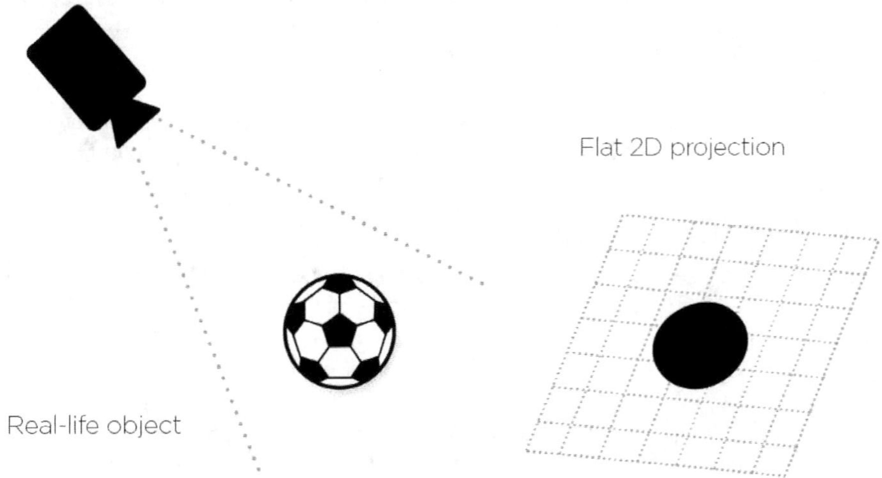

Figure 2. Two dimensional image for machine vision

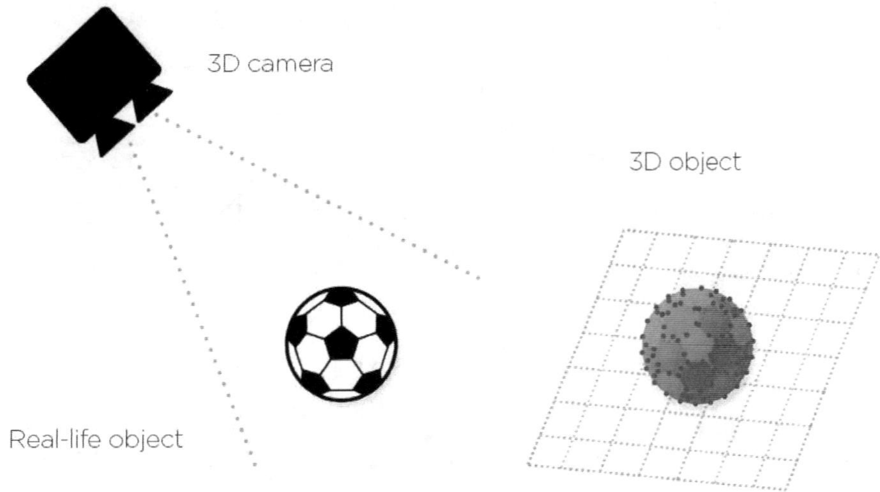

Figure 3. Three dimensional image for machine vision

6. Machine Vision for Quality Control and rapid manufacturing

Machine Vision stands behind many great advances in industrial automation. One of the most prominent achievements is the field of quality control. By adding Machine vision based automated systems offer an effective solution for quality and process control, adding values to inspection operations to enhance it productivity, the accuracy of the manufacturing process, reduction of operational cost . these automated systems includes a camera looking at a production line that capture images which are then are algorithmically compared do a predefined image in order to detect defective objects please stop the application has found major adoption for the detention of imperfections, joint metric inspection geometric inspections, packing control, product classifications, surface finish inspection, colour under texture analysis [12].

These can be analysis through the computer vision based working also:
Accuracy
It helps to ensure a higher grade of accuracy within the accepted tolerance in every manufacturing process. Even on a specific equipment such as a magnifying glass, delicate materials even in the detection of bacteria for medical equipment [12].
Repeatability
Machine Vision driven systems conduct monotonous tasks more effectively. Implementation of fully automated system definitely speed up the production time as the machine needs no time for think its accuracy and repeatability is for great [12].

Reduced costs
An automatic machine vision system provides tangible economic benefits play stop manufacturing companies do not require working personnel to manually perform control of manufacturing products, allowing them to concentrate on more important work. The cost of small human mistakes can sometimes be valuable at billions of dollars and mission vision helps to avoid it [12].
Method of machine vision:
In the machine vision there are different methods one of that is stereo vision system in this after collecting of images from two different cameras (one from Reference camera other is side camera) the pattern of its particular image is matched. Two Logitech C310 USB webcam is used to measurement of distance as shown in figure 4. It is used to identify the distance measurement of a person is worked by a cameras. With the computers after making disparity maps from the visible patterns of the camera it may helped to identify the matching of pattern [13].

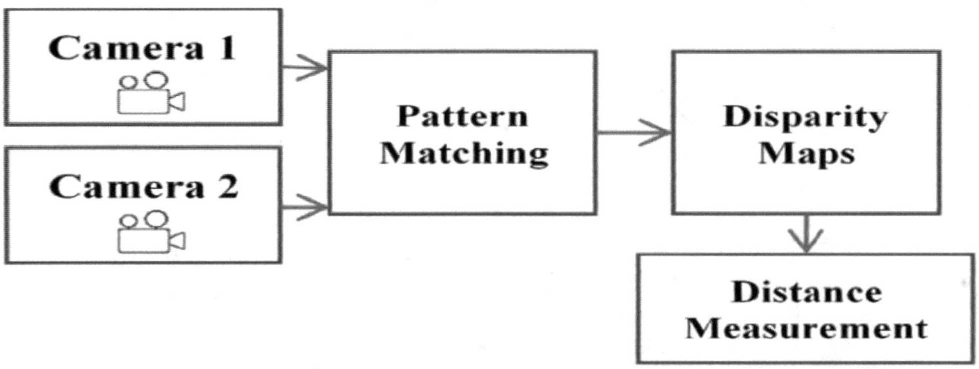

Figure 4. Collecting of two image pattern

It Show the inner components of a standard stereo vision system. It observed that the axes and image planes are composed of a parallel system when a stereo- camera whose camera planes are properly aligned and lens disturbances are completely measured and analysed [14]. The estimation of the height of tress and vegetation near the base poles to the depth map is inversely proportional to the disparity map [15].

7. Convolution Neural Network used at machine vision

The convolution neural network (CNN) is work more accurately and depth of success in field of high level applications. It is known as convolution neurons with in it. The convolutional is a normal neural networks this used to convolution instead normal multiplication of matrices in one of the layers. CNN is work like step by step process to defined number of convolutions are made at same time to batch of linear activations. In the second stage each and every activation is expected with an activation function is not linear. The step is called polling stage for the objective of modifying output of each layer [16] is shown in figure 5. In every case the grouping helps to make the representation almost invariant for small translations of the entry, if we translate the entry by a small amount, the values of most of the grouped outputs do not change [17].

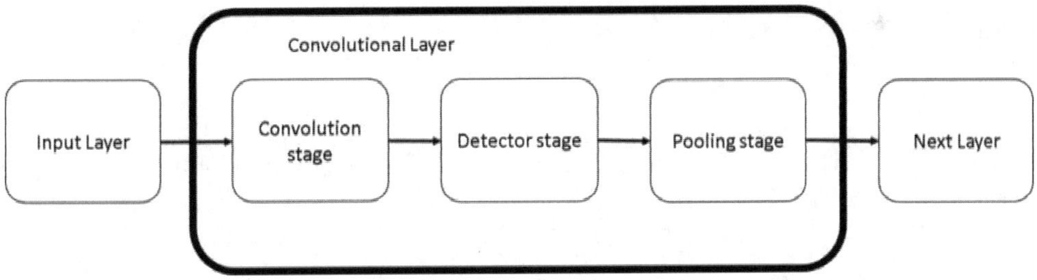

Figure 5. Image convolution layers

8. Report of Machine vision

The global machine vision market size validate at 12.29 billion in 2020 and is expected and expanded at a compound annual growth rate (CAGR) of 6.9% from 2021-2028. The increasing demand for quality inspection and automation in different industrial verticals is likely to drive the market for machine vision [18] are shown in figure 6.

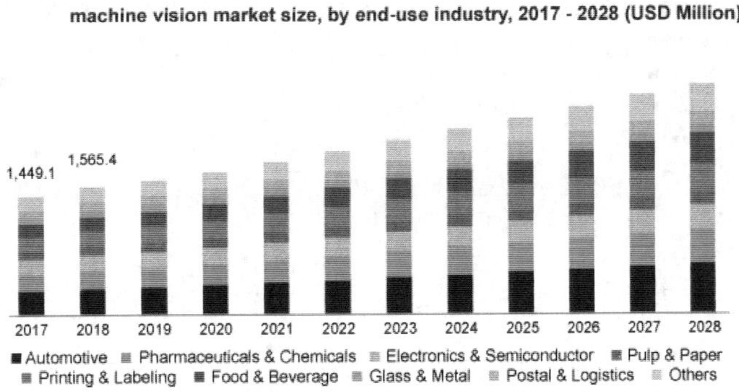

Figure 6. Machine vision market size mapping (Source:www.grandviewresearch)

9. Conclusion

Machine vision is used for industrial and non-industrial applications which will gives us accuracy, repeatability, and reduced costs etc. It also helps us to give exact and best productivity, by using latest technology like industry 4.0. By using of that technology and image classification we can detect the particular object in a different application of machine vision. The data will be classified in a particular pattern, it is used to validate the objects in the industries, it helps to determine and detect the product and validate the rapid manufacturing; the CNN (convolutional neural network) acts a major role to define the image data set from input layer to next layer. The process helps to define the different image from different directions of the conveyer line. The technology of using machine vision is rapidly growing as per the grand view research from 2017-2028. The machine vision is going to implement at different industries like automotive, pharmaceuticals & chemicals, electronics & semiconductors pulp and paper, printing and labelling , food and beverages , glass and metal , postal and logistics , others. Through this review, we try to reveal various confounding factors that have become barriers to the prevention of musculoskeletal diseases, and try to systematically solve these barriers. Various factors that have a significant impact on causing musculoskeletal diseases have been discovered, such as posture, strength, frequency, and working hours.

Conflict of Interest

None to report.

References

[1] Introduction to machine vision (A guide to automating process & quality improvements)/ pg.no:3, written by Vision Drive Natick.
https://www.assemblymag.com/ext/resources/White_Papers/Sep16/Introduction-to-Machine-Vision.pdf ,
[2] S. Sathiyamoorthy. Project scientist IV, NHHID, Anna University, Tamil Nadu, India/ International Journal of Research in Engineering and Technology. ISSN:2321-7308.
[3] Machine Vision, H. Golnabi, A. Asadpour. Institute of Water and Energy, Sharif University of Technology, Tehran, Iran Plasma Physics Research Center, Islamic Azad University, Tehran, Iran
[4] Lim DC, Yun SY, Jung BY, Hong CK (2001) Optical inspection method of lead frame using mathematical morphology. Opt mechatronic systems II, Proc SPIE. 4564:107–114
[5] Writer/Full Stack Developer/STEM Ambassador/ Articles on 10 Examples of using Machine Vision in Manufacturing.
https://www.devteam.space/blog/10-examples-of-using-machine-vision-in-manufacturing/
[6] Raghava Kashyapa. 7 common machine vision Applications in manufacturing.
https://qualitastech.com/7-common-applications-of-machine-vision-in-manufacturing/
[7] AI Multiple/ Machine vision in 2021: In Depth Guide by Cem Dilmegani on January 1, 2021 in what are machine vision use cases/applications areas? https://research.aimultiple.com/machine-vision/amp/
[8] L/A/C CONVEYORS & AUTOMATION News, Robotics By: Gemma/December 11, 2019.
https://www.lacconveyors.co.uk/machine-vision-systems-provided/
[9] AZO MATERIALS/ Methods used to provide Imaging-Based Automatic Inspection and Analysis by Reginald Davey/ Feb 14, 2020. https://www.azom.com/article.aspx?ArticleID=18988
[10] STEMMER IMAGEING/ Member of Primepulse/ Introduction to Machine Vision- Aguid to automating process& quality.
https://www.stemmer-imaging.com/en-gb/knowledge-base/cameras-line-scan-technology/
[11] blog of ZiVID/Machine Vision Written by Henrik Schumann-Olsen/2021-10-27/ https://blog.zivid.com/3d-machine-vision-vs-2d
[12] blog of Softengi/ computer Vision for Quality Control written by: Sergiy Tikhonov(Head of BA)/ 2021 https://softengi.com/blog/computer-vision-for-quality-control/
[13] Kerim kursat cevik/ Akdeniz University/36 publications/ 126citations/ Conference :2019 3rd International symposium on multidisciplinary studies and innovation Technologies/ At: Ankara, Turkey / DoI:- 10/1109 /IsmsIT.2019.8932817/
[14] M. Soyaslan. Stereo kamera sisteminde aykirilik haritalari yardimiyla nesne vzakhklarinin tespit edilmesi", Sakarya university journal of science, vol. 20, pp.111-119, 2016.
[15] L. Chung- Hee and K. Dongyoung. Dense Disparity map-based pedestrian Detection for intelligent vehicle, IEEE international conference on Intelligent Transportation Engineering, pp.1015 - 1018, 2017
[16]Universidad Autnoma de Ciudad Jurez @alumnos.uacj.mx http://www.uacj.mx/Paginas/Default.aspx
[17] I. Good fellow, Y. Bengio and A. Courville. Deep learning MIT press.
http://www.deeplearningbook.org,2016
[18] Grand view Research/ Machine vision market size/ published Date: Apr, 2021/Base year for Estimate:2020, Repot ID:GVR-1-68038-842-8/ Format: Electronic(pdf)/ Historical Data: 2017-2019/ Number of pages:154, https://www.grandviewresearch.com/industry-analysis/machine-vision-market

AUTHOR INDEX